W0037790

**WORLD BANK TECHNICAL PAPER NO. 461**

*Europe and Central Asia Environmentally and Socially Sustainable Development Series*

# Social Assessment and Agricultural Reform in Central Asia and Turkey

*Edited by*
*Ayse Kudat*
*Stan Peabody*
*Caglar Keyder*

*The World Bank*
*Washington, D.C.*

ISBN: 0-8213-4678-4
ISSN: 0253-7494

Cover photo by Stan Peabody.

Ayse Kudat is lead social specialist in the Environmentally and Socially Sustainable Development Sector Unit of the World Bank's Europe and Central Asia Region. Stan Peabody is principal social scientist in the Environmentally and Socially Sustainable Development Sector Unit of the World Bank's Europe and Central Asia Region. Caglar Keyder is adjunct professor in the sociology department at Bogazici University in Istanbul, Turkey.

**Library of Congress Cataloging-in-Publication Data**

Social assessment and agricultural reform in Central Asia and Turkey / edited by Ayse Kudat, Stan Peabody, Caglar Keyder.
    p.   cm. — (World Bank technical paper ; no. 461. Europe and Central Asia
    environmentally and socially sustainable development series)
    ISBN 0-8213-4678-4
    1. Land reform—Asia, Central. 2. Land reform—Turkey.   3. Agriculture and state—Asia,
Central.  4. Agricultural and state—Turkey. 5. Asia, Central—Social conditions—1991– . 6.
Turkey—Social conditions—1960– .  I. Kudat, Ayse.  II. Peabody, Stan, 1944– .  III. Keyder,
Caglar.  IV. World Bank technical paper ; no. 461.  V. World Bank technical paper.
Europe and Central Asia environmentally and socially sustainable development series.

HD1333.A783 S62   2000
338.1'8561—dc21                        00-022839

# Contents

## Chapter One
## Social Dimensions of Agrarian Transformation

## Chapter Two
## Social Assessment: A Comprehensive Framework for Development Initiatives

## Chapter Three
## Highlights of the Case Studies

## Chapter Six
## Uzbekistan Agricultural Enterprise Restructuring and Development Program .............. 219

## Chapter Seven
## Sanliurfa-Harran Plains On-Farm and Village Development Project ................................ 255

# Foreword

## Kevin M. Cleaver[1]

The mission of the World Bank is to reduce poverty and to support socially and environmentally sustainable economic growth through financing development projects and providing advice on the framework for reform. The present volume reflects this objective by illustrating an innovative way — the process of social assessment and stakeholder consultation — in which the Bank partners with key actors in the development effort in the former Soviet Union (FSU), Eastern Europe and Turkey (ECA).

The social assessment process is relatively new in the World Bank. Because its use is particularly relevant to countries in transition — politically, socially, economically and ethnically — where much is still unknown, in 1997 the World Bank published a book of case studies of the use of social assessment in general in the FSU.[2] This second volume focuses on the social assessment process in the rural sector and includes non FSU countries. It has several objectives: first, to disseminate our experience with social assessment for the use of development practitioners around the world; second, to initiate a dialogue about how the social assessment process can be improved in the future; and third, to demonstrate the usefulness of social assessment in an area of the world where dealing with rapid change and an uncertain knowledge basis is common.

People living in the rural areas of ECA have had their lives transformed several times in the twentieth century, and not always for the better, as forced collectivization and massive investments in agricultural infrastructure have demonstrated. Today, the aim of governments and the international community is to improve people's lives in ways that are relevant, effective, and coherent. Still, the basic principles that have guided the transformation of agrarian life have been imposed largely from above: the liberalization of the market, land reform, privatization and investments in agricultural modernization and expansion. If today's interventions are more successful than yesterday's, it may be because the people whose lives we are trying to improve are increasingly seen more as partners than as beneficiaries. Rather than seeing them as passive recipients of whatever comes their way we now expect them to be intimately involved in planning and implementing the policies and investments of change. They are the actors in the drama of reform. Integral to this new approach is the process of social assessment and stakeholder consultation, the subject of this book. This book is not just an exercise in theory but it also presents case studies of how the World Bank, governments, and civil society have come together to empower people to speak for themselves. The book also investigates whether the people are heard, and if so, whether their voices have an impact on how the designers of reform and the providers of financing behave.

---

[1]     Kevin Cleaver is the Director for Environmentally and Socially Sustainable Development in the Europe and Central Asia Region of the World Bank.

[2]     Michael M. Cernea and Ayse Kudat, editors (1997). *Social Assessment for Better Development—Case Studies in Russia and Central Asia*. World Bank.

The voices of the people, as heard through the process of social assessment and stakeholder participation, are not always easy for governments and donors to hear. But in this book we see examples of how social scientists have facilitated this process in one region of the World Bank's operations through the involvement of local social scientists with the assistance of funding from the Swiss Government. The social assessment and stakeholder consultation process is based on four pillars: (i) identifying social development issues; (ii) establishing a participatory framework for key stakeholders; (iii) clarifying the institutional and social organizational challenges; and (iv) setting up an ongoing process of monitoring and social impact assessment. The purpose of this effort is not simply to avoid adverse impacts, as we attempt to "do good," but rather to mobilize beneficiaries so that "doing good" becomes "doing better" through people's own involvement, allowing them to reap the rich harvest of social capital.

# Acknowledgments

In addition to those colleagues cited in different chapters, a number of people made important contributions to this work. Kevin Cleaver and Csaba Csaki made direct contributions, as well as providing support and critical advice throughout its preparation. The Task Team Leaders of the four projects for which the social assessments in this volume were prepared – Roy Southworth, Nedret Durutan, and Salem Gafsi – also inspired, challenged, and critiqued different parts of the work.

Ataman Aksoy, Arbi Ben-Achour, Cyoprian Fisiy, Hermine De Soto, and Bruce Harris reviewed earlier drafts and offered many useful recommendations. Peter Gordon and Meg Wilder edited different parts of the volume. Beaulah Noble, Christin Cogley, and Daphne Sawyer-Dunn skillfully processed, edited and formatted different drafts, and Alan Zuschlag prepared the report for publication.

The authors alone are responsible for any omissions, oversights, and mistakes in interpretation.

# Abstract

The social assessment process is relatively new in the World Bank. Because its use is particularly relevant to countries in transition — politically, socially, economically and ethnically — where much is still unknown, in 1997 the World Bank published a book of case studies of the use of social assessment in general in the former Soviet Union. This second volume focuses on the social assessment process in the rural sector and includes Central Asian countries and Turkey which were not included in the first volume. This volume has several objectives: first, to disseminate our experience with social assessment for the use of development practitioners around the world; second, to initiate a dialogue about how the social assessment process can be improved in the future; and third, to demonstrate the usefulness of social assessment in an area of the world where dealing with rapid change and an uncertain knowledge basis is common.

# Preface

# Achievements and Dilemmas of Agricultural Transition in Former Socialist Countries in Europe and Central Asia

## Csaba Csaki[1]

The transformation of Central and Eastern European agriculture started almost a decade ago. Looking back, it can be seen that the countries concerned made the right choice in setting their overall goals and policies for transition to a market economy, but the social costs have been high. Under the present economic and political conditions in the region, there is no alternative to the creation of a market economy based on private ownership. However, given the developments of the past eight years, it is clear that the initial expectations for transformation were overly optimistic and the transition process is far more complicated and complex than anyone imagined in 1991. The region's agrarian economy is still struggling to adjust to economic reality.

This volume presents a very fascinating overview of social assessments carried out in Central Asia and Turkey as they relate to the preparation of World Bank agricultural projects designed to support the transition of agriculture from a socialist, or semi-feudal, system to a modern market-conforming system. Taking a retrospective look at the agricultural transition in the region, one could come to the conclusion that the social aspects of the transition have been given short-shrift. Increased social tensions created by the delayed reforms have brought these problems to the forefront of agricultural policy decision making and have put pressure on the governments to solve them. This compendium not only provides a comprehensive review of the social problems emerging out of the transition in the regional rural sector, but it also identifies some of their root causes.

The social aspects of the agricultural transition in the Central Asian region have obvious importance, and therefore should be taken into account both by policy makers and analysts. There is however, a danger that policy makers give wrong answers to these real problems, and the analysts address these issues separately from the whole complexity of the transition process. Solutions to these problems must always be crafted with the political and economic dynamics of the transition, and the historical background of the problems in mind.

The political changes that affected the entire region in the late 1980s and early 1990s, created new conditions for reforms in agriculture. The former goal of reforming the socialist agriculture system was replaced throughout the region by the new goal of changing the regime and creating a market-type agrarian economy based on private ownership. The current ongoing agrarian reforms in the former socialist countries are aimed at realizing the latter objective. The creation of a market economy and privatization in agriculture involves the following actions:

---

[1]   Csaba Csaki is the Lead Agricultural Advisor for Environmentally and Socially Sustainable Development in the Europe and Central Asia Region of the World Bank.

- The creation of a new macroeconomic framework for agriculture, in terms of price and market liberalization.
- Land reform and creation of a new farming structure based on private ownership of land and productive assets.
- The creation of a competitive environment for agriculture in terms of privatization in agro-processing, input supply, and trade.
- The creation of a rural financial system, serving the needs of privatized agriculture and agricultural services.
- Institutional reform creating institutions and public services required by a market economy.

In theory, the former socialist countries have the possibility of shaping a legislative and regulatory system which, combined with adequate capital input, can bring about the creation of the farming structure most suitable for the market circumstances. At present, the region has only partly taken advantage of this possibility. What is the reason for this? One of the most important features of the past few years has been the predominance of politics over economic rationality. In all the former socialist countries, agricultural transformation is taking place fundamentally as a process guided by political, often short-term, motivations in which the economic consideration and the efforts to create an efficient new structure, as rapidly as possible, play a secondary role. The beliefs inherited from the communist period, distorted and often illogical views of the market economy have played a significant role in the former socialist countries. As a result of the ideologically charged prejudices against the inherited structure and the over-emphasis on seeking justice and reprivatization, the output has fallen by a greater extent than necessary, the loss of fixed assets has been greater than justifiable, land ownership has become fragmented, and people not engaged in farming have acquired a substantial share of land ownership. On the other hand, the political conservatism and the mistrust of private production and market relations often demand the continuation of subsidies in general, and the preservation of the remaining large-scale farms.

Beyond the common overall objectives, the picture is far from being uniform in the region when it comes to the concrete forms and implementation of the various tasks of transition. There are substantial differences among countries in practically all areas of reforms listed above.

In the leading Central European countries, the reform process is close to completion. The possibility of EU membership has accelerated reforms in those countries that were lagging somewhat behind the leading EU accession candidates, most notably in Lithuania and Latvia. Romania and Bulgaria, unfortunately, have shown only modest advancement. The agriculture policy agenda in Central Europe is characterized by efforts to complete the transition, to cope with the increased social problems in rural areas, and to adjust to the evolving CAP. Unfortunately the task of facilitating increased competitiveness has often been stymied by farm lobby demands to provide immediate protection in the agricultural sector and to provide income transfers to farming populations;

There has not been satisfactory progress in agricultural reform in the core countries of the CIS. In Russia, Ukraine, and Kazakhstan, this standstill in the reform process has contributed to the further decline in agricultural output. Some of the smaller countries in the

CIS such as the Kyrgyz Republic, Moldova, and Azerbaijan, however, have taken significant steps in 1998 to continue reforms in the agricultural sector. In 1998, Tajikistan, Turkmenistan, Uzbekistan, and Belarus had still not started any significant reforms.

Land reform and land ownership continues to be the subject of heated debates in practically all countries of the region. In the countries of Central-Eastern Europe the privatization of land based on some form of compensation is largely approaching completion. The new farm structure is characterized by a varied mix of small and large units. The remaining pieces of the state owned units from the socialist period are also increasingly undergoing change and adapting to market economy conditions. The legal settlement of land ownership relations is not yet completed, and the land register and the emergence of a market for land are still in the initial stages. In a few countries a heated debate is being conducted on the ownership of land by companies and foreign nationals.

While land has formally passed into private ownership in the decisive part of the CIS countries (Russia, Ukraine, Belarus), the large-unit sector remains practically untouched. The co-operatives and state farms have been transformed into share companies. However, in practice there has been no change in the real questions of substance. These large units continue be controlled by a central management structure and operate with low efficiency and face increasing financial problems. The role of independent private farming is marginal, not least of all because of the deterrent effect of the undeveloped market relations. There has been remarkable progress in Moldova and Azerbaijan in 1998 in the transformation of the Kolkhoz system. Radical land reforms have been carried out in only a few countries of the former Soviet Union, as a consequence of their special political and economic situations. This is the case for Armenia and Georgia where independent private farming now dominates. Here, the distribution of land carried out on the basis of family size resulted in very small farm sizes and this has gone together with a steep decline in agricultural production for the market. Conversely, in some CIS countries (Uzbekistan and Tajikistan) private ownership of land is prohibited by the constitution and the lease hold arrangements are an added uncertainty.

There are a variety of reasons why the accomplishments of land reform in the CIS countries, including Central Asia, have been fewer than originally anticipated:

- **Political and legal uncertainty.** Controversy over the general reform program and frequent changes of legislation have created a sense of uncertainty about the future course of reforms among the population. A sense of insecure tenure and a fear of the impermanence of the recent changes are not conducive to willing participation by individuals in the reform.

- **Lack of supportive environment.** Functioning markets for farm inputs and products have not yet emerged everywhere, impeding successful operation of the new privatized agriculture. Land markets, which are essential for creation of efficient farm structures, do not function at present because sales of privately owned land are subject to moratoria and other restrictions. While the government credit allocation system is discriminatory and not transparent, the financial sector is not geared to lend to the new privatized farms. Mortgage finance, an important instrument of agricultural lending for private investment, is impossible without legal title to land that can be used as collateral and without land markets that give objective valuation of land. Agricultural policies perpetuate heavy taxation of the sector and leave large powers of decision in the hands of government bureaucrats, thus making investment in farming unattractive.

- **Inadequate mechanisms for restructuring and exit.** Procedures for further restructuring at the farm level are inadequately developed, and individuals and farm managers lack basic information about the options open to them. Mechanisms by which a group of shareholders can present a proposal of separation, including specification of the land and assets they would like to take with them, must be designed. Procedures for adjudication of disputes that arise when the remaining shareholders do not approve a separation proposal will have to be devised.

- **High risk and lack of instruments for risk management.** Political uncertainty, lack of clarity in design of programs, and macroeconomic instability create a risky environment for private farming, while such basic instruments as secure savings and insurance are missing. Instead of establishing independent private farms, participants in land reform and farm restructuring are likely to choose to remain within larger units, where cooperative arrangements provide a measure of insurance against risk.

It is now clear that, contrary to original expectations, the transformation of large-scale agriculture will be a lengthy process. To foresee what type of units will emerge as dominant is extremely difficult. Future agriculture in the region probably will be characterized by a variety of forms of farming operations. Private farms, restructured cooperatives, commercial farms of various sizes, and small part-time subsistence farms producing mainly for their own consumption will coexist side by side. Applying the major lessons of the first phase of the transition to assess the prospects for the future (without attempting to make more concrete predictions), the following issues deserve attention.

- The region's agriculture must adjust to basically new internal and external circumstances. Rapid expansion of internal demand cannot be expected, and capturing a permanent share of external markets requires a basic improvement in efficiency, quality, and competitiveness. In this situation, the level of output will remain for some time lower than that reached in previous years, which means that agricultural production cannot become the engine for general economic recovery. What is even more important, private production is emerging under conditions of strict efficiency requirements, hard budget constraints, and intense market competition, so that this process will have losers as well as winners.

- A relatively large number of private farms may not be able to operate profitably, despite the startup support that many of them receive. It would be a serious mistake to keep these farms alive with further injections from the state budget. State support, if it is given at all, should serve to improve the general income conditions of agriculture and to enhance international competitiveness. A stratum of private farmers dependent on central government support could easily become an economic and political burden.

- The transition to a profitable and efficient agricultural structure should be shortened as much as possible. What is required above all is the establishment of unequivocal and clear land ownership relations. In the future, the development of the farming structure should be basically shaped by market relations and market tools, principally through an efficiently functioning land market. This means minimizing the restrictions on land ownership and land use, as well as on ownership and lease rights. The transfer of ownership and lease rights should be simple and cheap. The greater the government intervention in the operation of the land market, the slower and more costly the shaping of the new farming structure will be. The leasing of land will have a very important role in shaping an efficient farming structure, and the conditions for leasing must be determined fundamentally by the relations of demand and supply.

- It is unlikely that family farming of the Western European type and scale will become the generally accepted mode, even over a longer period. The countries in the region cannot afford to provide state subsidies over a long time to keep alive farms that are not viable under free market conditions. For this reason, the desirable farm size and form of farming must be capable of generating self-sustainable profits without subsidies or with relatively little support. These farm sizes almost certainly will be many times larger than the average size of independent, family-based private farms in Central Europe today (2 to 20 hectares), and possibly larger than farm sizes in some of the Western European countries.

- The traditional dual structure of former socialist agriculture and the emphasis on household plots in the process of reform makes it likely that the distinction between subsistence farming and commercial farming will persist throughout the region for some time to come. The success of the ongoing process of land reform and farm restructuring will be judged by a gradual increase in the number of commercially viable small- and medium-sized farms and continued downsizing of the traditional large-scale structures. Persistence of a dual structure of small subsistence plots and large subsidy-dependent capital-intensive farms will surely signal failure of the process.

- In international practice, agricultural production cooperatives have proven less efficient than farms based on individual ownership, especially in environments with adequately functioning markets. The development of market relations will definitely lead to further disintegration and transformation of the many agricultural production cooperatives that are still in existence in these countries. Precisely because of the social and political implications, it is advisable to guide this process not by administrative intervention but through market and efficiency instruments, in a manner comparable to the desired differentiation of independent private farms.

- The strengthening and spread of processing, marketing, purchasing, and other service cooperatives is a desirable development which can be expected to gain momentum. These service cooperatives will provide a viable support framework for the growth of private farms separating from the collective sector in the process of transition. The development of service cooperatives calls for government attention and support.

- Rural finance is one of the components of the farm support system that is totally undeveloped in the region at this stage. The traditional financial system in the socialist countries is not geared to the needs of new private agriculture. Debt write-offs and interest rate subsidies will be eliminated as agriculture moves toward the market. Both banks and borrowers will have to pay attention to creditworthiness and to take loan repayment seriously. Mortgage finance will have to be developed once private ownership of land and land titles are instituted, and legal restrictions on using land as collateral are removed. In the medium term, creative financial solutions may emerge as an alternative to the undeveloped banking system, including credit from suppliers and marketers linked to product delivery by farmers. Development of village-level credit societies and mutual credit arrangements is also a possibility for meeting the financial needs of a small private farmer.

- Finally, special attention should be paid to the social aspects of the transformation in agriculture. The decline in production and the change of ownership structure are obvious sources of social problems and tensions. Agriculture will be able to contribute to the solution of these problems only within the framework of efficient and profitable production. The fundamental objective of agrarian policy thus should be the achievement of an efficient and profitable farming structure based on private ownership.

There is a host of problems, but the basic final message is not one of despair by any means. Most of the countries concerned are not poor. Most can be self-sustaining, creditworthy, and have market access. We must, however, recognize the complexity of the changes taking place; the social costs, the political strains, the investments needed in nation-building. The point is to be realistic about the task ahead, the problems to be faced, and the time required. In particular, the following points should be emphasized.

- Dedicated people are at work trying to change the system and make the new approaches work. However, there is fear of the future.

- This is not a matter of merely revealing a pool of latent entrepreneurial talents. Generations of experience have taught risk avoidance, suspicion, and survival techniques aimed at circumventing the system.

- The scale of the problem is unprecedented.

- All previous discussions about sequencing have become academic.

- Speed is essential; but speed is also difficult. Efforts need to be focused on actions that will show results quickly (retail privatization, small and medium enterprise development in opportunities) and actions that will offer opportunity to more people and involve them in the process.

No one today is in a position to project where the agriculture sector in the former socialist countries in the region will be at the end of the decade. What we can say with some certainty is that:

- They will not all have evolved in the same way, nor at the same pace;
- Not all transformations that have been started will be fully completed;.
- Progress will not be linear, there will be setbacks; and
- Institutional weaknesses, social structures, and attitudinal changes will slow down structural change and growth.

To avoid disappointment, it is best to be realistic at the outset. The early euphoria envisioned a quick transformation, followed by an equally quick supply response. But developments thus far (and the experience of both Central and East Europe and former East Germany) suggest otherwise. Gradually, the perception is growing that the process will be slower than anticipated and that, consequently, the social and political strains on very fragile systems greater. However, one also has to believe that the whole task is not impossible. The region has all the natural, economic, and human resources to become a fully integrated and prosperous part of the developed world in the foreseeable future.

## Chapter One

## Social Dimensions of Agrarian Transformation

### Caglar Keyder and Ayse Kudat

## Introduction

Agrarian structures in the countries and regions studied in this volume are in the process of rapid transformation. The fundamental changes were often instigated from above: liberalization of the market, land reform and privatization in the former socialist countries, and the construction of a massive irrigation project in southeast Turkey. A decade after the start of the new "reform" programs, experience shows highly variable results on the ground, with each country and social group following paths determined by their needs, circumstances and opportunities. The transformations are redefined as stakeholders shape the outcomes through various negotiations. As a result, "reform" experiments differ from one another with emerging "paths" defined by preexisting institutions and relationships between the state and society.[1] Attempts to impose uniform "reform" plans meet with resistance and the impacts on people's lives diverge from what was anticipated.[2]

The resilience of pre-existing institutions, the diversity of needs and capabilities of social groups, sub-regions and communities, the dynamic relationship of traditional and emerging power holders, the varying availability of social, local political and administrative interventions all imply divergent social impacts and modes of coping with attempts to privatize the agricultural sector. But the complexity of forces that affect peoples' lives and the heterogeneity of institutions are often neglected by policy makers. The western institutions that support these uniform privatization efforts are often disappointed by the varying capacity of countries to ensure a rapid transition to efficient market conditions. They often insist on their own prescriptions, rather than attempt to understand the correlates of differential local responses and unsystematic and often superficial. Indeed, one of the most common errors of western observers is to start from the assumption that market systems are in place and to ask why they fail to function properly. Instead, they should start by asking whether or not the elements of a market system are in place, then identify both missing elements and other factors that currently impede development in rural areas.

The cases included in this volume show that there is high pay-off to understanding the desires and behavior of the people affected by development initiatives. Those on the receiving end mobilize their own resources, capacities, collective memories and social capital to cope with changes in their social and economic lives. The relationships of people in the rural society with the state, with representatives of the old regime, with new entrepreneurs and with those carving new

---

[1] David Start and Gernot Grabher. 1997. "Organizing Diversity: Evolutionary Theory, Network Analysis, and Post-Socialist Transformations." Mimeo.

[2] Vedat Milor, ed. 1994. *Changing Political Economies: Privatization in Post-Communist and Reforming Communist States*. Lynne Rienner Publishers.

spaces for themselves within a mosaic of changing conditions—all form complicated patterns and take unexpected turns. The studies presented in this volume are examples of attempts to forge partnerships between policy makers, especially those who provide financial assistance for reform, and other stakeholders, by providing glimpses of peoples' aspirations, attitudes and behaviors

The cases of transition presented here, although vastly different in scope, illustrate a wide range of examples of structural transformation and processes of social adjustment. They pose a challenge from the point of view of both the immediate study of ongoing change and the attempt to identify the directions of transformation. These two concerns are obviously related, especially if the purpose of studying current transformations is to alleviate the costs of adjustment and to help those who are being economically and socially disadvantaged. Without a perspective on the direction of change it is difficult to identify target groups, their needs, and the modes of formulating policies and interventions that should be directed to these needs.

This chapter first discusses the process of land reform as it was implemented in the Commonwealth of Independent States (CIS). While the case studies pertain to Central Asia (CA) and Turkey, the first two introductory chapters refer to Central and Eastern Europe and other FSU countries in order to provide a broad comparative perspective. The intention is to highlight the specific characteristics of this land reform by evaluating it within the perspective of its social objectives and economic rationale. A discussion of the legal framework within which reform and privatization were conducted seeks to emphasize some of the institutional problems confronting reform and shaping its impact. The following section attempts to identify structural dynamics in the rural society. A substantial literature exists that describes the transformation of agrarian structures following the reforms. There is a rich variety of paths, accommodations, and combinations of forms of organization; the task now is to draw some conclusions regarding the social dimensions of the various directions of change in this complex picture. One common characteristic throughout the region, however, is the increased importance of household-based production. This "peasantization" brings with it new economic and social requirements. The last section deals with social services, arguably the most dismal component of the entire tableau of economic and social crisis. The disruption in the delivery of services has itself become one of the factors promoting subsistence agriculture and shaping the specific mode in which households cope with the transition.

### *Changes in Land Ownership—Restructured Social Relations*

*Radical and rapid changes in land ownership are extraordinary events in most countries, but the countries of Eastern Europe and the former Soviet Union have experienced these changes twice in the twentieth century. In Turkey, ownership changes have been more evolutionary.*

Radical and rapid change in the relations of ownership and property in the countryside is a rare occurrence. The ex-socialist world experienced it twice over the course of the century. For the CIS, the revolution of 1917 ushered in an era of nationalization of the agricultural land which culminated in the formation of state and collective farms under Stalin. In Eastern Europe, the socialist regimes followed suit by collectivizing agriculture during the decade after the end of World War II. Before the collapse of the socialist regimes in the late 1980s, the agricultural sector in much of Eastern Europe and the Soviet Union was characterized by the overwhelming predominance of large collective and state farms, where the rural population was *de facto* employed as workers, rather than peasants in the European sense. The large farms were integrated into a planned network of sales and purchases run by state agencies. Hence, the non-agricultural rural economy of services was non-existent. At the same time, families owned or had use rights to what was variously called household, garden, or auxiliary plots. On these plots, generally more labor-intensive products were cultivated for home consumption and, in some cases, for sale. Household plots and the collective farms existed in a complementary symbiosis; in addition to producing grains and industrial crops, the large units provided cheap or free inputs to families for the production of potatoes, fresh produce and animals. The production of household plots increased the amount of consumables available to the families themselves, supplied nearby urban markets and added income to family budgets. The ratio of consumption to marketing from household plots varied widely from one area to another, however.

The second wave of change took place since the early 1990s. With the growing crisis of socialist economies, various attempts were made to reform the agricultural sector even before the political transformation of the regimes. Regime change brought with it a faith in the advantages of private ownership and the market system, thus defining the task of reform as making agriculture work efficiently in a market economy. Once again, the change was politically motivated (but necessary given the dismantling of the planning system) and, in most countries of the CIS, it initially enjoyed almost no support from the rural population itself. The basic vision underlying the reform effort has been that rural production would become efficient within the market system if individual farmers and agricultural workers were given new incentives. It was thought that a land reform establishing private ownership, changing the structure of ownership and tenure, would provide the conditions under which producers would be motivated. They would thus (accepting the assumptions of the individualist utilitarian model) have a stake in increasing production. It was through this reasoning that the reform was conceived primarily as the dismantling of the large state or collective farms and distribution of land and farming resources to individuals and families, or to voluntarily-constituted groups of rural dwellers. In many cases, urban dwellers were also able to gain access to land and/or to enlarge their existing plots.

Compared to this example of political upheaval and the sudden imposition from above of momentous changes in agrarian structures, the experience of Turkey has been closer to the historical norm in the rest of the world. Here, changes in property structure evolved in response to the slower movements of technology, population, changing factor endowments, and market prices, as reflected through government policy. Starting from a background of peasant agriculture without much landlessness, the post-World War II years in Turkey witnessed the consolidation of family farming. With mechanization in agriculture new land was brought into production during the ten years after the end of the War. Populist politicians made sure that the newly reclaimed (previously state-owned) land was acquired mostly by small owners. Meanwhile, the growth of manufacturing created incentives for migration out of the countryside, thus relieving the pressure on land.

There was significant regional variation, however. The Southeast, the subject of the social assessment included in this volume, had always been characterized by large landed property, landlessness of sizable segments of the population and personal dependency on tribal and religious leaders. This semi-feudal structure was reinforced by the advent of agricultural mechanization as landlords became able to replace mutual dependency relationships with those based on the labor market. In this new social balance, the landless were forced to find employment as workers and to supplement their income with seasonal migration to other regions. In the 1970s, the Turkish government initiated a vast development and irrigation project in this region (the Southeastern Anatolia Project) designed to increase productivity and establish a more labor-intensive agriculture which would increase employment and the incomes of the landless population. Due to political constraints, however, the project was not accompanied by a meaningful land reform: hence, its impact on the distribution of income and social relations in general is indirect, albeit substantial.

## Land Reform

*Variations in ownership patterns had different implications for social problems and policy interventions.*

Much of the collectivization experience earlier this century, in both the Soviet Union and in Eastern Europe, was in accessible to outside observers and analysts. Consequently, most accounts of the experience available in western literature were are reconstructed from indirect and anecdotal evidence, and in retrospect. With the current situation, however, we are witnessing a transformation that must be one of the most intensively studied social phenomena in history. Governments, banks, funding agencies, NGOs and social scientists are in the field, observing, monitoring and conducting research. This is not to say that all dimensions of a complex transition are documented or understood. The transition is an unfolding process whose course is influenced by developments in economic, political and social fields. Furthermore, as in all transitions when the direction of change is uncertain, most research is fueled by a desire to intervene and influence their direction. It is therefore important to try to document the multifarious dimensions of transformation as they are experienced by the immediate participants in this great experiment.

In the case of the former Soviet bloc, the most important dimension from which everything else follows is the change in property relations. State and collective property in agriculture is being transformed into private property, although this shift does not necessarily entail the distribution of land to individuals or households or even the privatization of land itself. As is well known, the reform was conducted in different ways in different countries. In Central Europe, there was an attempt to restore property rights to those who ownerd the land prior to collectivization. This was a response to moral and ideological concerns which aimed not only to restitute property, but also, in some symbolic sense, to recreate the society in its pre-socialist form.

In most of the CIS, however, land reform consisted of allocating land shares of the large collective or state farms to the members of these farms. Only exceptionally did the allocation of shares mean the actual exit of the recipients from the large farms. In these cases, individual households or groups of households took the land corresponding to their shares and set up independent farms. In other cases, larger groupings separated, resulting in the division of a large farm into a number of smaller units. In still other cases, there was little visible change at all. In Russia, Ukraine, and the Central Asian republics, for example, the new property structures were based on the transfer of ownership to individuals, but they generally did not result in the dismantling of large farms. Instead collective and state farms were restructured and re-organized into the legal forms of companies or cooperatives whose shares are held by the former workers, but managed by essentially the same managers or agronomists. In some countries, such as Kazakhstan, the pattern varied from one region to another. Nonetheless, in all cases, the new farms must adapt to the emerging market environment.

The differences in the experience of privatization are due to various factors: the political balances at the time of reform, different legal practices, cultures, risk perception, institutional endowment, previous existence of peasant economies, availability of marketing networks, social memory of pre-socialist land ownership, etc. As a result there has developed a spectrum of outcomes from a previous order of relatively homogeneous rural sectors under the socialist regime. The agrarian structures emerging over the region vary from one pole, in which a large majority of the rural population becomes family farmers of differing size and wealth, to another pole, in which large farms are transformed into commercial estates on which new "owners" continue to be employed nominally as workers, but may spend most of their time as subsistence farmers on their individual household plots. In between—and this describes the majority of cases—large corporate units, recreated as production cooperatives, joint-stock companies or other legal entities, coexist with new smaller family farms. Different points on the spectrum of ownership and management, have different implications for associated social problems and require different policy interventions.

### *The Social Development Objectives of Land Reforms Were Varied and Inconsistent*

*The Soviet experiment forcibly converted peasants into workers. Consequently many people living in rural areas after decades of Soviet experience did not have the range of skills to transform themselves into independent farmers and operate in a competitive environment without reliance on the state and the collective system .*

Historically, land reforms generally aim to establish rural societies dominated by family farms; they divide up the land and make previously landless peasants into owners. They aim to appease a landless peasantry and to prevent political discontent or rapid outmigration. Ownership of the land by those who actually have to work on it is socially and politically stabilizing. When previously landless peasants become owners, they become political allies of those who instituted the reform. Furthermore, propertied peasants are reluctant to abandon their farmsteads for uncertain prospects in the city; this factor slows down outmigration. In other words, if the distribution of land is successful it creates the legal and social background for an agrarian sector based on the use of family labor for the production, management and marketing of agricultural output. Land reforms generally aim at this outcome; successful examples of land reforms in history have all achieved the goal of establishing rural societies dominated by family farmers.

The political and social goals of reform in many CIS countries were not all clearly elucidated, and not necessarily consistent, however. One objective was to dismantle the entire structure of central planning, which had sought to supplant the market, but the social implications of this radical change were not clearly identified. This is not surprising because the rural population of the socialist bloc could not be characterized as a landless peasantry which expected to become owners of the means of production. Rural people were workers, not peasants, and an autonomous demand for land distribution was certainly not on the agenda. With the exception of some Central Eastern European states, the private memory of being a peasant did not exist; and without the village institutions to reinforce it, the collective memory had in all likelihood disappeared.[3] In some areas there was no previous peasant tradition at all, thus nothing to remember. From this point of view, the Soviet inspired social transformation in the countryside had been unique in providing a singular example of modernization of the peasantry and rural institutions.[4]

Historical experience elsewhere suggests that the modernization of the rural population has occurred through two concurrent transformations:

- Agriculture became economically integrated into national and world economies, thus inducing changes in economic and social behavior and attitudes. This is generally referred to as the transition of the peasantry from subsistence orientation to commodity production which, in many cases, funded the process of industrialization.

- Commercialization, mechanization and integration into the larger economy and society have resulted in migration from the rural areas to the cities. Peasants either became farmers and embarked on a process of social change as they continue to live in the countryside, and/or they migrate out to urban areas, where they transformed themselves into workers or were otherwise employed.

In the case of Turkey, these trends have been observed throughout the country since the early 1950s. Had land reform been carried out in Southeastern Turkey, the chances for modernization on the basis of market integration would have been high. In its absence, the Southeast emerged during the last decade as the region yielding the highest number of rural migrants to the bigger cities. This trend stabilized only recently in certain parts of the region, due to new economic opportunities brought in by irrigation. The process of social change in the Southeast is a particularly accelerated example of transformation where the irrigation technology changed the parameters of production and introduced the possibility of more labor-intensive commercialization. This mitigated the need for massive outmigration and reversed the trend.

By forcibly converting peasants into workers, the socialist model provided a unique experience in the modernization of the rural population. In this model, there was neither integration into market networks on the basis of household economies, nor modernization of the rural population by becoming workers in the urban economy. Instead, the rural population become regulated, modern workers in the countryside, within the plan and the confines of the model envisaged by the state. This move at once transformed social and economic lives, separated the spheres of production and social life, disjoined economic decision-making in the household, and devalued the importance of the family and the village by designating the government as the provider of social services. Thus, without having to migrate to the cities, peasants became workers in a highly capital-intensive agriculture; they also enjoyed the economic and social amenities of

---

[3]       Gavin Kitching. 1998. "The Revenge of the Peasant? The Collapse of Large-Scale Russian Agriculture and the Role of the Peasant 'Private Plot' in that Collapse, 1991-97." *Journal of Peasant Studies* 26(1):43-81.

[4]       David Kideckel, ed. 1995. *East European Communities: Poland – The Struggle for Balance in Turbulent Times.* Westview Press.

urban workers. They were guaranteed security; education, health services and retirement benefits; and stable employment with regular hours. In terms of social indicators, differences between urban and rural areas were much less than in the non-socialist developing world. These achievements were unusual compared to the gradual modernization of the peasantry in the rest of the world, which did not occur fully unless peasants migrated to the cities. It is not surprising, therefore, that when the Soviet system collapsed, there was low initial demand by these workers to become independent farmers where they would have to rely on individual skills, family structures and divisions of labor, and on social capital appropriate to peasant commodity production. Not only was this alternative undesirable for most of the rural population, but it was also obvious that the skills, demographic make up, and rural social structures within which a peasant economy would be embedded did not exist.

### Social Inequity Increased with Economic Reforms

*Agrarian reform provided mixed lessons about large and small-scale agriculture and the package of support measures that were necessary for each.*

The justifications for land reform in the ex-socialist bloc were generally in terms of economic efficiency, rather than social goals. Aside from the fact that a collective agriculture relying on central planning could not survive if the entire socialist system was dismantled, the economic reasons that were proffered in favor of land reform stressed the efficiency of private ownership and market regulation, rather than concerns for equity, landlessness, social stability and rural-to-urban migration, for example. While the arguments for the superiority of individual incentives may be valid, the literature about the economic efficiency of scale in agriculture is equivocal. Soviet collectivization was based, at least in part, on the assumption that there were increasing returns to scale in agricultural production. There are still arguments in favor of larger scale agriculture based both on technology and market considerations. The majority of the contributions to the debate, however, point to problems of labor management in large-scale agriculture, particularly when labor-intensive technologies are employed.

In the real world, the conclusive argument in favor of smaller scale agriculture derives from the organization of the labor process. Small producers are generally family farmers, one characteristic of which is the willingness and ability to self-exploit to a degree hired workers will not tolerate, unless they are bound by extra-economic coercion. Thus, family farmers survive in spite of adverse conditions because they are willing to work at rates of remuneration and under conditions that would not be acceptable to wage-earning employees. The argument in favor of small scale is bolstered by the observation that agricultural technology has evolved in a way that permits application in smaller doses. Most of the modern inputs—irrigation, fertilizer, insecticides, and high-yielding varieties of seeds—are divisible and do not accord any technological advantage to larger units. In the case of machinery, the practice in most peasant economies suggests that the bulkiness of inputs such as harvesters may be overcome through the functioning of machinery rental markets as long as markets permit reasonable participation for small producers. Thus, a land reform dividing up the land does have the potential of releasing 'peasant energies' without jeopardizing productivity. Such decentralization of production decisions would be rational in the overall scheme of things if large-scale organizations are believed to be inefficient due to the deficiency of the market environment and decentralization of production would mobilize the surplus labor potential in the countryside.[5]

Given the labor argument and the machinery consideration, the conclusion follows that family farming will have more of an advantage over large-scale farming in labor-intensive crops and technologies. Any land reform that divides up large farms will thus be economically successful if the new smaller farms also switch to higher value crops, which are generally more labor-intensive. In other words, the historical, political and social structural arguments which are deployed in the explanation of agrarian structures should be complemented by a technological awareness of the relationship between crops and relations of production. This compatibility was observed in socialist agriculture where a division of labor existed between household plots and collective and state farms: the former specialized in labor-intensive garden crops and animal products, while the latter concentrated on grains and feed crops and commercial production of cotton, mostly conducted in a technology-intensive manner.[6]

The Turkish case is an even more revealing example. The historically and politically determined relations of production in the Southeastern region have been characterized by an unequal distribution of land. Large owners cultivated their land using machinery and with the help of a small number of permanent workers. They had recourse to a larger number of seasonal workers derived from the landless or the small-owner population, when needed. Hence, as in the socialist model, there was a complementarity between large and small units, except that the landless population and small farmers were not salaried workers in the large farm sector, thus the income distribution was severely skewed. The land distribution and the prevailing relations of production dictated that the dominant commercial sector would be the large farms. In any case, as long as the region's agriculture was based on rain-fed farming techniques, there was no opportunity to switch to more labor-intensive crops that would favor household-scale farming. When irrigation was introduced to the region, a shift to seasonally labor-intensive cotton production took place, maintaining the dominance of large farms. However, given the agronomic conditions, it would be economically rational to diversify and switch away from cotton to other labor-intensive crops, such as vegetables and fruits, and to activities such as animal husbandry. This envisaged shift, however, would require the owners of labor, the households, to gain access to land and resources. In other words, considerations of economic efficiency required a land reform to transfer the ownership of land to those who would cultivate it with labor-intensive techniques.

The Turkish government failed to carry out a land distribution, however, thus there is now a growing conflict between the relations of production and the potential for higher value added. Large landlords persist in cultivating cotton, which enjoys increased value added due to irrigation, but has the added advantage for landlords of a technology that perpetuates large units of operation. In order to preserve their economic domination, large owners have to continue growing cotton. If

---

[5]         David Turnock. 1998. "Introduction." In David Turnock, ed., *Privatization in Rural Eastern Europe: The Process of Restitution and Restructuring.* Cheltenham: Edward Elgar.

[6]         In the USSR almost all rural households had what we are calling a "subsidiary farm" consisting of a plot near the house and another away from the house. During the last decade they provided a fourth of the family income of collective workers. Csaba Csaki, Kenneth Gray, Zvi Lerman, and William Thiesenhusen. 1992. *Land Reform and the Restructuring of Kolkhozes and Sovkhozes.* The World Bank.

they switch to labor-intensive crops, such as vegetables, they will lose their ability to control the production process or to manage their farms; they may have to rent out their land and become rentiers. We might expect the result the eventual outcome in Southeastern Turkey to be a gradual move toward more lucrative labor-intensive crops, cultivated on the basis of family farming. It is not possible as yet to predict the exact manner in which families will gain access to more land—perhaps through an increase in the amount of leased land or through the workings of land and credit markets. In addition to increasing overall productivity, such a de facto distribution of land, will also be a factor retaining the population in the region—a tendency toward modernization of the rural population implemented through economic integration.

Economic rationality derives not only from the division of land into smaller household units, but also from the institution of a land market. From the point of view of economic efficiency, property rights have to be transferable so that inefficient or unwilling producers will sell or rent out their lands or their land shares. The existence of land markets allows part of the rural population to leave the countryside and those who want to remain in agriculture to increase their holdings. The accumulation of land in the hands of more efficient producers is a prerequisite for the capitalist path of concentration, which assumes that technological change, efficient production and modernization of agriculture become possible through increasing scale and elimination of inefficient producers. In other words, according to this argument, an internally stratified rural economy in which land is held by those who use it more efficiently is preferable to one in which equity is the major concern. Agriculture will then boast generally larger units of ownership, in which some former farm members are either reduced to agricultural workers or they migrate out. Competition in selling products and acquiring property would impose the rationality of the market on the agrarian economy. Entrepreneurs would behave according to market signals and efficient outcomes would result.

Because this argument conflicts with the social and political objectives that land reforms often pursue, the institution of land ownership for the rural population, land markets have been much disputed in the former socialist bloc. Even without the transfer of rights on land, however, rational economic behavior requires land tenure rights to be secure so that individual owners may feel confident to invest in, maintain and improve their property. For conventional credit markets to work, property rights have to be secure so that the owners may establish collateral. According to institutional analysis, enforceable property rights and secure tenure are preconditions for the efficient use of resources, even if there are prohibitions to the transfer of ownership through land markets. There is also, of course, an older tradition of political theory which argues that good property rights make good citizens, on the assumption that property rights protect individuals from the arbitrary impositions of the political authority and lead to other civil rights. This leads to the willingness of property owners to participate in decision-making regarding public matters.

### *Social Development Objectives Were Not Necessarily Reflected in Legal Frameworks*

*Land distribution took place on the basis of mixed criteria with differential gains for social groups.*

When the practice and the prospects of land reform in the CIS first came on the agenda, researchers concentrated on the legal aspects of the transfer of ownership to families. The privatization process has taken place in different ways in different countries, and the methodology of land distribution has led to varying consequences. In Eastern European countries restitution to the previous owners was accepted as the principle of privatization. In Latvia, for example, the attempt was to reestablish the pre-World War II farming structure, itself a result of land reforms between the World Wars that divided up large estates. Former owners of the land and their heirs claimed their shares, and by 1998, 85 percent of the agricultural land was in the hands of family farmers.[7]

One problem associated with restitution was its temporal inequity. Often, land was returned to elderly people with prior claims who were not in a position to engage actively in farming, or to urban-dwelling heirs of original owners.[8] Meanwhile, current residents of villages who were recent arrivals and other users who had no ownership rights dating from the earlier period were unable to claim land. For example, the Roma in Romania and Turks in Bulgaria were employed by the farms during the Soviet period, but could not claim prior land ownership. In some places, claims exceeded available land. In Latvia, claims amounted to 25 percent more land than was available for distribution. In Romania claims on land in the plains exceeded the amount of land available, while demand for mountainous fields remained low. Most places had no mechanism other than litigation to sort out the surplus claims. In addition, cadastral records were less than complete and old landmarks had disappeared thus attempts to sort out contested claims often resorted to complex negotiations and court cases.[9]

In the CIS, land distribution was generally based on allocation of shares to the members of the collective farms.[10] In some countries outside of Central Asia, land shares were distributed during the first round. These were ownership rights to a theoretical amount of land (weighted by the fertility of the soil) which could be used to obtain specific land parcels. Thus, members of the collective who decided to become independent farmers opted to convert their shares to land titles while others—pensioners, students and the non-entrepreneurial—chose to lease or sell their shares to other individuals or farm management. In addition, authorities in some countries encouraged individuals to identify leaders to form partnerships.

In such cases, members were polled to identify the likely leaders and the elected leaders then collected shares from other individuals with the intent of leasing or purchasing land and/or property from them after distribution. These leaders were expected to play a crucial role in the

---

[7]        William H. Meyers and Natalija Kazlauskiene. 1998. "Land Reform in Estonia, Latvia, and Lithuania: A Comparative Analysis." In Stephen K. Wegren, ed., *Land Reform in the Former Soviet Union and Eastern Europe.* London: Routledge.

[8]        David Turnock. 1998. "Conclusion." In David Turnock, ed., *Privatization in Rural Eastern Europe: The Process of Restitution and Restructuring.* Cheltenham: Edward Elgar.

[9]        K. Verdery. 1998. "Fuzzy Property: Rights and Power in Transylvania's Decollectivization." In J.M Nelson, Charles Tilly and Lee Walker, eds., *Transforming Post-Communist Political Economies.* National Academy Press, Washington, D.C.

[10]        Under Belorussion law most agricultural land is allocated to cooperatives, companies, and other large agricultural enterprises in permanent use. The land is not allocated to enterprise workers or other rural dwellers in any form resembling a land share system and only a small fraction has been allocated to citizens for use on private farms (Rolfes 1998:2).

breakup of the collective farm by providing a release for individuals who no longer wish to be involved in direct agricultural production. This created an atmosphere in which those who did not conform to the norm of entrepreneurial farmers (able-bodied males) shied away from independent enterprise. They were encouraged to sell or lease their shares to the leaders even before they converted them to claims on actual pieces of land. In fact, some pensioners and female-headed households did yield their shares in this manner. The alternative was to remain in the fold of the old organization, which became joint-stock companies, production cooperatives and the like. This alternative offered the possibility of retaining shares without having to become active or entrepreneurial.

There were inevitable inequities in this method of reform as well. The distribution of land only to workers on the state farms excluded some occupational groups.[11] In some countries, workers employed in social sectors—schools, hospitals, kindergartens and such—were not eligible to receive land certificates, or they received partial land shares without property shares, even if their lifelong careers had been tied to the farm. When social services collapsed, these employees not only lost their jobs, but were also denied access to other income opportunities that could be generated through landholdings. In other countries, such as Kazakhstan, employees of social services received land shares, but not property shares.

A greater inequality has concerned women, both in the laws and in their implementation, especially in Central Asia. In some countries, laws governing the distribution of property do not discriminate against women, but discrimination may result from local implementation. Under Russian law, individuals, not households, receive land shares, and their land share certificates must be registered individually. Such a legal provision provides formal protection for women shareholders. However, Russian law also requires peasant farms to be held in joint ownership and registered under the name of the head of household; the emerging gender inequities have been noted by many. In some other newly independent states, women did not enjoy equal rights under the law. Often land rights are recognized vis-à-vis the household unit despite the continuity of cultural constructs associated with parentage and marriage, which may leave women with few rights upon divorce and widowhood. Customary use rights have changed with restructuring and women's access to land also varied depending upon the quality of land.[12] In Tajikistan, women are systematically excluded from the land distribution process; thus an unmarried woman was left with no resources unless she was head of a household or belonged to a male-headed household.[13]

---

[11]     In Tajikistan, many people, particularly members of vulnerable groups, are still unable to access land due to them by law. Other groups who desire are prevented from doing so by law unless they were members of collective or state farms during the Soviet period. Melisa Bodenhamer et al. 1997. "Returning the Roots: Rocky Renascence of the Dekhhan Farmers in Tajikistan." The World Bank.

[12]     Jeanne Koopman et al. 1998. "Gender Issues in Farm Restructuring in Uzbekistan and Kyrgyzstan." ICRW. Washington, D.C.

[13]     It is reported that the "re-traditionalization" in Central Asia encouraged by the "official policy of indigenisation " is strengthening the role of traditional social fabric—the informal networks of solidarity based on clans, extended families and neighborhoods—and the emphases of the patriarchal system and the social control of male elders. This, in turn, may mean the further erosion of women's property rights although the basic elements of the traditional structures were largely unaltered among the ethnic Central Asians in rural areas. Olivier Roy. 1999. Re-traditionalization of Central Asian Societies: Implications for Privatization and Gender Issues. Mimeo.

Kyrgyz law provides for only one land share certificate to be issued per family; the law is unclear as to whether individual members have a right to withdraw, sell or lease their land share. In Kazakhstan, the household is the legal unit and the head of the household controls land transactions.

In the Turkish case, the project designed to increase productivity on irrigated land has had the unexpected consequence of strengthening women's claims to land. In a situation where land holdings are dispersed in several plots for each farmer, consolidation of these plots in larger holdings was considered to be an effective measure to facilitate the adoption of land leveling and irrigation technologies. Consolidation has indeed proven to be a cost-effective measure to increase productivity. An unexpected benefit has emerged from the willingness of some women, who heretofore had implicitly accepted the de facto transfer of their ownership rights to their husbands or brothers, to now claim their property shares. In other words, for women, cadastral registration has asserted a right that they had abdicated.

With some exceptions, the legal framework under socialism generally excluded private ownership of land. Thus, new laws were needed for land reforms to achieve the goals as defined by advocates of capitalist development. Despite extensive legal aid and pressure from western agencies, it was not easy to institute such legislative change. Political concerns about the transfer of land to non-agricultural use, to foreigners, about food self-sufficiency and the dispossession and expropriation of the rural population were reflected in legislation. Thus, in many countries, measures were taken to restrict markets in agricultural land either temporarily or permanently.

The legal arrangements which underpin the land reforms in each country have peculiarities deriving from history and from the balance of forces at the time of legislation. Some summary judgements can be made:

- Long-term leasing was easier to work into legal systems than outright sales by the state. As a result, most peasant farms have formed on such leased land.
- Governments have been concerned that urban people would speculate in agricultural land. Fear of this speculation prompted legislation that restricts eligibility to purchase, limits the sale of land shares to other members, and specifies the amount of time a property must be held before it can be resold.
- There has been a reluctance to sell agricultural land to companies, and especially foreign companies. Again, in many countries private property is not a newly established category, and legal systems restrict ownership of village land by foreigners or companies.

In Russia, for example, private ownership of farms was established through presidential decrees, rather than formal legislation. After several versions of the law regulating the market for agricultural land, a compromise was reached which seeks to ensure that agricultural land remains in farming, and that there are no speculative purchases. This 1997 agreement between the Parliament and the President's office prohibits foreign ownership and sets a ceiling on the size of holdings by individuals. Even without overt political opposition to land markets, the process of organizing the legal and administrative infrastructure is slow and difficult, especially in the case of new forms of ownership. Partly for this reason, long-term leasing from the state has become a prevalent mode of acquiring land and may be considered an efficient transitional compromise.[14] It seems that in Russia, the application of the legal framework for the land market will increasingly be determined at the local level.[15]

With the exception of Azerbaijan, the Central Asian republics have been particularly reluctant to institute private property in land. This reflects both a paternalistic attitude towards rural people and a convergence of self interest by local officials and old farm managers. Turkmenistan is the only country with a constitution that formally recognizes private ownership of land, yet this has done little to spur land reform or develop a land market. In fact, according to law, privately owned land in Turkmenistan may not be sold, given away as gift, or even exchanged. In the Kyrgyz Republic and Kazakhstan, private ownership of agricultural land is not allowed; however, up to ninety-nine year leases are allowed and these may be sold, purchased, bequeathed, gifted, or mortgaged.[16] Following a referendum in October 1998, however, Kyrgyz law is being changed to permit private ownership. Similar legislation was under consideration in Kazakhstan, but discussion was suspended recently due to the protests of women. Uzbekistan is unique in not instituting any scheme for privatization, even on paper. Here the concession to reforms has been through the restructuring and reorganization of large farms, and, of course, the expansion of household plots. Thus, Uzbek farmers have not received shares in the large farms, nor have they been able to form peasant farms.

In all the cases reviewed here, there was concern that the newly distributed shares would not carry sufficient legal status to grant new owners either security of tenure or the possibility of exit. In most part these concerns have not proved valid. During the initial stages of land reform there was a great deal of confusion as to how it was being implemented and the targeted population of former workers did complain about a lack of information. Had there been ways of disseminating information at that time many, of the recipients of shares might have made better choices about how to use them. Kazakhstan illustrates this. When the Ministry of Agriculture started a concerted information program to acquaint farmers with their rights and responsibilities, a huge wave of people left the old farms to become independent farmers. In retrospect, however, the much-deplored tendency of share holders to decide to remain within the fold of the large farms seems understandable given their social and material constraints. It was not only the lack of information, but also insecurity of tenure, graft and corruption of the authorities that constituted the principal obstacles to the establishment of an independent farm sector. Rather family conditions, pre-and-post transition forms of peasantry, lack of start-up capital and technical knowledge and the market environment have been the underlying reasons why independent farming has remained subordinate to large units.

Since the reforms, there has been some adaptation to the new conditions. The family, extended family and other social ties are reactivated to cope with the transition from socialist work

---

[14]     Roy Posterman and Tim Hanstad, eds. 1998. *Legal Impediments to Effective Rural Land Relations in ECA Countries: A Comparative Perspective.* The World Bank.

[15]     S.K. Wegren and Vladimir R. Belen'kiy. 1998. "The Political Economy of the Russian Land Market." In *Problems of Post-Communism.* 45(4):56-66.

[16]     Renee Giovarelli. 1998. "Land Ownership." In Roy Posterman and Tim Hanstad, ibid.

to peasant households.[17] At this junction of transition, for example, the elderly are sought after both because they receive pensions—regardless of how small the amount may be—and because, in some countries, they are entitled to larger property shares of the state farm. In Azerbaijan, rural households who are able to pool the assets of the elderly with younger generations fare better than other types of households. In Russia, men who are able to obtain "brides" have a better chance of surviving than those who live alone.[18]

### Unclear Common Property Rights Increased Social Vulnerability

*Agrarian reform put pressure on maintenance and sustainability of common assets such as grazing lands and irrigation systems.*

Agricultural land that is cultivated may exist under the ownership of an individual or a household, and this may indeed yield a greater efficiency in terms of economic outcomes. The same is not necessarily true in the case of natural resources which are in the form of commons or public goods. In fact, irrigation water, grazing land, and forests have posed problems in the land reform process. In some CIS countries, the process has focused on individualizing and/or privatizing rights to arable land, not on defining rules governing widely shared or so-called "common property" resources such as pastures or grazing land.[19] The transfer of common resources from collective farms to the state, which lacks funds for their maintenance, have increased the need for associational forms of commons management—users associations and cooperatives. However, several CIS countries lack clear legal ownership regimes or rules regarding common resources. A good example is animal husbandry in Central Asia.

Livestock production in Central Asia depends on natural pasture, and rainfall limits pasture production in any one place. Households or groups that engage in animal husbandry typically respond to these constraints by moving their herds within grazing territories—acting like pastoralists. Pastures can rarely be divided into viable lots for individual households; given variable rainfall conditions, the area available to each household would not be sufficient to sustain animals and farm family throughout the year. The maintenance of pastures and grazing lands might appear relatively less important in transition given the significant decline in livestock ownership throughout ECA. However, it is reported that the poorer households rely increasingly on livestock management for subsistence purposes. In the Kyrgyz Republic, for instance, uncertainties surrounding property rights in agricultural land and grazing resources created a situation in which independent farmers have started to seek security by increasing their livestock.[20]

---

[17]     The disruption and perhaps the permanent disappearance of social services in rural areas is another factor which strengthens reliance on the family and creates the need for a new household division of labor in tasks such as looking after children and the elderly, and even cooking, to substitute for the meals that the collective farm used to provide.

[18]     Sue Bridger, ibid.

[19]     Giovarelli, op. cit.

[20]     The absence of clear legal rights and responsibilities governing the use of lowland pasture lands, in the Kyrgyz Republic is becoming a problem. As a result of the increase in herd sizes and intensification of grazing in the lowland pastures conflicts over grazing rights are arising and the pastures are starting to degrade rapidly. The state has

With sharp falls in incomes and the restructuring of many livestock state farms, most households have difficulty getting commercial feed, thus increasing the reliance of the poor on pastures. Moreover, due to high fuel prices, the practice of transporting livestock to highland pastures during the summer has practically ceased. As a result, pressure on lower pastures and grazing lands surrounding the inhabited areas has increased dangerously since independence. Since privileged segments of the rural population and the collectives have better access to high quality feed, it is the poor who are most dependant on pastures and grazing lands. Therefore, there is a need to develop effective ways to maintain pastures and grazing lands, especially using participatory mechanisms, and to encourage associational forms of commons management, so as to ensure sustainability.

In other areas, however, the response to new realities is more ambiguous. For example, in some countries, independent farmers were located in one service unit so as not to disrupt large-scale automated sprinkler systems; the small units were refused irrigation service until they organize themselves to make use of the existing technology. In another larger system, in the same country, officials are making great efforts to develop complementary delivery systems using small pumps to be able to serve one-hectare outlets efficiently. Most countries have introduced some sort of charge for water, but payment and collection are problematic and maintenance of the systems is difficult. Also, only a few countries have transferred ownership of the infrastructure to users, who are less than eager to assume ownership and/or management of largely dysfunctional systems. Nevertheless, there are tendencies for cooperation inherent both in the pre-Soviet communal systems and in Soviet collective practices that provide social capital to build on.

In the Turkish case, large irrigation canals are owned by the state agency which is responsible for building and maintaining them. Local water user associations manage small canals and decide on the allocation of the water among users. Payment for the water is problematic and the prices charged do not fully cover the costs of building or maintaining the infrastructure. Where land distribution is uneven, the beneficiaries of the irrigation scheme are disproportionately richer landowners. If they had to pay more realistic prices and the revenues were used to improve the lot of the landless and the poor, there would be a net improvement in total welfare.

Inequity in access to irrigation appears along ethnic and gender lines. In countries where members of certain ethnic groups are concentrated in enclaves, their circumstances are improved or worsened according to the priority they are given in water distribution. In the Kyrgyz part of the Ferghana Valley, for example, differential access to irrigation water has exacerbated ethnic conflicts as problems of upstream/downstream access are interpreted in ethnic terms by residents. Similar situations are observed in other countries that share the valley. The reported concentration of female labor in garden plots may also be a factor that reduces the potential for the benefits of irrigation improvements to accrue to women. Although the role of these plots is recognized, there

---

maintained ownership of the land and allocated management responsibilities to the village governments, with few guidelines for appropriate practices or requirements to formalize rules.

is nevertheless a de facto distinction between the new peasant farms and private plots, with the former getting greater recognition in some countries.[21]


## Social and Structural Dynamics of an Ongoing Transition

*Agrarian reform in the former socialist bloc did not result in one type of farm but a range from family to corporate farming, calling for different policy interventions.*

Although it is crucial to gauge and understand the current situation, it must be remembered that the transition is still taking place; it is by no means complete. It is difficult to understand and theorize about the transition in agriculture because it consists of a number of moving targets which are inconsistent and contradictory. Many patterns exist, but is perhaps too early to ascertain which trends will ultimately dominate in different areas.

The 'agrarian question' was the label given to the debates at the turn of the century concerning the future of agriculture and prospects for the peasantry.[22] As the question was asked by scholars who were engaged in attempts to draw political lessons from the transformations underway in rural society, the corresponding analysis was often conducted in class terms. Were farmers evolving into capitalist landlords; were small peasants being expropriated and becoming landless workers; were ancien-regime landowners turning into commercial-minded landowners? Some participants in the debate welcomed a social transformation toward capitalist-like structures in which the rural society would be polarized between landlords and landless proletariat. They considered this to be progressive, both for purposes of political action and in terms of its potential for economic and technological change. Others, who were labeled populists by the first party, favored a rural society dominated by family farmers. They argued that agriculture had specific technical attributes which allowed family farmers not only to be as efficient as larger units, but also to adopt new technology more rapidly. Their primary concern, however, was to avoid social upheaval in the rural society, which would be caused by a transformation toward a polarized capitalist structure.

These two paths to a modernized rural society were labeled as the capitalist path and the peasant path. The history of agrarian transformations in the twentieth century has favored the peasant path. A combination of political economic factors, and a wave of land reforms in the developing world, led to a situation where the survival of large ownership with landless laborers has become uncommon. The transitional cases considered in this volume illustrate two tendencies in the countryside: in some places such as Azerbaijan and South Kazakhstan, often reflecting an historical tradition of sedentary agriculture. In other areas, large industrial farm units created

---

[21]     This is noted for Russia by Sue Bridger. 1999. "Rural Women and the Impact of Economic Change." Paper distributed in the Seminar on "Making the Transition Work for Women in ECA." The World Bank. This paper also quotes the head of the Agrarian Institute of Pskov to observe that "no renaissance, no reform is possible if we do not know what is going on at the level of the family...In the next four to five years...family farms will en masse become one of the most important producers in Russia, i.e., our major providers." (p. 53).

[22]     Karl Kautsky. 1988. *The Agrarian Question.* London: Zwan. (original 1899). Athar Hussain and Keith Tribe. 1981. *Marxism and the Agrarian Question.* Atlantic Highlands, NJ: Humanities Press.

during the Soviet period continue to dominate.  As already mentioned, the socialist regimes emulated the capitalist path in terms of forms of organization and size of operations, but on a huge scale.  To some extent, post-Soviet land reforms are supposed to change this situation and divide up the land, thus bringing the agrarian structure closer to the global norm.  In the Turkish case, technological change was also supposed to redress the anomaly of social backwardness associated with landlordism.  In neither case, however, has the transition advanced sufficiently to be able to predict with confidence that family farms will dominate in the near future.

The former socialist countries developed a spectrum of outcomes, in part related to the rural diversity that existed under the socialist regime.  Different types of farms coexist within the same economic universe.  The old bi-polar structure of large farms and household plots has yielded to a more continuous distribution as large farms divide up into smaller components, a new sector of intermediate scale family or peasant farms, and a growing number of household plots.  In fact, given the slow pace of privatization in most countries, it is remarkable that such a wide variety of farm types and production arrangements has emerged.[23]  Various new outcomes will surely develop, depending on the degree of development of land and credit markets, the opportunities that exist outside the farm sector and the pace of privatization and demonopolization in agricultural services.

Currently, the agrarian structures emerging over the region vary from one pole, where a majority of the rural population become family farmers of differing size and wealth, to another, where large farms continue to operate, albeit with different ownership patterns on paper and, in many cases, changed functions.  In Poland and Hungary, an agrarian structure similar to Western European norms of family establishments is evolving.  At the other extreme, people remain employed as workers in the same large estate with which they were associated before privatization, often accompanied by economic crisis and dire poverty.  In between, we find the creation of larg units (albeit smaller than the state or collective farms of the preceding era) as new cooperative farms or joint-stock companies of shareholders.[24]  Household plots also exist over the entire spectrum.  Where independent family farming becomes dominant, household plots and peasant farms exist in a continuum.  Within the household plots as well, there is a wide variety of orientation.[25]  Farmers in the more commercially advanced areas use these plots to produce directly for markets, almost as contract farming.  In the crisis-ridden areas far from the cities, the household plots are the primary source of subsistence for families.  In contrast, when large farms dominate, there is a bipolar distribution of scale between the industrial farm units and household plots, with an insignificant intermediary category of peasant farms.  These cases cover the spectrum of

---

[23]　　　Max Spoor. 1997. "Agrarian Transition in the Former Soviet Union: The Case of Central Asia." In Max Spoor, ed., *The 'Market Panacea', Agrarian Transformation in Developing Countries and Former Socialist Economies.* London: Intermediate Technology Publications. Pp. 95-111.

[24]　　　There are important sub-national or regional variations in this regard. In a recent analysis regional factors have been shown to determine income and income mobility as well as distinct patterns of adjustment. Tatiana Yu. Bogomolova and Vera S. Tapilina. 1999. "Income Mobility in Russia: Scale and Intensity." Mimeo.

[25]　　　The conditions of household production was improved in Russia when mandatory obligations to produce for the state and other forms of payments were abolished. E. Serova et al. 1999. "The Impact of Privatization and Farm Restructuring in Russian Agriculture." Mimeo.

ownership forms which each have different inequality implications and call for different policy interventions.[26]

### *Large Farms and Associations—Old Forms and New Social Roles*

*Large farms continue to exist because people often see no benefit in dismantling them and are reluctant to relinquish the fabric of support they appear to provide—either in material or psychological terms. Newly independent farmers also seek to form voluntary associations. The model of small, independent farms may run against tradition in some areas and may not be viable in the face of current economic realities in others.*

In many CIS states, there has been no large-scale exodus of people from collective structures who subsequently set up independent farms. This has been due in part to the subversion of reforms by regional and local administrations, peoples' own negative assessment of the results of privatization and the basically unfavorable administrative and economic environment for family farming. In many countries, during the first wave of privatization, starting in the late 1980s, a small number of entrepreneurs left state farms to set up independent units. In most cases, these people were the technical staff of state and collective farms. Since then, it has become more difficult and risky to leave the large units. The continuing decline in production and welfare, the decrease in institutional support or institutional opposition, the depletion of capital stock and the lack of financial and technical resources have all helped to dampen whatever latent entrepreneurism may exist. Many rural people are also convinced that the old regime was better. Of course, that does not necessarily mean that they expect to be able to return to the past, but it does cloud their concept of a future. While it is true that private farming has generally been found to be more efficient than larger industrial farms, it is also the case that setting up independent family-owned farms requires initial assets or available credit; particular household demographics; adequate information; access to trading networks; and willingness to take risks. Except where the population density is high—where farms are located near towns—and where crops tend to be more diversified and generally commercial in nature, the lack of these factors have been sufficiently daunting to cause much of the rural population to remain in larger management structures where they continue to function as shareholder-employees. Meanwhile, the overall situation has deteriorated in most rural areas of the CIS. In some cases, such as Kazakhstan, however, some farm employees are also prompted by the seriousness of the economic crisis to set up independent farms as a defensive measure. That is, once they understand that they indeed can exit from the privatized former collective and state farms which offer them virtually no prospects for future gainful employment, they escape before the total collapse of the enterprise.

It has been argued that privatization in the former Soviet Union has been slow because governments fear various possible outcomes. One factor that was much discussed during the early days of privatization, especially in the context of Central Asia, was the fear that ethnic conflict might arise as a result of differential access to the land. Another argument was the inertia and active opposition of the rural *nomenklatura*, who both nationally and locally undermined attempts

---

[26]         They also have different "social costs" as perceived by the people. In Russia, for instance polls indicate that over 60 percent of the rural people are frustrated with the reforms; only 10 percent believe that reform objectives were fully met. M. Stroyev. 1997. "The Conception of Russia's Agrarian Policy in 1997-2000." Mimeo.

at reform and tried to increase their own power through the reforms. There is no doubt that both these arguments are valid to some extent. At the same time, managers of the large farms use their own experience and skills to substitute for largely absent market mechanisms, at least during this crisis-ridden period of transition. All accounts suggest that farm workers would be in a much more desperate situation without the individual initiatives of managers, especially in the more remote provincial areas.[27] Nonetheless, there are also many cases in which the managers have bled the farms, siphoning off resources for their own enrichment.

Fundamentally, however, large farms continue to function. Some are operating successfully in the changing institutional environment; others have survived only marginally, often assuming new roles. No institutions have emerged to take over the roles the large farms play in the rural economy, including providing social services. Pensioners, who constitute a significant proportion of the shareholders in these farms, are reluctant to leave the large farms for understandable reasons. Even if they could rely on receiving pensions from the local government, their shares in the large farms provide supplementary income and a familiar social environment. This segment of the population, generally would not want to sell their shares in the farms if they could. In some ECA countries, one pattern for collective farm members has been to trade in (to the farm manager) land rights for a promise of continued wage employment. Similarly, some pensioners transfer their shares or lease them to managers. In other cases, pensioners transfer shares to their children residing in the cities. In turn, they lease the land to the farm and their parents receive the proceeds, if any. Many families maintain formal links to the privatized former state and collective farms to ensure income from multiple sources, in addition to other reasons.

Another critical reason for opting to remain within the former state and collective farms is the hope of maintaining pension rights. It is unclear whether pension benefits continue subsequent to the restructuring of state farms or collectives; nevertheless rural households try to maximize their coping potential by ensuring the continuity of wage employment in large farms. These arrangements underscore the urgent need to address the issue of rural pension systems. The uncertain situation with regard to rural pensions is more than a social problem concerning the elderly; it makes the establishment of independent farming difficult and, in some cases contributes to the concentration of control in the hands of the managerial elite.

Farm workers who trade their land shares for the promise of employment may end up disappointed. Most large farms are in a precarious financial situation. To become commercially viable, they need to reduce their labor force and social commitments. When this happens, farm members who have retained ownership of their shares or leased them to the farm enterprise have the opportunity to negotiate terms for dividends or fees that can result in an income stream. Those who have relinquished their shares, however, risk losing their incomes entirely and may thus need to rely more completely on subsistence farming on their home plots or migrate.

Even if employment and wage payments are not guaranteed, large farms and their management provide some security in an environment of commercial and financial instability and uncertainty and under the conditions of a deep crisis that has characterized the rural economies of

---

[27]    Gavin Kitching. 1998. "The Development of Agrarian Capitalism in Russia 1991-1997: Some Observations from Fieldwork." *Journal of Peasant Studies* 25(4).

CIS countries. The farms operate in ways that are similar to their previous functions: they provide essential services such as the production of costly inputs which the household sector purchases, they operate machinery and they supply transportation services. The household sector is thus able to concentrate, much as it did before the reforms, on higher value added tasks which are more immediately remunerative in the market environment. Hence, there is not much incentive to undermine the large farms since they do not threaten the household sector. Farming families seem content to go on with the symbiotic relationship that is the legacy of the socialist era. This may change if the broader economic environment changes so as to offer greater employment opportunities in the urban sector and making migration out of the countryside even more attractive than it is now. Then there might be incentives both for the rural population to leave and for a more entrepreneurial farming population to attempt to increase their holdings. In the present situation, there is neither a 'pull' from the urban sector, nor a 'push' from the rural; instead coping strategies dictate a conservation of the existing arrangements.

In the majority of the cases, privatization as destatization is a reality: former state and collective farms have been reorganized as cooperative ventures, joint-stock companies or shareholders' associations, even if the mode of operation remains much the same as before. Hence, the labels and the categories have changed along with new legal organization, and the larg structures appear to be likely to succeed in preserving their assets and therefore technological and marketing advantages. The reorganization may involved physical restructuring, as well. In some situations this restructuring takes the form of vertical disaggregation where, for instance, the machinery resources of a farm spin off and become a service cooperative, or the commercial and production management sides may be separated. In other cases, the farm land and assets are divided in to several separate units. Nonetheless, it is likely that farming operations in many countries of ECA will be large, multi-family farms co-existing with small household plots, with some independent peasant farms in between.

The willingness to form associations that work as cooperatives is high in Central Asia. After legally separating from the large farms, many farmers regroup into associations or companies and many group together before they separate. It seems that the larger-scale production systems hold a credible promise for the rehabilitation of agricultural production given concerns over technological and economic efficiency. In Moldova in 1996, for instance, these associations were not simply marketing or purchasing cooperatives; they also functioned as production cooperatives where the agricultural work was undertaken jointly; "an average association of peasant farms managed around 200 ha, implying group membership of 100 farmers per association."[28] In Romania, some 1.5 million families organized in associations work on 40 percent of the privatized land. It is expected that this share will climb higher.[29] In Bulgaria, the former state farms have been dismantled to a large extent; only 7 percent of the land is still under cultivation by the large

---

[28]         World Bank. 1997. *Land Reform and Private Farming in Moldova.*

[29]         David Turnock. 1998. "Romania." In David Turnock, ed., *Privatization in Rural Eastern Europe: The Process of Restitution and Restructuring* Cheltenham: Edward Elgar.

farms. Individual peasant farms and household plots occupy just over half of the land (52 percent) while various partnerships, cooperatives and other intermediate forms take up the rest.[30]

In Azerbaijan, many newly independent farmers also choose to remain in collective arrangements. A majority of them have formed groups along the lines of the work brigades under collective farms and they cooperate in cultivating their land in blocks. Villagers form associations with other households they can trust, often close relatives. Such family affiliations seem for many to become the only way to survive, especially when self-sufficiency and sustainability of an individual farm becomes problematic and almost impossible for the most farmers in conditions of poor financial and technical support.[31] In other cases they form associations based on expediency with farmers whose lands neighbor their own. Armenian private farmers cooperate with others in a variety of activities, the most prevalent of which is irrigation: "more than 60 percent of farms with irrigated land report that they cooperate with other farmers for irrigation and water management."[32] Of course, such cooperation is generally essential in surface irrigation systems, particularly where landholdings are fragmented. Other activities of cooperation include soil amelioration, and machinery use. In the two oblasts in Kazakhstan that were the focus of the SA in this book, only 37 percent of the independent farms consisted of a single family; the rest were of two or more families in joint operation. Some of these farms were created on the basis of former joint activity in the collective farms. In the Kyrgyz Republic, there are production cooperative farms with up to 10 workers who pool their land and livestock shares to form a single farm unit under collective management.[33]

Farmers in CIS countries have been part of large-scale farming organizations for at least two generations; indeed, much of the rural population in Central Asia was forced into state and collective farms without prior experience in sedentary agriculture. Consequently, most rural people in the CIS have virtually no direct experience in self-exploitation in the manner of the peasantry in much of the developing world and Europe. Except in areas of traditional high-value crops production, especially fruits and vegetables, generally on home plots, CIS farmers are not skilled in labor-intensive methods or farmstead management. Therefore, faced with the dissolution of collective agriculture, especially in the grain-producing flat lands of Russia, Ukraine and some Central Asian countries, individual farmers will not be able to maintain the level of capital intensity required for extensive farming. If they opt for independence, it is improbable that, as small independent units, they could compete in the world market at their level of technology, capital intensity and human capital, and without major state subsidies. Associations and cooperatives give them a chance to do so, however; increasing both their social and physical capital. One reasons for farmers to join such associations is to gain access to machines, equipment and other inputs. In this sense associations provide an incentive and an opportunity to increase the scale of operations to

---

[30]   Diana Kopeva. "Transition, Land Reform and Adjustment in Bulgaria, 1997." In Max Spoor, ed., *The 'Market Panacea', Agrarian Transformation in Developing Countries and Former Socialist Economies.* London: Intermediate Technology Publications.

[31]   Ayse Kudat and N. Egemberdi. "Azerbaijan Agricultural Development Project." Mimeo. The World Bank.

[32]   World Bank. 1995. *Land Reform and Private Farming in Armenia.*

[33]   World Bank. 1996. "Staff Appraisal Report: Kyrgyz Republic, Sheep Development Project."

conform to available technology. If the associations retain a participatory structure based on voluntary cooperation and do not develop into rigid managerial structures, they constitute an appropriate response to the pressures for change in the CIS and Turkey.

Cooperative ventures have a long history, and the new forms emerging in former socialist countries may well evolve into more advanced versions of co-operation. While it may be understandable in political and ideological terms that most observers demonstrate a preference for the western model of independent family farms, the different history and context of the CIS inevitably will lead to results other than a mere replication of the western model. Moreover, such biased observations do not necessarily provide appropriate guidance for policy proposals intended to address issues of social development in a social context that clings to the collective production model. Given the likely prospect that large management structures are here to stay, at least in the medium term of a decade or so, social development initiatives should start to focus on distribution, participation, and empowerment *within* the structure of the large collective farms.

Theoretically, participants in large collective units are shareholders—owners of property and land use rights who have chosen to remain within a co-operative structure or have willingly formed associations to re-create such structures. It is important for legal reality to conform to this model. In Ukraine and most Central Asian republics, land is still owned by the state and farmers are given long-term inheritable use rights. This is gradually changing, however. While this may not pose problems to the managers, the ambiguity of the ownership category, and much continuity in management of large scale enterprises, creates problems of participation in and review of management practices. In some countries the new agricultural units (especially joint-stock companies and production cooperatives) are actually kolkhozes or sovkhozes disguised with a new name. The old structure, wage standards, restricted access to information, and lack of participation in decision-making, remain the same. It is important to insure that tenure arrangements among members of co-operative farms are transparent, because members of the new collective groups can only pursue the demands for participation and equitable distribution that are likely to arise if they understand clearly that they are indeed legal owners. In far too many areas, local administrators have obfuscated this fact.

The major risk faced by the rural population who have opted to stay within large structures is the concentration of assets, income and power in the hands of the managerial hierarchy, consisting primarily of the previous managers of collective farms under the socialist system. How will it be possible to ensure that the knowledge and network assets of the managers which may potentially turn into ownership differentials, are in fact subject to checks of co-operative owners? This will require a learning process as well as clear legal provisions that protect ownership rights.

### *Family Farms—New Pioneers or Victims?*

*Large productivity increases have occurred in household plots, either to ensure the subsistence needs of families, to generate income, or both. Household farming peaked, however, as families secured access to as much land as they could cultivate using labor-intensive farming. Further expansion of household agriculture is constrained because of its coexistence with large farms, and the consequent limits to the supply of labor.*

Legal ambiguity as to land markets and private property in the CIS did not preclude the expansion of private household plots, which began under Perestroika. Since privatization, this expansion continued through long-term leasing and the sales and allocation of plots to individual families from the reserve lands controlled by local authorities. This has served to increase the size of already existing household plots. In some areas, this expansion served both to integrate rural families into the market and to provide subsistence to families in an environment where the larger farms were in a crisis.[34]

Such short-term changes do not necessarily point to an identifiable trend, but it is important to attempt to trace out some general lines of development behind the complexity of transition. One discernible line is the greater success of private farms, including household plots, in adapting to the emerging market environments. This outcome is not surprising. First of all, only those farmers who are comfortable with the requirements of commercial farming go into independent production, or significantly expand the size of their household plots. They are the people who have the needed family labor, the energy, the connections, and the entrepreneurial willingness to separate themselves from the collective group and attempt independent operations. Secondly, while larger farms have to rely on larger scale commercial networks for their continuous operation, smaller farms may thrive even on sporadic or informal markets and one-time deals. More importantly, the symbiotic relationship between household plots and large farms which used to prevail in socialist agriculture continues in the market era with the now larger private farms and the reconstituted large farms. Previously, household plots in some areas were thought to be more successful because they were allowed to specialize in high value products and were subsidized with inputs by the large farms. This especially applies to farms near major urban areas. The relationship remains the same: in Russia, while large-farm output has declined, private plot output increased in areas accessible to markets. Private plots concentrate on the production of cash crops such as potatoes, vegetables, animals, and animal products.[35]

In retrospect, it might be argued that an important contribution of private plots was to prepare farmers for post-socialist agriculture. In the transition period, private plots have expanded in size and acquired much importance because they indisputably belong to the families and, because of their small size and labor-intensive nature, it was possible to adapt them quickly to the new conditions. In Moldova, as in other CIS countries, "the first stage of the land reform aimed at the distribution of land for household plots."[36] It more than doubled the total area of household plots: they now occupy 13 percent of the agricultural land by area. The average household plot in 1996 was 0.41 ha, compared to 0.27 ha in 1990, an increase of more than 50 percent. Where privatization has successfully led to the creation of family farms, home plots may merge with the newly created independent farms; elsewhere their size has increased—usually to around three times their original size.[37] Home plots in countries where family farming has not become

---

[34]    In Altay, a major agricultural region in Russia, half the households did not produce for the market in 1995 but 80 percent of the rural households provided food aid for relatives in urban areas (Serova et al., ibid).

[35]    Kitching 1998: 56, ibid.

[36]    World Bank. 1996. "Staff Appraisal Report: Moldova – First Agriculture Project." P.5.

[37]    In cases where size has increased to one hectare or more the plots may even be sufficient for the reproduction of the household if there is a switch to labor-intensive crops and techniques. In Armenian agriculture where private

dominant have so far remained complementary to employment in the larger units, although their contribution to the household budget and their claim to the household labor is, in every case, much larger. In Kazakhstan 95 percent of rural households have garden plots attached to the house; in Tajikistan, they provide subsistence to families who have been left without employment as a result of restructuring. As shown by the case study in Uzbekistan, virtually all rural households—even those who are employed in non-agricultural activities—have home plots and many sell some surplus production in order to supplement their incomes. Therefore it is important to find ways and means to support garden (home) agriculture, including the support of small scale food processing activities. To the extent that women play a large role in garden agriculture, such a support would also imply a positive gender impact.

The vast majority of those who buy land to expand their household plots do so in order to assure subsistence production or to supplement family incomes. Incomes from household plots are subsidiary to other primary employment—employment in larger farms, in most cases--but plots generally are the principal source of family food. This does not imply that the income obtained from a household plot is less than the salary or the income share from the large farms, but merely that the household in question is not ready or willing to abandon the large collective unit or joint-stock venture. In this sense, rather than leading to a transformation of property relations, land markets permit families to expand their household plots and helps conserve the agrarian social order by making it possible for rural families to stay within the large farm structures.

For farmers who continue as shareholders in larger farms, particularly in areas surrounding large cities, the income from a household plot may outweigh a monthly salary.[38] It is the availability of family labor which constrains the expansion of household plots as a second activity, in addition to membership in the collective farm. The question then arises as to why these families do not fully exit from the larger structures, giving up their income from that source in order to expand household production. The answer lies in the continuation of the symbiotic relation. Large farms leave the production of lucrative crops and products to households; they assume the responsibility and costs of marketing products such as milk; they provide transportation and machinery free or at subsidized prices to the families, as well as fodder for animals. Without these transfers, household plot production would be less remunerative, if not impossible.

Given labor constraints, there is a potential trade-off between the success of household plots and the large farms. Since the large farms are unable to pay higher wages their members continue to operate their private plots. By doing so, they continue the symbiotic relationship whereby they are subsidized in many ways by the larger units. This relationship relegates the large farms to service units which provide essential inputs, but do not operate to maximize their own profitability. Thus, they become service cooperatives because their mode of operation is basically determined by social relations, benefiting the private operations of their members and of their managers. It has been suggested that this model of "large farm/private plot symbiosis" may indeed

---

farming has taken hold, "half the farms report between 0.6 ha and 1.8 ha of land" and in districts where there is irrigation and higher population density, the average size of the farm is 1.35 ha (World Bank 1996: 8)

[38]      Kitching 1998, argues that especially animal sales provide household farmers with incomes that easily equal wage earnings from the large farms.

be the peculiar way in which post-socialist Russian agriculture will evolve.[39] This evolution would be an intermediate trajectory that responds to some of the problems associated with the emergence of a pure peasant farming structure. The large farms would perform some of the economic and social functions that are not found in post-socialist agriculture because of the history of the market, the absence of a peasant tradition and the lack of appropriate demographic structure within the ranks of prospective peasants.

In addition to the growing share and importance of household plots, a new category of peasant farms has emerged as the result of restructuring, land sales by the state or successful restitution. These are integrated farms in the sense of being capable of existing on their own, rather than needing to be in a symbiotic relation with the large farms. Their success does not depend on the division of labor that household plots enter into with the larger farms; their incidence correlates with success in dismantling large farms. The countries of the Baltics and Eastern Europe where large farms have been eliminated to a greater extent—especially through restitution of land to the original owners—have the highest proportion of peasant farms. In Russia as a whole, however, family farms account for less than 10 percent of the agricultural area; in Ukraine the figure is 2 percent.[40]

In Central Asian republics the structure has not yet rigidified and the proportion of independent farmers in the rural population is expanding along with the proportion of land under household control. Land in Azerbaijan is undergoing a rapid privatization and many independent holdings have been formed. In Kazakhstan, privatization was implemented through the provision of land-use rights and asset ownership certificates, but large segments of the farming population opted to remain in the framework of large cooperative units. Individual family farms now cultivate 20 percent of the land. In Uzbekistan, the figure is 12 percent, peasant farms are located within the area of large state/collective farms because these are the only irrigated lands. This means managers of large farms and owners of peasant farms maintain a close relationship. These new peasant farmers are said to behave more like household farmers than independent operators. Their numbers as well as average size have grown steadily, and they seem to concentrate on livestock breeding for income.[41] In Turkmenistan peasant farms account for 10 percent of the agricultural land, and 9 percent in Tajikistan. In Russia private farms average 42 hectares per farm and employ three to four workers.[42] In Ukraine, despite legislation that creates private farms through sales to individuals from a special category of land, actual implementation depended on the good will of local administrators. By the end of 1996 there were only 35,000 such farms, averaging 23 hectares in size.[43]

---

[39]      Gavin Kitching. 1998. "The Development of Agrarian Capitalism in Russia 1991-1997: Some Observations from Fieldwork." *Journal of Peasant Studies* 25(4).

[40]      C. Csaki and T. Nash. 1998. *Agrarian Economies of Central Eastern Europe and the CIS: Situation and Perspectives.* The World Bank

[41]      Zvi Lerman. 1998. "Land Reform in Uzbekistan." In Stephen K. Wegren, ed., *Land Reform in the Former Soviet Union and Eastern Europe.* London: Routledge.

[42]      S.K. Wegren. 1998. "The Conduct and Impact of Land Reform in Russia." In Stephen K. Wegren, ed., *Land Reform in the Former Soviet Union and Eastern Europe,* London: Routledge. P. 26.

There are many examples of successful formation of family-farming sectors. By 1998, Poland had 82 percent of the land under private farms and the rest was in the process of restructuring. In Slovenia family farms cover 90 percent of the arable area; all of the agricultural land in Albania is in family farms; and in Armenia, private farms constitute 65 percent of the total arable land. In these countries the majority of the rural population have become household producers. Particular circumstances explain the preponderance of family farms. In Poland, private holdings continued to dominate under socialism; in Albania village communities were quickly recreated after independence; in Armenia, land reform was carried out swiftly and completely, shortly after independence, turning most rural people into landowners virtually overnight. In other countries such as Hungary and Romania, where land under family cultivation is 49 percent and 54 percent, respectively, independent family farming has taken hold in part because of the previous significance of household plots. Under the old regime, household plots covered around 15 percent of the cultivated land in Romania and 20 percent in Hungary.[44] Furthermore, the economic system in Hungary had been revised to accommodate household plots which enjoyed marketing freedoms and the active cooperation of the state farms.[45] This experience provided the requisite background for the success of peasant farms in the 1990s.

The diversification of activity and income sources requires that a market, transportation, communications and other support networks be available to the rural sector. In Eastern Europe, parts of the Caucasus and in areas close to large urban centers elsewhere, these networks exist. In Poland, for example, many agricultural workers have been combining farm work with other economic activity, especially those located near large urban centers. For 23 percent of Armenian farming families, the farm provides less than 25 percent of family income; for 50 percent of the families surveyed, farm income is less than half of household income. Elsewhere, however, diversification has been more limited and slow, although it will accelerate if, for instance, credit becomes available for non-traditional economic activities. Generally, policies supporting off-farm activities in the countryside will contribute to rural development and equity by permitting family enterprises to employ their assets more efficiently. In addition, assisting in the development of small-scale industry in the rural areas will create employment for those with no access to land and reduce migration from the countryside. In particular, agro-processing activities—large-scale or on-farm—and other developments in farm diversification, such as rural tourism, have in fact gained an important momentum in Eastern Europe.[46] It is even speculated that some local industry in the form of traditional crafts and other ancillary activities, may revive in response to a demand from rich urban centers.

---

[43]      Timothy N. Ash. 1998. "Land and Agricultural Reform in Ukraine." In Stephen K. Wegren, ed., *Land Reform in the Former Soviet Union and Eastern Europe,* London: Routledge.

[44]      Z. Lerman et al. 1994. "Self-Sustainability of Subsidiary Household Plots: Lessons for Privatization of Agriculture in Former Socialist Countries." *Post-Soviet Geography* 35(9).

[45]      David Turnock. 1998. "Introduction." In David Turnock, ed., *Privatization in Rural Eastern Europe: The Process of Restitution and Restructuring.* Cheltenham: Edward Elgar.

[46]      Floarea Bordanc, Stanislaw Grykien, Nicolae Muica and David Turnock. 1998. "Aspects of Farm Diversification." In David Turnock, ed., *Privatization in Rural Eastern Europe: The Process of Restitution and Restructuring.* Cheltenham: Edward Elgar.

### *Family Farms and the Village Framework—Emerging Social Structures and Solidarity*

*The village community, which in many parts of the world provides a framework for social relationships and solidarity among households, is still evolving in the former Soviet Union. In Turkey, the village is a stable community with well-articulated internal and external linkages.*

Despite significant variations in the conduct of land reform, there is one consistent outcome in the ECA region: the share of privately managed plots has increased. In one way or another, a sector of independently operated family farms has emerged both through the enlargement of private plots that previously existed alongside the collective sector, and as an outcome of newly created categories of peasant farming. In certain countries in Eastern Europe and in the Caucasus, independent family farms are so dominant that they determine the direction of development of the entire agricultural sector. Such is not the case in the CIS, however. Nevertheless, even in countries where privatization has not been significant and where large farms of the Soviet era survive under different names, there is a growing number of peasant households in which a large proportion of the available labor time is devoted to economic activities organized and carried out by the household itself. The household also derives a larger proportion of its livelihood from such activities. From the point of view of the social universe of the rural population, there are now two simultaneous spheres within which people operate: the transforming collective sector and the growing private sector. New complementarities have developed between these two spheres, as well as new divisions of labor.

Family farms cover a wide spectrum worldwide. At one extreme are the technologically and commercially sophisticated family corporations in the American Midwest and Western Europe, which are served by developed commercial and financial service sectors. At the other extreme are the poor peasants of the Third World who subsist thanks to various family and village social structures which embed their economic activity. To survive, these complex, mixed-farming units require both a diversified family labor force and a favorable commercial environment.

The array of tasks on these farms requires a family division of labor among men, women, the elderly, adults and children. It is generally the case that households which lack one or the other component of this supply of labor are unsuccessful as independent farmers. The demographic imbalance in Russia and some of the other ECA countries will make it difficult to sustain an agrarian sector based on independent family farms. This is why solutions that are specific to the actual conditions become more plausible. It is likely, for example, that many families will remain members of the collective enterprises employed as aging workers, because they can not supply the able-bodied family labor required for tasks on independent farms. In their case, peasantization is not an option. As in all family farming sectors, one result of the demographic structure on the production patterns will be the redistribution of land in units that will seem more equal in terms of households, but not individuals. In other words, in private independent farms, aggregate land ownership or possession ultimately correlates with household size. Since labor is the constraining factor in private farms, the larger the labor supply of the household, the larger will be its holdings of land or animals.

Farmers in the developing world have recourse to family and village networks, reciprocity in the village context, and other non-market forms of social capital. These social relations facilitate networks and provide access to resources. They may also cushion the adverse impacts of economic transition. Without the village social structure, its networks, and solidarity among households, the survival of independent farming in the developing world would be difficult. Similarly, in much of the CIS, certain conditions external to the production unit itself have to exist for there to be successful and competitive agriculture based on family farming.

In the CIS, a market environment barely exists for rental, credit and sale of inputs and farm production. Similarly, social relations are weakly structured. Family and the village social life in the CIS evolved in a different way than the peasant societies of developing countries and Europe. Most economic links were provided through the large farm management, thus farm workers themselves did not participate in vertical networks with merchants or creditors. Since there was a highly developed system of state-provided services, horizontal links with other households in the farm were also not as crucial as they would be in other peasant societies. Members of the farm lived in the same locality, but this locality did not possess the density of social relationships that might be expected of a village in a comparable context. The absence of informal social safety nets poses important problems for the success of land reforms that aim to establish independent family farming in the CIS. This deficiency is one reason why many farmers in Central Asian states opt for various forms of collective farming, rather than the alternative of independence.

The existence of village communities is also important from the point of view of resource management. In the Turkish case, water user associations which manage irrigation are constituted on the basis of the traditional social hierarchy of village society. The headman of the village is an ex officio member of the association. While this arrangement perpetuates the inequalities already present in the village by transplanting the village social structure into the association. Nonetheless, the existing village structure confers stability to the irrigation scheme that it might otherwise lack, as the accustomed channels within the village provide the means to deal with conflicts and help build new alliances.

Another illustration of the importance of the village context comes from a micro-credit project in Albania. There is a ready social context within which to operate a group credit scheme when the borrowing households all are in the same village community. The village community can supply crucial assets of solidarity and social control in a situation where conventional credit schemes could not work because normal banking and credit institutions would not find the potential borrowers creditworthy. A similar "social collateral" system is used in a rural credit scheme in Kyrgyz Republic. Just as in the case of the village money lender, whose principal coercive mechanism is the disapproval of the village community, the micro-credit scheme in Albania uses the existing village administrative structure to vet borrowers and guarantee repayment, to achieve a high repayment rate. Existing village structures have been employed similarly in credit extended for irrigation schemes and other larger scale investment projects in different settings. For purposes of social development, the use of already existing village structure in project implementation is a major asset, although it carries the risk of perpetuating existing inequalities by vesting it with additional functions. In addition, social funds and other community based projects can be used to strengthen new institutions in both the private and public sector.

### Cooperatives and Associations—New and Old Forms of Social Capital

*Voluntary cooperatives and associations began to compensate for the withdrawal of the state support and the absence of effective local-level institutions to replace those that are being restructured.*

In the former Soviet Union, households are situated within the periphery of larger cooperative farms, thus there usually is no village community independent from the collective farms that can substitute for more formal institutions. The situation is even worse in Azerbaijan and other countries suffering the effects of internal wars where a dislocated population is being settled in rural areas. Here, to reach the peasants with assistance or credit, it may be necessary to create a new institution to substitute for the absent village community. One alternative is to encourage the creation of voluntary associations comprised of 50 to 100 independent farmers who have exited the cooperative structures and owners of household plots, whether or not they remain on collective farms. Such associations could potentially be mechanisms for targeting micro-credit to individual households or it could serve as the appropriate unit for extension work, organization of irrigation assistance and the like. In fact, cooperatives may evolve in this direction—as a civic equivalent of the absent village structure. If they succeed, their independence becomes a major asset.

One major outcome of purchasing or marketing cooperatives is the building of social capital through the participation of individuals in collective decision-making and operations. This is much needed in the present situation of social disarray characterized by the withdrawal of the state and the absence of local-level institutions originating from below. When the collective farm structure was dismantled, opportunities for collaboration and group functioning decreased greatly. The emergence of new social divisions, and the conflicts and competition these may engender, contribute further to the erosion of existing solidarities and trust. When farmers voluntarily choose to associate in production or marketing cooperatives, they tend to re-create previous networks, such as production brigades in state farms.

In fact, new owners do express a desire to solve different population, financial and social problems. In the Kazakhstan SA included in this volume, 55 percent of those questioned would like to unite with other farms in peasant farm associations. In the Kyrgyz Republic, individual sheep-farmers live in remote areas. Associations may be the only way to promote participation for profitable cooperation in marketing and production and to manage and sustain common resources.

Where independent farms constitute a small proportion of the rural economy a major concern is to establish their viability. One policy would be to secure their access to markets while protecting them from the power both of larger farms and larger merchants. To achieve a level of income commensurate with their effort in production, farmers must have the ability to sell products and to access markets that are not prejudicial to them. Marketing and purchasing cooperatives would be one means to gain a bargaining strength. Individual farmers do not have the resources to purchase machinery and equipment, fuel and other inputs. Associations that function as purchasing cooperatives would make it possible for them to enter the market.

In many countries, such as Turkey, marketing cooperatives are quasi-state bodies. Trade councils also assume the role of providing the farmers with technical services, seed development, quality control and storage facilities. In fact, one of the criticisms of water user associations in Southeastern Turkey is their slowness in developing in this direction. It is clear that in terms of providing technical and marketing information, as well as for purposes of market access, such associations perform an important function. The small farmers in the irrigated lands of southeastern Turkey would like to diversify out of cotton, but they lack both the knowledge and the commercial wherewithal to do so. Such quasi-official undertakings might be encouraged in order to provide producers with knowledge of alternatives so they can diversify and become less dependent on traditional crops and traditional commercial networks.

In the case of remote and sparsely populated areas of extensive farming, such as sheep-farming in Central Asia, commercial networks and other advantages of geographical proximity do not exist. In the Kyrgyz Republic, farmers' associations are established under the EU (TACIS) Livestock project as grassroots multi-purpose service cooperatives. These provide both purchasing (fuel and oil) and marketing services. They also organize machinery services, credit, water and pasture regulation. The Bank has proposed a project to expand the operation of such associations to develop an export-marketing infrastructure in lamb and wool markets, manage common pastures on a sustainable basis and improve sheep breeding practices. It is expected that village-based producers' associations will play an important role in taking over the tasks that were previously performed by state farms and enterprises, with the added advantage that they would empower the producers themselves through their participation. The intention would be for the cooperatives to become fully self-governing after a period of five years.

### Credit and Diversification—Social and Institutional Mismatches

*Agrarian families face severe liquidity problems that can be ameliorated by provision of credit on appropriate terms and by diversifying sources of income. Easy access to markets and to transportation is essential for families to improve their economic and social conditions. In most areas, however, the needs of current farms and the institutions that should serve them are acutely mismatched.*

The extension of credit to small producers, especially in agriculture, usually poses problems because of the institutional biases built into the banking system. Family farmers are rarely in a position to provide the collateral, security, and accounting that lending institutions demand to extend credit. The recognition of this difference has led countries to establish specific agricultural credit mechanisms oriented to poor farmers. These have had varying success, but have generally been unsuccessful in avoiding the bias toward larger and more commercial (therefore more mortgageable and accountable) operations. Consequently, small farmers in developing countries have remained outside the formal market of bank credit and have to rely on local merchants, family members or moneylenders for loans. These loans are generally short-term and carry high interest rates. Against this background, recent experiments with micro-credit schemes offer a genuinely novel alternative. As mentioned earlier, small projects in Albania and Kyrgyz Republic seem to replicate the success of the method elsewhere.

In most of rural ECA, farms have severe liquidity problems. This applies to large farms as well as small independent ones. In Russia, there are reports that farmers have problems with the concept of repayment; in Azerbaijan, farmers are willing to borrow where they are certain to be able to pay back within an agricultural season and where they have some control over the factors of success, especially livestock. On the supply side, rural finance institutions are largely lacking. In Russia, for instance, there are no attempts in the financial community to access agricultural markets. Some institutions are interested in agribusiness, but they are few and they only cater to large enterprises. The newly emerging private financial sector is inexperienced in assessing the risks associated with farm credits. Thus there is very little credit available to large farms and virtually none available to small holders. If farmers borrow to meet their needs, it is often from relatives and friends. Loans, usually in the form of seeds, feed and other inputs, are extended to independent farmers from the former state farms and paid back shortly after the harvest. Currently, advances or inputs, whether made by state farms or friends, must be repaid in kind at harvest. As in all developing countries, such loans have a tendency to penalize small farmers who face a disadvantage because of the way prices are reckoned and whose repayment generally occurs when prices are lowest. As opportunities for rural credit improve, differential access to land, farm equipment and information will be reflected in inequitable access to credit. Similarly, the lack of credit experience of small holders means that those who are better endowed—ex-managers of state farms—express a higher demand on financial institutions. At the same time, however, credit institutions are starting to seek group based mechanisms for small-scale lending. This is so in Azerbaijan and Kyrgyz Republic. Inequitable access to credit is already a problem where credit is available; independent farmers have the least access to credit while larger farms and those controlled by former managers of state farms have the greatest access.

Rural populations in the developing world are diversifiers. It is common to see some household members engaged in grain cultivation while others work in labor-intensive activities such as dairy farming or poultry keeping, and some involved in trade while others work in a nearby town or migrate for seasonal work. Occasionally, secondary activities capture a market niche and become commercially viable. This is the direction of movement when seasonal migration becomes permanent and former peasants move out of agriculture altogether—a path which has characterized the Turkish experience in the southeastern region.[47] In socialist agriculture there was limited potential for diversification, framed in the complementarity between the collective farm and household plots. Nonetheless, some subsidiary activities did gain permanence over time, responding to market demand. The production of strawberries in Georgia is one famous example and onions in Kazakhstan is another. However, under the impact of liberalization in the trade of agricultural goods, some of the rural population in the former socialist bloc have already begun to diversify. Families cannot meet their needs from a single source of income, whether pension, wages from the collectives, or farming the private family plot. In Uzbekistan, the share of income from non-agricultural work amounts to 45 percent of the total. Regional differences indicate, however that the range is between 30 percent and 54 percent, depending on proximity to large cities. As such, a rapid transition is occurring from a rural economy in which all families depended primarily on wage employment to one in which families rely on multiple sources of income. At the same time, family labor is becoming more specialized and interdependent

---

[47]     Caglar Keyder. 1983. "Paths of Rural Transformation in Turkey." *Journal of Peasant Studies* 13(1).

One aspect of diversification relates to opportunities for employment in non-agricultural activities in the rural economy. Policies which support off-farm activities in the countryside will contribute to rural development and equity by permitting family enterprises to employ their assets more efficiently. In addition, assisting in the development of small-scale industry in the rural areas (especially agribusiness) will create employment for those with no access to land. Due to the elaborate division of labor fashioned under the Soviet system, only a small share of locally produced raw materials were processed locally. There is, therefore, much potential to develope local agro-industry, thereby creating employment and contributing to the prosperity of the countryside.

The existence of marketing opportunities is a precondition of diversification. The disappearance of elaborate exchange and marketing systems that existed in the Soviet era has affected much of the rural landscape. As a result, neither the restructured farms nor independent farmers have access to an adequate marketing infrastructure. The restructuring or closure of agribusiness has also removed key outlets for marketing farm inputs and farm produce. Both input and output markets were affected and independent small holders appear to suffer disproportionately. In Eastern European countries small-scale, rural agro-processing in dairy products and distilleries has achieved some success. Elsewhere, the development is slow.

In-depth interviews in Russia indicate that organized groups block farmers' attempts to market their produce. There is growing evidence that this occurs in Central Asia, as well. Attempts to take a small load of produce to a nearby urban market or to sell by the road side are said to meet the opposition of local groups, referred to as the Mafia, who demand tribute in kind or in cash. In many instances the demands of these criminal elements are excessive and many farmers simply give up before they even try to market their produce. In Tajikistan, for instance, the absence of markets is a major constraint to the establishing viable private farms and the high cost of transport to markets is prohibitive. "A major constraint in marketing is checkpoints located on transport routes, which demand informal payments in cash or produce. The higher the number of checkpoints, the higher the cost, and the lower is the security of the transported goods."[48] To get around the problem, some farmers travel on buses and pay the drivers, who then pay the gatekeepers at checkpoints.

Anecdotal evidence from other ECA countries indicates that farmers feel threatened by illegal elements and even the local police. This is one reason the rural population prefers to remain attached to the large farms, maintaining themselves at a subsistence level on their home gardens. In Azerbaijan, people say that the difficulties compounded by marketing inefficiencies result in less than half of the independent farmers being able to sell a portion of their produce; the rest only produce for their own family needs. In Kazakhstan as well, household cash income is obtained from animal sales, rather than the sale of crops. In part at least, this indicates how difficult it is to avoid road side harassment by rent seekers. In rural areas closer to urban markets and along populated borders, it is more common to see a diversification based on market advantage. Both structural and informal marketing constraints constitute key issues for independent farmers to be

---

[48]        M. Bodenhamer et al. 1997. "Returning the Roots: Rocky Renascence of the Dekhkan Farmers in Tajikistan." The World Bank. P. 15.

successful. Country specific differences and sub-regional variations are substantial. These difficulties are more pronounced for some crops than for others.

The limited number of social assessments carried out for Bank-financed projects identified three solutions that rural people perceive to be appropriate for them. First, develop local agribusiness where farmers can directly take produce, enabling small holders to market their goods safely with low transport costs. These would also produce some wage employment for people whose farm size is too small to use all the family labor. The second, is to make well-organized and regulated local markets available. Third, people would like to see local corruption reduced.

Credit policy is also important in facilitating diversification of the rural economic activity. There are successful examples of using credit incentives to initiate self-sustaining activities, such as dairy farming in a project in a poor interior region of Turkey.[49] The introduction of new activities is important to increase the incomes of the rural population, make the rural economy more self-sustaining and prosperous and provide employment opportunities for those who would otherwise leave the rural areas to work in the cities. Credit projects tailored for the purpose, operating according to the specific conditions of the area, always have a high chance of success. The proposals for labor-intensive production in the newly irrigated areas in Southeastern Turkey, when accompanied by credit, have a chance of realizing all these goals in addition to contributing to an improvement in income distribution in the area.

Successful diversification requires a wide range of information about markets and techniques, and the acquisition of new skills. Managers of existing joint stock companies, production cooperatives and collectives must also learn new approaches to farming in order to develop enterprises that are commercially viable and environmentally sustainable. Consequently, there is a high demand for advisory and extension services. However, where extension services are available they are geared to deal with collectives. Thus there is a need to redefine the content of extension services and to reorganize and reorient them to reach small independent farmers. These services need to equip farmers with technical knowledge about agricultural production, as well as financial, marketing, legal and management aspects. Since some farmers are independent while others prefer to operate in groups, it is important for the extension services to meet the needs of various categories of producers, based on an empirical assessment of their needs.

The design and implementation of advisory/extension services require systematic consideration of key social development issues and a specific assessment of the knowledge requirements of key social groups. Different ethnic populations in ECA countries from one region to another participate in different aspects of agriculture; likewise participation rates differ with important social implications. In Bulgaria, for instance, a higher percentage of ethnic Turks and Roma engage in agriculture and, compared to Bulgarians, a much higher percentage of them have lower levels of education. Cropping patterns, living conditions, social institutions as well as patterns of information/communication are all likely to differ on ethnic lines, which must be factored into the design and implementation of extension services. Likewise, in Uzbekistan, farmers of family plots, independent farmers, those in the newly established collectives, and those employed in state farms all have distinct knowledge and management needs. Their ability to

---

[49]      Known as the Corum-Cankiri Project and implemented in the late 1970s.

receive the relevant knowledge depends on a range of socioeconomic, institutional and situational factors; thus specific considerations in knowledge management efforts are required.

### *Urban/Rural Inequity on the Rise*

*Rural families in the CIS enjoyed more equal conditions under socialist regimes. The withdrawal of the state from many areas of life is increasing inequality between rural and urban people. Subnational inequalities within the rural sector are also growing.*

The ideological and political interrelationship of the state and the family manifests itself in many forms in rural societies. For example, often the need for national self-sufficiency in food is an explicit policy of governments. At a deeper level, family farmers operating on their own are associated with the ideal of rugged individualism; their disappearance would seem to be a betrayal of the very foundation of the social and economic culture of the market society. For these reasons family farmers enjoy a privileged relationship with governments, especially in more developed countries. In many countries, they are subsidized through price support, income guarantees or technological externalities provided free by government institutions. They also often have the political power to resist policy changes that might threaten their livelihoods. This last point is evident in the case of the European Union where half of the EU budget goes to farm subsidies, and farmers, especially in France, periodically strike against unpopular measures.

Agricultural subsidies have been instrumental in increasing the volume of output in more developed countries, thus their cessation would affect world market in a significant way. Some Eastern European countries have low income and wage levels and proximity to rich markets, thus market-regulated agriculture may expand trade and output. Conversely, it may be argued that once agricultural producers of the ex-socialist bloc are out of the division of labor and central planning of the previous era, and more and more subject to market regulation, they will have to confront competitive pressures they may not be able to withstand. This is because, even the oil-rich states of the CIS will not have the fiscal resources to extend the same favorable treatment to their farmers that European and American farmers enjoy, with their already higher technology and productivity. In other words, market pressures on farmers in the ECA region can be expected to accelerate the weeding-out process with its attendant outmigration and abandonment of marginal lands, especially if they have to compete against subsidized western agriculture. One of the significant variables that influence the direction of change and its social consequences is the degree of liberalization in agricultural markets, both in terms of the reflection of world prices, and in terms of the continuation or cessation of subsidies in the developed world.

Poorer countries elsewhere in the world have tried to squeeze agriculture in an attempt to transfer accumulated resources to the cities and the manufacturing sector, with a resultant inequality between rural and urban incomes.[50] By contrast this urban bias was much less evident the socialist bloc. Rural families in the CIS and Eastern Europe enjoyed a standard of social services and consumption which, in comparison to developing countries in the rest of the world,

---

[50]      Michael Lipton. 1974. *Why Poor People Stay Poor: Urban Bias in Underdeveloped Countries.* London: McMillan.

was closer to that of the urban population. The social consequences of urban bias are well known. In addition to lack of investment and low production in agriculture, which leads to food shortages and loss of self-sufficiency, it also results in widespread rural poverty. Such a bias against agriculture and the rural society may be attractive to less developed countries within the CIS which depend heavily on agricultural exports, such as Turkmenistan and Uzbekistan which depend on cotton, and the Kyrgyz Republic with its dependence on wool. The Uzbek government is extremely reluctant to give up its cotton-purchasing monopoly or to privatize the trade, and it offers farmers prices that are substantially below world-market levels. Such monopoly over purchasing and agro-processing, resulting in adverse terms-of-trade for farmers, is obviously detrimental to the development of the rural economy. It undermines the possibility of the autonomous development of the rural areas, draining resources funds that could be invested, leaving farmers poorer than they would otherwise be. Governments should be cautioned against this temptation, which is self-defeating in the longer run.

The transition to private farming and the emergence of peasants require a careful design of policies that govern state support to agriculture. These policies often pertain to regulation of the price inputs and outputs, the determination of subsidies, the establishment or restructuring of the regulatory functions of the government, etc. In this context, there is still a large role for the state. Through subsidies and import taxes, the state can regulate the quality of input supply—seeds—to influence prices. This affects output and cropping patterns—by carrying out facilitating research and extension—and enhances the development of a competitive private market for marketing—purchase, production and sale—of seeds, fertilizer and pesticide. In the design of these policies it is important to take into consideration the particular problems faced by small holder private farms—as opposed to collectives large farm enterprises. This is particularly critical in countries where private land shares are small and a more intensive use of land may have high returns.

## Rural Social Services—The Growing Crisis

*Social services provided by the collective farms and the state in rural areas have deteriorated in the CIS. It will take time for new local organizations, both government and non-government, to replace them. Where local resources are insufficient, national governments will have to play a role. In the meantime, the burden for providing social services rests increasingly on the family, particularly on women who care for the young and the elderly.*

The cases presented in this volume concern agricultural development; they do not deal extensively with rural social services and infrastructure. Nevertheless, it is important to end this chapter with a short review of these broader concerns. The momentous transformation of the past decade in socialist agriculture has had an overwhelmingly adverse effect on the social welfare of the rural population. This negative judgement, however, has to be seen against considerable geographical variation. In fact, the restructuring of the agricultural sector has met with a degree of success in some Eastern European countries. The score sheet is less ambivalent in the case of the former Soviet Union. There, against a background of overall economic crisis, agricultural output has suffered drastically. Incomes in the countryside have fallen and distribution has worsened.

The disruption to the lives of the men and women who live in the countryside in the CIS has been massive. All the certainties of the previous period have crumbled—production practices, social relations and services expected from the government and local authorities. As the government recedes, essential services of health, education, child and elderly care have suffered. Under the Soviet system, social protection was based on subsidized commodities, cash transfers and employment related services. With reforms designed to liberalize markets and prices, subsidies ceased; transfer payments became arbitrary and eroded with inflation.

Throughout the CIS, social infrastructure in the rural areas has deteriorated. Community centers and kindergartens have been closed, clinics and schools depend on local contributions to stay open, and the quality of remaining services is deteriorating rapidly. The current generation of rural youth is being exposed to a declining quality of education, increased irrelevance of textbooks and teaching materials, declining quality and commitment of teachers and administrators, and shortened school years to reduce energy costs. Families are called upon to pay part of the hidden costs of education out of declining incomes. In addition, their ability to contribute to educational expenses is inversely related to the size of the family; family size is relatively higher in rural areas and among certain ethnic groups. Consequently, today's rural youth are already disadvantaged relative to their urban counterparts and this disadvantage is likely to increase over time. Larger family size of certain ethnic groups and sub-regions will also augment inequities and create structural poverty in rural areas, particularly pronounced in some sub-regions.

Previously, social services such as schools and clinics were mostly delivered through the collective farms, which accommodated them within service centers, and assumed responsibility for or greatly supplemented state-dispensed social services. The farms also acted as local extensions of the government in dispensing pensions and elderly benefits. Farms no longer enjoy the secure incomes deriving from the procurement and purchases system; they have to operate as autonomous enterprises. This translates into an inability to guarantee services which require money wages for the personnel who provide them. In most cases, these services have been transferred to the responsibility of the local governments along with the social assets in education and health services. The governmental unit varies from one country to another: for example, in Kazakhstan and Kyrgyz Republic, *rayons,* are responsible for social assets and services; in Turkey and Azerbaijan, village councils are responsible; and in Uzbekistan, the *mahalla* is the unit responsible. In each situation, given the fiscal crisis, and their own exceedingly limited authority to generate revenue, however, local governments have generally found it impossible to provide the same level of benefits and services. Indeed, in many places, farm managers are reluctant to transfer social assets because they know that the local governments have neither the financial nor the human resources to maintain them.

With the disruption in the delivery of social services, those needs that were previously met by the state (especially child and elderly care) are now in the process of being privatized and/or are reverting back to the family. The short-run consequence has been to create a disincentive to leave state farms, because even if wages are reduced through inflation and in arrears, the larger units provide some services in kind, and some transfers to pensioners and the unemployed. More crucially, however, the erosion of social services implies that the vacuum has to be filled by families taking on new responsibilities with respect to the young and the elderly. It also suggests changes in the family division of labor where women may have to spend more time in child or elderly care and thus withdraw from other tasks and community participation. These changes would amount to a re-invention of the family in the peasant mode.

The burden of social services on the family budget has become onerous and makes it difficult to set aside funds for farm investments, particularly when there is little cash at all in the countryside. In Uzbekistan, households are now responsible for three key social services which previously were supplied by the state: health, child and elderly care, and education, or they now pay for them. The impact of .the erosion of social services is widespread, but some of the population suffers relatively more—such as the elderly who are not part of larger households and are not in a position to set up private farms or expand their garden plots. This is a big problem, especially because the rural population is older and rapidly aging. A recent World Bank mission to Bulgaria notes that the elderly who live alone, particularly the single elderly women, represent a high degree of vulnerability.[51] It is crucial to note that countries in Eastern and Central Europe, Russian, Ukraine and Belarus face uniquely high levels of aging, especially in rural areas. This is not yet an important social concern in Central Asia and Turkey. In these areas, however, more villages and certain ethnic groups also display disproportionately high levels of aging.

The deterioration of social services, coinciding with a transitional period during which there was pressure to find sources of income to supplement pensions, appears to have taken a high toll, both on the elderly and on those who care for them. The Uzbekistan SA included in this volume notes that many key social services are no longer being provided by the state and some, most notably education and medical care, although nominally free-of-charge, have become very expensive. Medical supplies are seldom available through the public health care system, and are purchased at high prices on the market or from doctors. The practice of paying medical personnel in order to receive quality medical attention, which existed to some degree in Soviet times, has become more widespread in the last few years, especially since wages in the medical professions have not kept pace with the rise in the cost of living.

Children are adversely affected by the inaccessibility of both health services and educational facilities. Women have cushioned some of these adverse impacts with greater time and effort for the care of children. Likewise, while the elderly might have suffered disproportionately from the deterioration of health services at a time when they need it most, women have to spend greater time and effort cushioning these impacts by providing greater support for the care of the elderly. The loss of support mechanisms makes women's tasks more difficult, but this negative impact is exacerbated because it is women who lose employment and income through the erosion of the social services.[52] Women were employed as teachers, kindergarten employees, doctors, nurses, administrators, etc., and when social services ceased to

---

[51]     Opposite results are reported for Moldova (David L. Lindauer. 1997. "Labor and Poverty in the Republic of Moldova." Draft report. The World Bank), whereby elderly are noted to be better off than children and working age unemployed and where vulnerability of the elderly living alone appear to be lower than those who live with their families.

[52]     A study carried out by UNICEF in Kyrgyzstan found much higher rates of female unemployment than male unemployment in the country. It noted that job losses meant the loss of access to many important benefits, including family benefit, child care support, in-kind contributions of food and fuel, etc. The study also notes that nomadic women and women on privatized rural farms are extremely vulnerable to significantly worsening poverty and many are suffering shortages of medicine, fuel, food and clothing.

function, there was a large scale loss of employment, income and benefits. For these reasons, female-headed households are the worst off. Women find it more difficult to manage without kindergartens, childcare, and deteriorating schools, especially when their employment base is also shrinking. It would be desirable to target female-headed households and households consisting of the elderly in any kind of differentiated social assistance program.

The restoration of pensions to their pre-inflation level, as well as the restitution of other social services, is conditional on the restructuring of the government and the likelihood of attaining fiscal health. The prevailing economic crisis in the CIS makes it unlikely that fiscal health will be regained in the near future. Re-peasantization and/or re-traditionalization, in the sense of reliance on family and community for social services, are therefore realities in most other countries of the CIS. This is more visible where pre-socialist village structures were revived, as in Albania. Similar developments are observed in Azerbaijan and other countries in Central Asia.

It should also be remembered that public expenditures on education, training, health, and welfare services are part of the social wage. They add to the incomes of the recipients in addition to contributing to their social development. Conversely, when social services are privatized or have to be substituted for by the family or the community, the real incomes of the individuals and the households suffer a decline. Therefore, it is not only the provision of social services to the rural population that is in question, but also equity between the urban and the rural sectors and distribution of the national income. The higher the social wage available to the rural population, the more attractive is the rural sector. This is true not only because greater equity between the rural and the urban sectors tends to keep people in the countryside, but also because equity within the rural sector itself is a factor that tends to make the countryside more attractive. Widespread availability of social services tends to create a less unequal society. Pronounced inequality in the countryside tends to push the wealthy to towns in search of better opportunities in education, consumption and investment, while the poor leave villages in order to find work in marginal sectors in the cities.

Social services and the provision of public goods have to be maintained to promote rural development, to prevent rural poverty and to avoid the less desirable aspects of reliance on the community and the family, especially in terms of the gender division of labor. Restoration of social services is a priority for reasons of equity, health, education and maintaining levels of income—in short, for social development in the rural sector. As in all policies which improve the quality of life in the rural sector, maintaining social services will make the population less eager to leave the countryside. Services or transfers (as in the case of pensions) have to be provided independent of employment in larger farm units; the state has to be urged to assume the responsibility for social services, and to deliver them effectively in rural areas.

In order to design projects aiming to restore the level of social services to which the rural population was accustomed, more has to be known about the functioning of local governments. They have undergone change during the reform process—their sources of revenue are not the same and their responsibilities have expanded. It is entirely possible that local governments (district or municipality) would be capable of delivering the social services they are asked to provide if they are assigned certain sources of revenue. It would seem, however, that in a situation where their revenue depends on local conditions, the quality of these social services would undoubtedly vary greatly. There are some interesting practices emerging in this regard, such as in Azerbaijan where 30 to 40 percent of the land of the collective farms is set aside to provide income to the village council. More universalistic assistance schemes are also needed through social safety net and infrastructure projects, to equalize spending levels across regions, rayons and villages. To achieve this, it may be necessary to remove some of the responsibility of the village councils and think about establishing an administrative extension of the central government which would take on the role formerly played by collective farms. The functioning of local governments, their potential and limits in delivering social services, and the alternative of complementing their operation with a new or revived arm of the central government should all be the object of a new research agenda.

## Conclusions

The review of the changes that has taken place during the last decade in the countries of Eastern Europe and the CIS, as well as those southeastern Turkey, point to a large number of important observations that cannot all be summarized in a short paragraph. However, some of them are important to emphasize. The changes have been multi dimensional and rapid, with enormous regional and sub-regional variations. The target for change has been constantly moving. As such, what *transition* is or was, for large number of ordinary citizens, was a relatively slow and difficult change in an often ambivalent direction. The speed and content of these changes were largely determined by the pre-transition configurations at the national level, and especially at the local level. The lack of detailed understanding about the social structures that were expected to absorb the changes often led to national or more local expectations for rapid change that did not materialize. In addition, it is clear that many of the most important stakeholders in the sector were either poorly informed about the process or opposed it.

The focus on farms, rather than farming communities and rural communities, caused major distortions in expectations and outcomes. First, this focus prompted both governments and external funding and support agencies to give primary concern to the nature of state and collective farm enterprises, rather than with communities. Thus, agricultural crop production received great attention, while the issues of common property ownership and management, and the management of rural infrastructure and social assets were largely neglected. Secondly, it resulted in a strong bias toward farming, as such, rather dealing also with off-farm activities, such as processing, which are also essential for rural livelihoods. Third, by focusing narrowly on farms as enterprises, and assuming that the role of such farms would continue as before, most analysts remained oblivious to perhaps the most important result of the transition, particularly in Central Asia. That is, to survive the collapse of the sector, the majority of rural people became subsistence farmers, almost totally dependent upon home gardens, while the role of many large farms changed from being industrial/commercial agricultural enterprises, to doing little more than providing fodder and other inputs to the household plots or their employees, at ever-declining levels. Fourth, the focus on the reform of farm structures, rather than the whole agriculture sector, caused a growing mismatch between the needs of new farm units, both large and small-scale, and the scale and structure of the markets, input distributors, suppliers of information and other institutions that were available to serve them. Finally, the official focus on the amount of agricultural production, similar to the Soviet-era concern with targets, virtually ignored other critical elements of the rural sector, making it easy for both formal and informal rent-seeking opportunities to develop, such as in transport and transit, which exploited the rural population and undermined the social and economic objectives of the reforms.

The focus on rural transition or change must be shifted from enterprises to communities of people, from farm production to the structure and function of all elements of the agricultural sector. When this shift in focus occurs, the current social organization and demographic characteristics of communities become as crucial as their recent history in the design of strategies to facilitate change and to cushion adverse impacts. An equally important element of change is growing inequity, especially in those areas where change has been rapid. Although the shift from a planned economy to a market economy inevitably generates inequities, the level of inequity developing in Central Asia, for example, is much higher than necessary. We can gain better understanding of how inequities grow and with what consequence for the poor and vulnerable by sharpening our focus on local communities and the relationships between rural and urban areas. The issues involved are important and the ability of policy makers to address them depends, among other factors, on a better understanding of communities, their differences and similarities, and the patterns through which they are changing. Given the nature of changes underway, their magnitude and historical uniqueness, interventions should be well-grounded empirically. Social assessments are thus crucially important to help design effective policies/projects and social impact monitoring processes to enable us to learn from experience and to redesign development interventions accordingly. In the following chapter follow, we discuss the conceptual underpinnings of the social assessment process and some of its specific features as reflected in the transition economies of Central Asia and in Turkey.

# Chapter Two

## Social Assessment:
## A Comprehensive Framework for Development Initiatives

### Ayse Kudat

## Background

Over the past three decades the World Bank has made substantial progress in using social analysis as an integral part of the definition and implementation of Bank-financed development initiatives. Indeed, under the leadership of Michael Cernea, social analysis has been used in many different forms in projects.[1] Social analysis focusing on mitigation of potential adverse impacts of Bank-financed projects have since become mainstream practice throughout the institution; there has also been substantial success in establishing dialogues with borrower countries in legislating social policies for mitigation. The systematic use of social analysis in parallel with economic, financial, and technical analysis in the preparation of projects has had less success. Nevertheless, over the years the Bank has increased the number of social scientists among its staff, and social science inputs have become more central in shaping development initiatives of the organization.

Theoretical developments in a broad range of disciplines, ranging from development anthropology, clinical sociology, social psychology, participation, organizational theory, management sciences, and many others have all been incorporated into the Bank's work. Indeed, the Bank has been instrumental in disseminating the thinking generated by these disciplines in a large number of countries by using researchers and practitioners from a wide range of disciplines in its work, and it has imported new ideas from the academic and development communities through seminars and other knowledge management activities. In particular, ideas generated in applied anthropology and applied sociology have found a home in the Bank's work; the relevance of social engineering or social architecture, clinical sociology and development anthropology can easily be traced in the contributions to this book.[2] Similarly, by providing consulting opportunities for a large number of academicians and practitioners in the social sciences, the Bank may have had an impact on developments in the respective disciplines. The parallel and mutually reinforcing developments in the use of

---

[1]    Michael Cernea, ed. 1975. *Putting People First: Sociological Variables in Development.* New York: Oxford University Press.

[2]    A. Podgorecki, ed. 1975. "Sociotechnics: A Trend Report and Bibliography." *Current Sociology* 23(1). The Hague: Mouton; J.M. Fritz. 1989. "The History of Clinical Sociology." *Sociological Practice* 7:72-95; Howard M. Reback and John G. Bruhn. 1991. *Handbook of Clinical Sociology.* New York and London: Plenum Press; Michael Cernea. 1995. *Social Organization and Development Anthropology: The 1995 Malinowski Award Lecture.* The World Bank, Washington, D.C.

applied social sciences in other international and bilateral organizations have further contributed to the use of social sciences in development projects and policies.

Many of the issues and historical developments specifically relevant for social assessment were analyzed in *Social Assessments for Better Development,* a collection prepared for Bank lending activities in Central Asia.[3] To recapitulate, as poverty alleviation policies took hold in the mid-and late 1970s and as lessons from failures caused by socially and culturally inadequate projects kept accumulating, the need for social analysis became more marked. The Bank adopted guidelines for population displacement and resettlement in 1980, and guidelines for project affected indigenous populations became effective in 1982. In 1984 the Bank introduced guidelines for social appraisal of projects so as to broaden the use of social analysis in projects which did not involve resettlement or indigenous populations. These guidelines involved the sociocultural and demographic characteristics of the populations, the social organization of productive activities and of social services, the cultural acceptability of the project, and the social strategy for project implementation. Many of the elements basic to social assessment as defined below were incorporated in the 1984 guidelines, but few projects followed the guidelines.

A concern with participation spread throughout the Bank in the early 1990s. It became increasingly important to listen not only to the direct borrower but also to the client (often defined as the poor) and to a range of other stakeholders whose interests might be promoted or hindered through Bank-financed activities. Sourcebooks were prepared, special funds were created to promote participation, and many projects were prepared in a participatory manner. The Bank's understanding of institutions broadened substantially; the earlier emphasis on the borrowing organization as the institution shifted, and a broader understanding of the institutional framework included the state, community based institutions, the rules of the game, and the law. More recently, the Bank's focus on development effectiveness has been sharpened with a better understanding of the people, their needs, the mechanisms through which those needs could be met, and the sociocultural context. This has required a closer monitoring and evaluation of projects, including social impact monitoring. Social analysis, institutional and social organizational analysis, participation, and monitoring of development effectiveness of projects have come to play equally important roles in the design and implementation of Bank-financed development initiatives. It is these four elements that are brought together within a framework that we refer to as social assessment.[4]

While good examples of social assessment are being generated in all regions and departments of the Bank, social scientists working on these projects have little time and opportunity to document and publish their experiences. In this regard, perhaps one of the more important contributions of the collection edited by Cernea and Kudat was to document

---

[3]        Michael Cernea and Ayse Kudat, eds. 1997. *Social Assessments for Better Development.* The World Bank, Washington, D.C.

[4]        This conceptualization of social assessment is more comprehensive than the one used in the previous volume (Cernea and Kudat 1997).

some of the diverse contribution the World Bank makes to applied social science. The compendium noted that some of the methods used in applied work are well known but others are invented on the job, elaborated more by creative happenstance than by cold inference from textbooks on methodology. This creativity comes in response to the pressing demands raised by change programs. The drawback, however, is that such contributions of applied researchers risk being forgotten and unreplicated.

Since the publication of this earlier collection, the Bank has established a Social Assessment Thematic Team[5] and created a Web page to disseminate knowledge of good social assessment practice. However, there is a general misconception that socioeconomic surveys carried out at the outset of projects constitute a social assessment (SA). Even today one often hears task team leaders talking about commissioning SAs, referring basically to qualitative and/or quantitative surveys. Such surveys provide relevant and useful information for the SA process and can be considered as consultative tools to facilitate listening to some of the stakeholder groups. However, they are not synonymous with a comprehensive SA.

In what follows we shall first provide a conceptual framework for social assessment; its poverty specific objectives; its integrative functions with respect to social, institutional and participatory analyses; as well as an evaluation of its development effectiveness. To this end, four pillars of SA will be identified and discussed.[6] Special emphasis will be put on the interrelationship of these pillars and the advantages of informing development initiatives with one participatory instrument. We will show that SA, defined as a process, provides a roof under which social development concerns can be integrated into development initiatives during their design and implementation, as well as being useful in monitoring the social outcomes of these initiatives.

## Objectives of Social Assessment

The overarching objective of social assessment is to ensure that development initiatives contribute to poverty alleviation and at the same time enhance social inclusion, strengthen social cohesion, increase social capital, build ownership, and reduce the potential adverse social impacts of economic development. Bank-financed projects aim at changing behavior; the traditional use of economic, financial, technical, and environmental criteria do not always facilitate identification of the required behavioral changes or show how this can be achieved. As an input to induced development, SA provides information about social organization and cultural systems in order to ensure quality at entry for Bank-financed projects and success during their implementation. It is an iterative and participatory process to

---

[5]    The core Thematic Team consisted of Ayse Kudat, Stan Peabody, Arbi Ben-Achour, David Marsden, Cyprian Fisiy, Estanislao Gacitua-Mario, and Bruce Harris.

[6]    The four pillar approach was adopted by the Social Assessment Thematic Team in October 1998; it is not yet an official Bank policy. However, a large number of Bank staff uses this conceptual framework to guide the integration of social development concerns into the design and implementation of their projects.

prioritize, gather, analyze, and use operationally relevant information about social development and institutions. The compatibility of development initiatives with the needs of affected stakeholders is also a crucial concern. SA delineates the social forms and processes that affect all aspects of development projects—their architecture and shape, their size, their direction, and ultimately, their outcomes. It helps task teams incorporate social development concerns and participatory processes into design, implementation, and monitoring of projects and analytical work and assists clients in reaching the poor and vulnerable.

SA facilitates the participation of key stakeholders—especially, the poor, low-income, vulnerable, and excluded social groups. It requires commitment to sharing knowledge among groups and agencies in order to incorporate the views of distinct clients; thus it enhances ownership. Further, by articulating and advancing the interests of the poor and vulnerable groups, SA mobilizes broader support for the project from a wider range of stakeholders. The broader group of stakeholders includes the private sector, civil society, government and non-governmental organizations and their members, and others who facilitate or hinder the ability of the poor to have equitable access to the goods and services offered by the development initiative.

Understanding the social fabric—the social context—of a development initiative is thus a basic element in appropriate project design and implementation. In this view, SA is one type of feasibility analysis, and it complements economic, financial, technical, and environmental analyses. The knowledge of the social fabric, community institutions, social capital and trust, social diversity, and other social variables based on a solid and systematic field investigation and consultations with a key stakeholders is as essential for the design of development initiatives as other types of assessments. It maps out the asymmetries among social groups, brings out patterns of inclusion and thus allows the identification of institutional arrangements required, creating more equitable and participatory societies. It also facilitates the assessment of the impacts of development interventions.

SA broadly focuses on the social context but is very specifically designed for a particular project or program. It also relates to the project or program cycle from initial design to implementation and the measurement of social impact. Although a given SA may produce one or more written products, it is fundamentally a process rather than a specific product. Consequently, much of the following discussion will refer to the SA process as it elaborates the components and objectives of SA. However, the cases presented in this volume will cover the analytical aspects of the SA process rather than describing how the process was launched and carried forward.

## The Social Assessment Process: Integration of Four Pillars

SA incorporates four principal analytical elements, or pillars, each of which is consistently revisited during the life of the project. The pillars are:

*Identification of key social development issues.* This pillar addresses social issues relevant to the project or program. Issues are identified within a narrowing context, generally going from a national and sectoral perspective to specific project sites. Depending on the situation, the issues may include poverty, inequity, social diversity, gender, social capital, social exclusion or others.

*Stakeholder identification and the formulation of the participation framework.* This pillar identifies major stakeholders in the project and their particular interests or stakes. Stakeholders may range from individuals to specific social groups, authorities and local, national and international organizations. A major product of the analysis is the development of a participation framework designed to ensure the active participation of key stakeholders in project design, implementation and evaluation.

*Institutional and social organizational analysis.* This pillar concerns the structure of social relationships and behavior. The analysis identifies institutions, both formal and informal, which establish the "rules of the game" in the project context, and the incentive structures which affect the extent to which the rules are either followed, undermined or ignored. The product of the analysis is a series of recommendations on the institutional development or reorientation that is needed to achieve project objectives.

*Establishment of a monitoring and evaluation framework (M&E).* This pillar develops a monitoring and evaluation framework for project implementation, focusing on aspects relevant to the social development objectives of the project. The purpose of this analysis is to establish mechanisms to measure social changes and social impacts during implementation to inform stakeholders about mid-course corrections needed to ensure that social development objectives are achieved.

Where appropriate, each element would also address the need to mitigate any negative social impacts. This will mean: identifying the adverse social impacts pertaining to resettlement, indigenous populations, or cultural heritage; assessing the institutional capacity to mitigate them; ensuring stakeholder participation in the mitigation plan; and integrating monitoring of the mitigation measures into the overall project monitoring framework. Indeed, if development initiatives are routinely subjected to social assessment, mitigation issues can be systematically identified and dealt with, whereas a focus on mitigation alone will not always incorporate an understanding of broader social development and participation concerns. Since the SAs included in this volume do not address the adverse impacts as defined by Bank's operational policies, they do not provide an illustration of the use of the four pillar approach with respect to social safeguard or mitigation practices. It is nevertheless important to stress that concern with social mitigation is an integral part of the social assessment process.

The social assessment process has an analytical focus on social development, participation, and institutional issues (Figure 2.1). This process contributes to the design of the institutional and participation framework of a project so as to facilitate inclusionary practices and the targeting of project benefits. Social impact monitoring ensures that these benefits are actually received. A SA process may be considered "good" if all four pillars are included in it. Because this conceptualization of the SA is new, most of the existing SAs are partial, often focusing on stakeholder and social issues identification. The SAs reported in this volume are also uneven in their treatment of all four pillars. When there is insufficient knowledge about the project area, the affected populations, and their institutions, the SA may concentrate more heavily on information gathering than on other elements. The SA on Uzbekistan did not include considerations of social impact assessment.

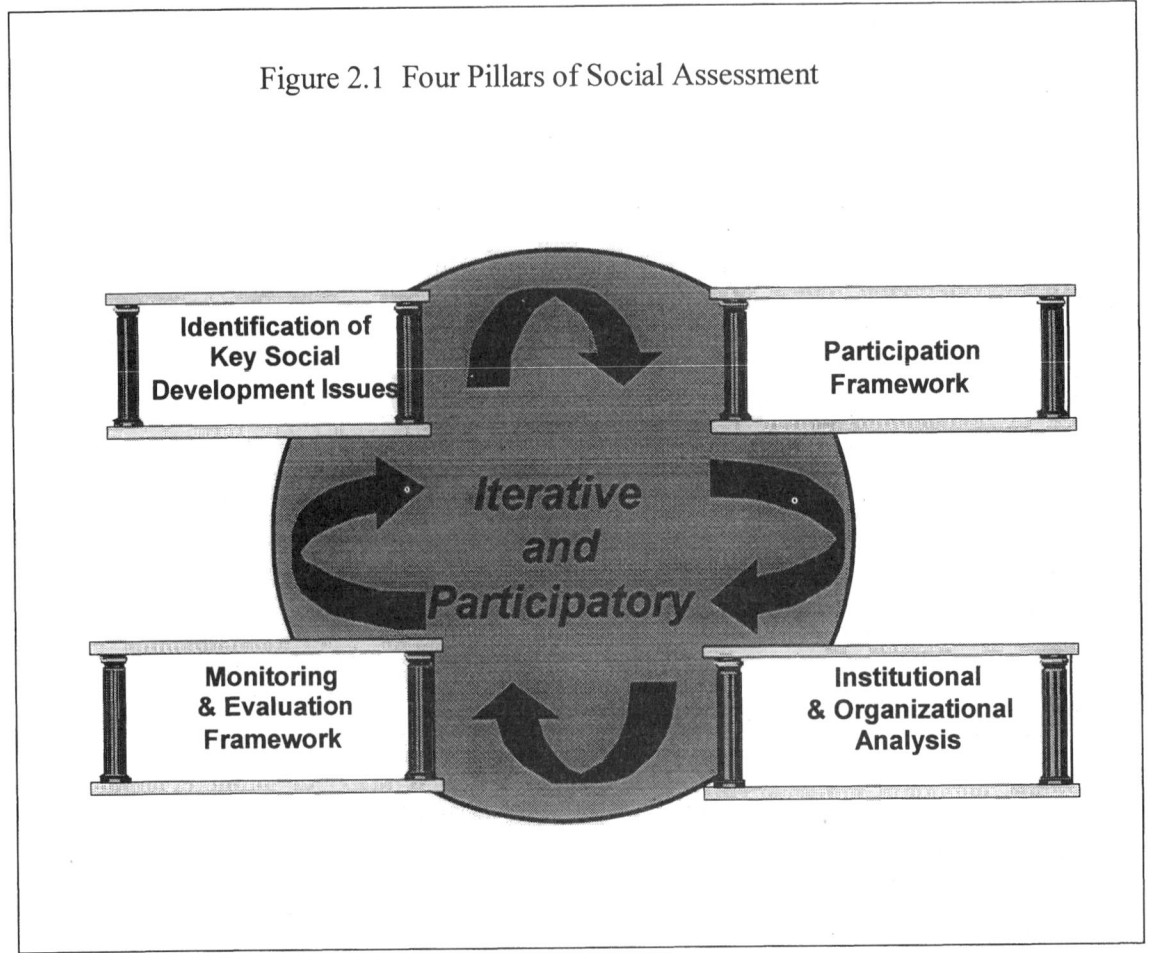

Figure 2.1  Four Pillars of Social Assessment

The SA case for Turkey reports the first phase of the process, focusing on qualitative and quantitative social investigations and stakeholder consultations. For the subsequent phases, it raises institutional and participation issues and provides information that would help design the M&E component of the project. However, the SA case for Turkey is important in its attempt to trace causes of outcomes of development interventions similar to those envisioned for Bank financing. By attempting to "net out" the effects of interventions through use of "control groups" of sub-regions that previously benefited from them, the SA pointed to corrective action.[7]

---

[7]       Kene Ezemenari et al. 1999. "Impact Evaluation: A Note on Concepts of Methods." Mimeo. The World Bank.

SA is a participatory **process** and is not merely confined to social analysis. Also, SA is not only a socio-economic survey or qualitative investigation although it is conceived as such by most, including the social scientists. The strength of the SA process derives from the integration of all four of its pillars in an interrelated manner into development initiatives. Social analysis is key to the establishment and monitoring of social development objectives specific to a project. Such an analysis is also critical for institutional and participation analyses. Institutions cannot be understood without taking into account the diversity of cultural patterns, norms, and values. Stakeholders cannot be identified without a full understanding of social organization, tribal and family structures, indigenous groups, gender, and inter-generation issues. A project's effectiveness cannot be evaluated unless the changes required in rules and incentive structures are specified and monitored. In many development initiatives, participation concerns are expressed during the start of a project's design, but mechanisms to ensure stakeholder participation in project implementation and monitoring are not always established. The inadequacy of social and institutional analyses is part of the problem.

The **art, science, and relevance of SA** are all important. The art is in converting social science knowledge into a participatory tool for building and maintaining consensus for change. The science is in creating this knowledge in a systematic and verifiable manner. The relevance is in making a difference. Although a "good" SA is often measured by its specific impacts on the development initiative for which it is are carried out, social assessments have had different effectiveness on the design and implementation of these initiatives. Among many other factors that would enhance SA linkages to projects/policies is their ability to move away from being "products" or "reports" to being processes through which stakeholder perspectives are taken into account in creating institutional mechanisms for sustained participation of the poor and the vulnerable among development beneficiaries.

### First Pillar: Identifying Social Development Issues

Social development issues are diverse, and no body of social theory, methodology, or knowledge is irrelevant to their identification. Social anthropology, sociology, ethnography, geography, political sciences, social psychology, and many other social science disciplines are brought together with a mix of qualitative and quantitative methodologies to launch and complete a social assessment process. Issues of poverty, social exclusion, and social capital have been researched by every social science discipline and are universally included in the SA process.[8] Therefore, the social assessment process aims to integrate inputs from these disciplines into the design, implementation, and evaluation of development initiatives. Issue identification and analysis often starts with broad sectoral and country-specific information and then focuses on the project context. The usefulness of the SA process derives mostly from its concrete and situation-specific nature.

---

[8]    Putnam defines social capital as those features of social organization such as networks, norms and trust that facilitate coordination and cooperation for mutual benefit. Robert Putnam. 1993. *Making Democracy Work: Civic Traditions in Modern Italy*. Princeton: Princeton University Press.

Preliminary identification of the key social development issues is based on secondary information if available. This may include:

- Country social science studies, other secondary literature, demographic data, relevant socioeconomic statistics, historical research, political and institutional information;
- Research on broad regional issues that may have applicability for countries contained within the region;
- Social development literature/studies relevant to the sector of concern;
- Existing social development profiles or other background information prepared for the country assistance strategy (CAS) and economic and sector work (ESW);
- Social impact monitoring studies for similar projects in the same or other countries;
- Consultations with knowledgeable local and international experts.

Poverty and inequity and social diversity are highlighted in the following sections. Depending on the project context and content, however, this pillar could cover other major social issues, such as social inclusion and exclusion, governance and corruption.

### Poverty and Inequity

Three of the cases included in this volume focus on Central Asia and existing knowledge on poverty in the region provides an important underpinning for analyses of this key social issue within each of the specific SA context. The existing knowledge about poverty trends within the Central Asia region show that the presence of poverty and the threat of inequity are of particular concern for rural areas and can effectively be addressed through agricultural development inputs. In reflecting upon the region wide knowledge, Husain[9] notes that "unlike other poorer regions of the world there is very little awareness among outsiders about more than 50 million people living in the Central Asian Republics and Caucuses, the majority of whom are below the poverty line. Central Asia republics have at least three characteristics distinct from the other republics of the former Soviet Union constituent republics. The proportion of the people living below the poverty line is exceedingly high— ranging from 80 percent in Tajikistan, to 35 percent in Kazakhstan. These countries face twin challenges of transition from the centrally planned economies to market economy and transition from poverty to decent living standards for the majority. Initial resource endowments are also highly differentiated with Azerbaijan, and Kazakhstan at one extreme; Tajikistan and the Kyrgyz Republic at the other."

Husain also notes that the majority of the poor live in the rural areas and derive their livelihood from agriculture, thus placing emphasis on raising productivity and incomes will

---

[9]     The poverty analyses provided in the next five paragraphs are taken entirely from an internal memoranda written in 1999 by Ishrat Husein, country director for Azerbaijan, Uzbekistan and Tadjikistan.

have a positive impact on the poor. He raises six key issues which are critical to poverty reduction several of these are addressed in the case studies included in this volume:

- The privatization of state and collective farms while avoiding concentration of assets in a few large hands and ensuring equitable distribution;
- The judicious use of irrigation water while protecting the small farmers and the poor and reducing massive state subsidies;
- Re-orienting the existing local level institutions and creating new institutions to respond to the emerging market-based opportunities;
- Revamping the social protection system so that it is efficiently targeted and cost effective;
- Promoting cross-border trade; and
- Creating the enabling environment for the non-governmental organizations to reach out the poor.

Husain draws attention to the fact that "the presence of local governmental apparatus at the Oblast and Raion levels and the institution of collective farming provide excellent opportunities for decentralization, delivery of social and economic services and mobilization of resources at the local level. For example, voluntary association of water users to manage irrigation water can be easily carved out of the old collective and state farm structures. Credit Unions for the provision of rural credit services and savings by members can be established to serve the new privatized farm families. Farm Machinery service centers can operate as commercial entities before they can be privatized. Village level enterprises for providing farm inputs and processing agriculture produce can be encouraged. The challenge therefore is how to reorient the old local level institutions in order to adapt and respond to the emerging market based opportunities and to create missing institutions." It is also noted that "each of these countries has inherited a system of social protection which consumes a significant proportion of the national and local budgets but does not reach those who deserve these payments on the basis of needs. Uzbekistan is reported to have developed a well-functioning, cost effective and decentralized Mahalla system for delivering Social Protection payments", but a systematic evaluation of these systems has not yet been carried out by the development community.

Let us take the SA for Turkey to examine how availability of country wide and sub-regional data help in the identification of social development issues. At the country level, the recent Bank studies identify growing inequity in living standards as the most significant trend of the 1990s. The work shows that "Turkey does not face a problem of absolute poverty by the standards of a developing country. Poverty in Turkey affects mostly specific groups of the population."[10] Large differences in poverty incidence are established between regions. East and Southeast Anatolia have the poverty risk that is 50 percent above the national average however, even the richest regions have groups of poor. Because different poverty lines are applied to rural and urban areas, small differences are found in vulnerability and poverty

---

[10]     Ruslan Yemtsov. 1999. "Living Standards and Economic Vulnerability in Turkey between 1987 and 1994." In "Turkey – Economic Reforms, Living Standards and Social Welfare Study." Draft Mimeo. April 1999. The World Bank. Page 1.

between the urban and rural poor. However, separating provinces on the basis of their degree of "rurality", the Bank study shows that nearly nine million people in Turkey live in regions ranked low by human development indicators; 60 percent of them are in predominantly rural and 36 percent of them are in significantly rural districts.[11] The study shows that rural poverty in East and Southeast Turkey is persistent and has geographic and socioeconomic dimensions. Thus, low living standards, low levels of human development and high inequity are among the key social development issues identified by the existing literature on Turkey.

A modest amount of social science literature was available for the proposal Project area, pointing to the presence of large inequities and deeply rooted tribal and lineage structures.[12] Demographic, socioeconomic, and other data collected by the State Institute of Statistics, as well as small-scale studies focusing on girls' education and migration, all pointed to the complex social fabric of the area. The large number of immigrants from neighboring regions experiencing conflict, the dynamic nature of lineage relationships across national borders, and the visible changes in income levels that took place since the start of the construction of the Ataturk Dam further enriched this fabric. This rich body of literature and the extensive familiarity of the manager of the SA process with the peoples of the area facilitated the formulation of the SA information strategy. The knowledge of past trends also helped in the interpretation of the results. However, the social fabric of the specific Project area was very different from the broader southeast region described in previous studies. Unexpectedly, first phase of the SA process launched showed that the project area population was predominantly Arabic speaking. However, little or no previous social science research was available to guide the SA team concerning the implications of this finding for the Project.

The context of other SAs is rather different. In many cases, the SA is undertaken where there is little, if any relevant information available about current conditions in the affected areas and/or agriculture sector activities. Consequently, many SAs are groundbreaking empirical studies in their own right. As was the case with the first collection of social assessments prepared for the ECA region, this volume presents a set of unusual social studies—unusual both for the countries where they took place and for the World Bank which initiated them. Except for Turkey, the countries covered in this volume established their independence less than a decade ago and little was known about the rural landscape. Getting to understand the socioeconomic context, rural human relations, people's aspirations and culture, and deciphering the unfamiliar social map were all important. Thus, the SAs prepared in transition economy countries are important contributions to knowledge of the rural landscape of the countries concerned as well as contributing to the design of relevant development initiatives.[13]

---

[11]     A. Halis Akder. 1999. "Dimensions of Rural Poverty in Turkey." Mimeo. The World Bank.

[12]     Ayse Kudat. 1974. *Kirvelik*. Ankara, Ayyildiz Matbaasi. Bahattin Aksin (1985) produced a number of unpublished qualitative and quantitative research papers for the Southeast Anatolia Project (GAP) administration, the primary client for the proposed agriculture project.

[13]     The disastrous effects of collectivization of agriculture with millions dying of hunger in the Soviet Union are well known. Also known is the fact that socialist countries had poorer living standards as compared to the western nations with significant country differences. Eastern Germany, for instance, was relatively *well off*

In the transition economies, where a major change from a traditional to a modern society is taking place, there is little knowledge of social capital and its impact on development. Social capital, the norms and social relations embedded in social structure including the family, tribes, firms, and formal and informal institutions, influences development processes. Although, sharp decline in trust for many formal institutions is evident throughout the FSU, less is known about the interaction of formal and informal institutions and the extent to which social networks reinforce "anti-modern" institutions.[14] Until 1928, rural production was carried out by individual households and although in 1917 land ownership was vested in the state, "the peasants viewed it as their property.[15] By 1940s early all farms were worked collectively, but household plots individually held by rural and urban families contributed a quarter of the total production. During the process of collectivizing Soviet agriculture, the "production brigades" organized labor for both state farms (sovkhozy) and collective farms (kolkhozy).[16] Although these were supposed to supplant family ties among workers and destroy peasant communial solidarity by re-molding peasants into industrial workers in practice the residential and the work group largely overlapped. Although unrealistically low state purchase prices caused the disappearance of incentives for collective work, by 1979 a new emphasis was put on "collective contracts" and attempts were made to ensure that family ties reinforce work group solidarity.[17]

Several SAs show that social capital based on tribal and lineage ties is strong and important for community mobilization and social safety net functions. However, the extent to which kinship continues to be an organizing principle for the functioning of formal institutions and associations has not been researched. The case from Turkey draws attention to the knowledge gap concerning the civil society and its relevancy for reducing potential

---

while "Rumanian citizens lived in utmost misery" Leszek Kolakowski. 1992. "Minds and Body." In Kazimerz Z. Poznanski. *Constructing Capitalism.* Westview Press.

[14]  Partly because of demographic factors and partly because of the Soviet ideology, many components of social capital inherent in transition economies were destroyed. For instance, one of the legacies of the command economy was the reliance on residential institutions and lack of community-based care for vulnerable groups, including the elderly. "Almost a million …vulnerable or disabled children in the 26 countries of Eastern Europe and the Former Soviet Union live in early years of their lives isolated in 4,500 regimented, large residential institutions that stunted their growth." David Tobis. 1998. "The Transition from Residential Institutions to Community-based Services in EE and the FSU." Mimeo. The World Bank. To re-create community institutions to provide social safety nets, however, will take more time and effort than that required for their destruction. On situational aspects of social capital, see Richard Rose. 1998. "Getting Things Done in an Anti-Modern Society: Social Capital Networks In Russia." The World Bank.

[15]  Csaba Csaki, Kenneth Gray, Zvi Lerman, and William Thiesenhusen. 1992. "Land Reform and the Restructuring of Kolkhozes and Sovkhozes." Working Paper 3. The World Bank.

[16]  The differences between these two forms were effectively eliminated by 1960s.

[17]  Don Van Atta. 1990. "Toward a Soviet Responsibility System? Recent Developments in the Agriculture Collective." In Kenneth Gray. 1990. *Soviet Agriculture.* Iowa State University Press. Ames. 130-154.

adverse impact of social capital based on tribal relationships.[18] It recommends in-depth studies of social capital and of the interrelationship of formal and social organizations. In other cases, such as Uzbekistan, the qualitative nature of community bonds is not well researched in relation to the proposed Project. How these affect the creation and functioning of new institutions such as the free market, the private sector, water user associations, and others, is not fully examined by the SAs, but attention is drawn to the need for further comprehensive work.

Trust in the institutions and members of a society is also an important indicator of social capital and reflects the ability of individuals to establish bonds and confidence in institutions and groups other than the nuclear family. "Singlestranded" and "multistranded" coalitions have characterized many peasant societies and while some peasant societies organize coalitions along kinship lines, others more stratified are organized through patron-client relationships.[19] Ties of extended kinship, friendship, neighborliness, and mutual aid within the community often cushions the shocks of dislocation or cyclical poverty. The cross cutting alignments also strengthen community relations. "Rich and poor peasant may be kinfolk, or a peasant may be at one and the same time owner, renter, sharecropper, laborer for his neighbors and seasonal hand on a nearby plantation. Each different involvement aligns him differently with his fellows and with the outside world."[20] This may enhance trust at the community level. Yet, the dichotomy of the indigenous local institutions and the local administrative structure superimposed from outside to rural societies as well as their greater exclusion from national politics leads to distrust. At time of externally induced change as that observed in the transition economies further augments distrust and requires that policies and projects aiming to induce such change make a special effort to inform rural people of these.

The SA in Kazakhstan provides a good example as to how through the SA process lack of trust can be identified as an issue and how a development initiative can be designed to enhance trust. It shows that lack of trust is widespread and governs relationships to local and national formal institutions as well as some of the non-governmental bodies. Therefore, the SA suggests strengthening social capital through systematic dissemination of information and provides specific input to the formulation of the impact monitoring procedures for the Project. It maintains that sharing information based on the monitoring of the farm privatization process can enhance transparency in agrarian reform implementation and assure stakeholder support. It also shows that a business plan and loan based on a narrow assessment of the potential commercial operations of newly restructured farms, without assessing institutional concerns regarding landholdings and leasing provisions, could both exacerbate social inequities and give a false picture of financial viability.

---

[18]      Ashutosh Varshney. 1998. "Ethnic Conflict and the Structure of Civic Life." Mimeo.

[19]      Eric Wolf. 1970. "Social Aspects of Peasantry." In Marshall Sahlins. ed., *Foundations of Modern Anthropology*.

[20]      Eric Wolf. 1969. "On Peasant Rebellions." *International Social Science Journal* 21. P. 368.

### Social Diversity

In the past, operationally relevant knowledge on social diversity has been a weak feature of Bank-financed development initiatives. Indeed, in 1993 many in the Bank considered these issues "controversial"; the operational implications of ethnic diversity are only recently considered.[21] The Bank's accelerated efforts to assist reconstruction of countries in ethnic or religious conflict have enhanced the focus on social diversity, and research has multiplied over the past five years. The contributions to this volume analyze some of the diversity issues with respect to development outcomes. This is an important concern of SAs since issues of social diversity, social capital, and social equity are often closely interrelated. The ethnic factor was an important dimension of adjustment to the privatization process in Kazakhstan; families who were there temporarily and wished that at least their children would return to their homeland were less motivated to become private farmers and succeed.

Sub-regional and community specific heterogeneity of a national society is likewise an often ignored but a critical dimension of social diversity.[22] Attempts to understand variability and commonality of local responses to general processes, including changes in property regimes are few but important in structuring institutional responses.[23] Likewise, the role of the regional leaders and their local/national networks, their ability to forge ties at various political levels to stabilize their power require in-depth knowledge that is often difficult to obtain during the SA process, primarily because of the time and resource constrains. Although sub-national variability is identified as a key social issue in Azerbaijan, an adequate understanding of the root causes of observed inequities is not provided. However, this concern is incorporated into the social monitoring of the Project so that after a period of implementation a deeper understanding of sub-national elements of development can emerge.

Gender is another aspect of social diversity that is considered in most SAs. An adequate consideration of gender would require that exclusions based on gender are documented and the contributions of various institutions are analyzed. Whether these are based on traditional patriarchal systems, or caused by formal legal framework, or induced through local institutions determine the ability of policy makers to affect changes in existing relationships. Inadequate understanding of the inter-institutional dynamics in shaping gender relations could result in unrealistic expectations about changes that can be affected through project specific interventions or in attempts to bring these changes about through inappropriate institutional arrangements. A focus on gender within the SA context would also mean a search for an equitable participation of women and men in a) the SA process; b)

---

[21] Ayse Kudat. 1999. "Ethnicity in Central Asia: An Update of Trends and Implications." Draft Mimeo. The World Bank.

[22] Assumptions of unitary national culture can be extremely misleading as shown by Julian Steward, etc. al. 1956. The People of Puerto Rica. Urbana: University of Illinois Press. See also Stephen Knack. *Social Capital, Growth and Poverty: A Survey of Cross-Country Evidence.* The World Bank.

[23] Eric Wolf. 1989. "Distinguished Lecture: 'Facing Power—Old Insights, New Questions.'" 88th Annual Meeting of the American Anthropological Association.

among the beneficiaries; and c) the contributors to policies/projects. However, gender issues are often raised with respect to benefit distribution. A focus on how women could be mobilized for the implementation of development initiatives is often ignored. There are growing number of projects where analyses of women's constraints, priorities and needs have helped maximize the productive contributions women can make to agricultural development provided that gender specificity in the delivery of project inputs (e.g., information, credit, etc.) is achieved.

Typically, Bank-financed rural development or agriculture projects systematically focus on gender issues, recognizing the key role of women in the productive process and the need for effective targeting of women. In the transition economies, however, gender analyses of agricultural or rural development projects have so far been weaker than for similar projects in other geographical regions. While some of this relative oversight can be explained by the fact that many of these FSU projects aimed at system wide land reforms and farm restructuring efforts through policy based lending, there still remains a relatively visible lack of gender focus in their design and implementation. This is despite the fact that de jure and de facto inequities with respect to land ownership and/or land registration is documented for several countries as noted in Chapter 1. Even in the case of the restructuring of the livestock farms, equity of distribution varied from one area to another, often associated with the position in the administrative hierarchy, gender and age of the beneficiaries.[24] It appears that the criteria of property share distribution discriminate against women who tend to have shorter years of work experience (being eligible for retirement five years earlier than men). The concentration of women in labor intensive non-mechanized agriculture also meant lesser likelihood for women to gain access to property shares in farm restructuring. In Central Asia, women hold fewer full time positions as agricultural wage earners than men and are less likely to work outside the collective farms. They are also over represented in the social sectors that have been hard hit by the economic crisis of the 1990s.[25]

While several dimensions of gender are captured by the SAs included in this compendium, there appears to be dramatically little documentation of the roles women play in rural communities of the FSU.[26] It is argued that the Soviet legacy implies similar influences on women in the different transition countries.[27] Some believe that the concentration of women in the livestock farms and the restructuring of this sector would create disproportionately adverse impacts on them. Similar results are expected from the concentration of women in occupations such as manual field labor and bookkeeping. The rapid loss of social infrastructure that was once enjoyed by rural women is also expected to

---

[24]     Marnia Lazreg. 1998. "Review of the Status of Rural Women in Transition Economies." Mimeo.

[25]     Jeanne Koppman. "Gender Issues in Farm Restructuring in Uzbekistan and Kyrgyzstan." ICRW.

[26]     G.T. Robinson. 1972. *Rural Russia under the Old Regime.* Berkeley; Thomas S. Pearson. 1986. "Authority and Self-Government in Russian Peasant Administration." Mimeo.

[27]     Sharon L. Holt. 1995. "Gender and Property Rights: Women and Agrarian Reform in Russia and Moldova." Mimeo. The World Bank.

cause adverse impacts across the FSU countries.[28] It is also argued for the Central Asian republics that "re-traditionalization" has weakened the status of women. Starting in the late Soviet period, girls started dropping out of schools in Uzbekistan, re-allowing polygamy was discussed in Tadjikistan, and the social control of the elders (aksaqal) on women strengthened through the re-emergence of the system of mahalle. The events surrounding the "re-islamization" appear to add yet another blow to women.[29] While there is merit in these observations, the empirical reality appears to be far more diverse and complex both in the past and in what is yet to come.[30]

There seems to be an implicit assumption that under socialism women gained a more equitable place in society and that both in the private (household plots) and in the collective sector they stood on equal footing as men as producers and wage earners. In a sense, the availability of social infrastructure was the means through which women's productive capacity for agriculture was captured. Again, the reality on the ground not only indicates a historically marginal presence of women within the village assembly and the traditional commune management based on patriarchal lineages, but a visible current presence of gender specific social hierarchy and continuing exclusion of women from leadership. Even under the collective system and the livestock farms where women are said to have heavily concentrated, top management roles for women are exceptions rather than the rule.

Gender was shown to be of critical importance in Turkey and the SA documented the presence of system-wide gender inequities. It showed how land consolidation, for instance, may expedite the process of women's access to property despite strong traditions that continue to hinder this. In Azerbaijan, the Bank-financed Project was prepared in partnership, among others, with one of the key NGOs concerned with women-in-development and specifically addressed the gender dimensions of Project components. It dealt, in detail, with gender targeting of the credit and advisory services incorporated in the Project.

Other elements of social diversity are more often neglected in project analyses and SAs. For instance, although closely related to gender inter-generation issues and population aging have been ignored primarily because of the overwhelmingly young composition of populations in developing countries. The countries of Eastern and Central Europe, and the Former Soviet Union are however dramatically different. A recent study of Kudat and

---

[28]    Holt shows that in Armenia day care centers did not decline with privatization because few state or collective farms provided these services to start with. Sharon Holt. 1995. "Using Land as a System of Social Protection: An Analysis of Rural Poverty in Armenia in the Aftermath of Land Privatization." Mimeo. The World Bank.

[29]    Roy, ibid.

[30]    A quantitative study of wage and price liberalization in Slovenia indicates higher returns of transition to women because women occupy sectors less adversely affected by the transition. Peter Orazem and Milan Vodopivec. 1994. "Winners and Losers in Transition." Mimeo. The World Bank.

Youssef[31] shows that in these countries the elderly constitute nearly a fifth of the population.[32] Moreover, in many countries, including Ukraine, Russia and Belarus, older people are disproportionately represented in rural areas. In Belarus, for instance, nearly a third of the rural population is over 60 years of age (Figure 2.2).

Most of the Central and Eastern European countries have a relatively large part of their population living in rural communities with a small number of inhabitants, dispersed settlement patterns, and a low population density. Not only the majority of the elderly are women, but also in some countries older women are more likely to reside in rural areas than are older men. Rural pensions are lower than urban pensions, but in many countries age has enabled the elderly to receive a slightly larger share of the distributed assets of the restructured farms. Despite their limited ability to cultivate the land and the pressure from many, including the ex-managers of state farms, to forego their entitlements, the elderly continue to contribute to agricultural production with help of their kin. Thus regional issues such as population aging are key social development issues for the SA.

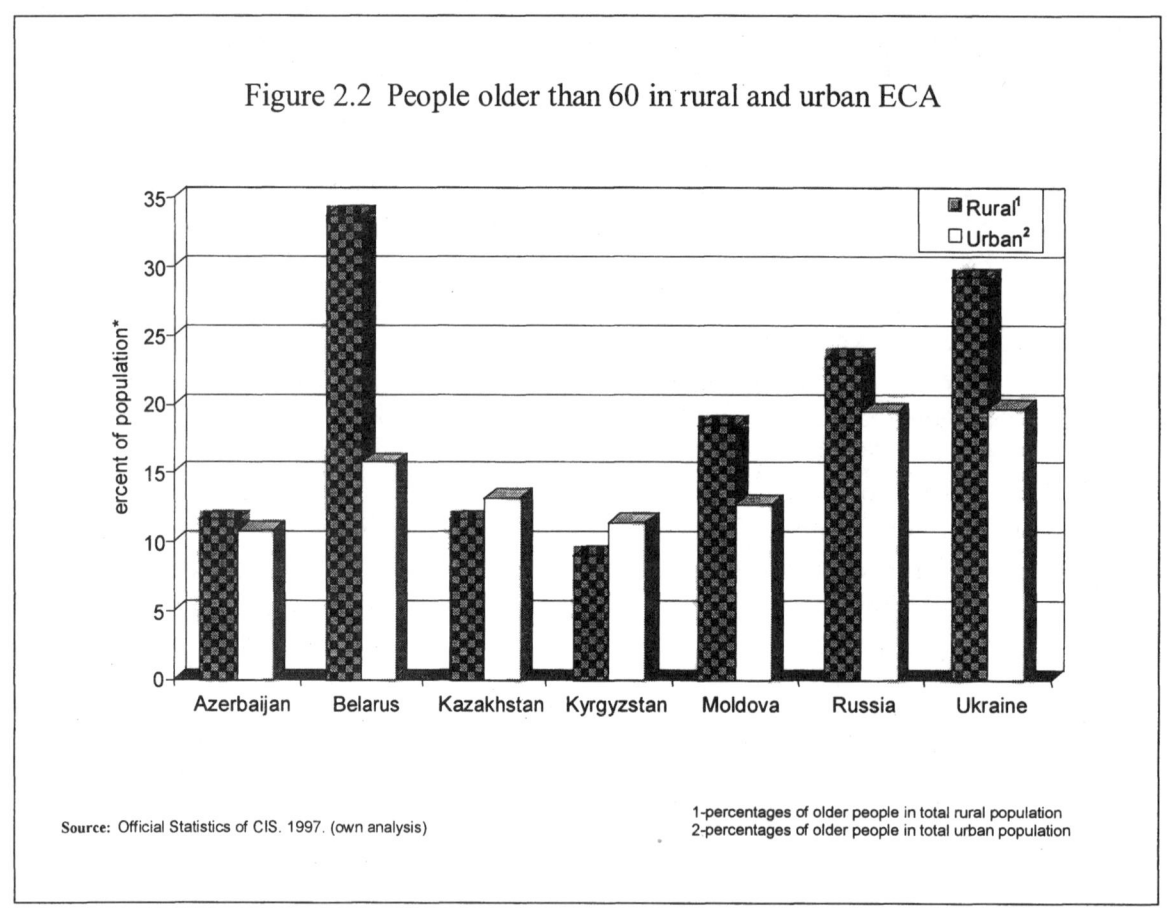

Figure 2.2 People older than 60 in rural and urban ECA

Source: Official Statistics of CIS. 1997. (own analysis)

1-percentages of older people in total rural population
2-percentages of older people in total urban population

---

[31]     Ayse Kudat and Nadia Youssef. 1999. "Older People in Transition Economies: An Overview of their Plight in the Europe, Central Asia Region." The World Bank.

[32]     Recent evidence also points to a disproportional concentration of older people in the remote rural areas of Turkey. Ayse Kudat. 1999. "Social Assessment of the Turkish Forest Sector." The World Bank.

### Identifying Social Development Issues in the Specific Project Context

We have provided some illustration as to how the broader knowledge of social development issues can help identify issues of potentially relevance for a specific project. By knowing the nature of poverty, or of social structure, social capital, and social diversity within a geographical region or a country and by having an understanding of sector specific and country wide issues useful working hypotheses can be formulated on social issues to be considered by the project. As the number of SAs undertaken in a given set of countries or sectors increases, the ease with which new ones can be launched also increases. The Kazakhstan SA, for example, pursued many issues identified in the SA for an irrigation project in the same country, which described inequities in the farm privatization process and analyzed the role of managers of former collective and state farms in subverting the restructuring process. The Azerbaijan case was built on lessons learned from Kazakhstan and Moldova and on earlier SAs carried out for different projects in Azerbaijan as well as from the stakeholder feedback received during the supervision of a Bank-financed Pilot rural Project.

Once social scientists have carried out SAs for several different projects in a country, their ability to do project-specific social and institutional analyses and stakeholder identification increases. Often, the empirical knowledge and analytical approaches developed for one SA provide useful inputs for others, building, at the same time, capacity for social assessment within the local social science community. Capacity building is an important by-product of the SA process in ECA where the Bank has established Social Science Networks and help strengthen them through their contributions to a series of social assessments carried out for various development initiatives.[33]

After reviewing and clarifying broad social development issues, the SA proceeds to focus on how the participation of the poor and vulnerable groups may be affected by the proposed project and how their participation may impact project ownership and sustainability. The SA presents analysis of potential gains and adverse impacts with a view toward understanding how specific stakeholder groups may facilitate or hinder the participation of the poor in the development initiative. More general issues such as social cohesion, equity, social capital, social diversity, social organization, and social exclusion may also be identified as they pertain to the project or the economic sector work. The needs and interactions of significant social groups are considered, including gender. These concerns were particularly relevant for the cases included in this volume; for instance, understanding the composition of social capital, whether it leads to maintaining the status quo of exclusion, aggravates

---

[33] Social Science Networks are associations of social scientists that were established in four Central Asian Countries in 1994 with funding from the Swiss Government. Through this generous support, the Bank was able to support capacity building for applied social sciences in these countries.

institutional constraints, and blocks opportunities is important in the specific project context in recommending policies to foster equality of opportunity.[34]

General knowledge of the social map of an area is often insufficient to guide the formulation and implementation of a project. In the case of Turkey, there was ample knowledge about the presence of large asset inequities in the area prior to the start of the Project specific fieldwork. This did not help determine whether or not the proposed Project would exclusively benefit those who are already well off; this had to be shown to both the Bank and the borrower through systematic research. Because inputs similar to those proposed for the Bank-financed Project had been received in other sub-regions a "control group" analysis was carried out. Sub-regional comparison showed that while landowners would continue to receive a large share of the benefits, the landless would benefit disproportionately provided that labor intensive cropping accompanies irrigation investments. Comparing the project area with irrigated areas in other provinces also suggested that supporting the process of democratization of irrigation associations and strengthening other civil society organizations such as farmers' cooperatives would further help the benefits to trickle down from the landowners to the landless. It was shown that the income inequality that existed in the highly stratified project area communities would be somewhat reduced through the land registration process. Because some of the landless are direct heirs of the economically dominant groups, newly emerging wealth would also be shared.

### *Second Pillar: Stakeholder Identification and the Participation Framework*

#### **Stakeholder Identification**

Stakeholders in any project or program will include various social groups, formal and informal agencies in the public and private sectors, and civil sector and non-governmental organizations (NGOs). Stakeholders are groups of people, connected to one another through formally or informally defined ties, that have something to gain or lose from the proposed development initiative. Some aim to support the poverty focus and inclusionary objectives of Bank-financed activities, and it is therefore important to ensure their participation in projects. Others may aim to block social development efforts; it is therefore important to understand how to reduce or eliminate their opposition. A review of the available secondary literature and initial fieldwork will reveal which groups and agencies are most directly concerned about the initiative. A tentative listing must be complemented by consultations with policy-makers, representatives of central and local government, knowledgeable local and international social scientists, and local NGOs.

In many rural development projects, central and local government agencies, private sector institutions, and civil society groups all play important roles. When the rural

---

[34]     Sidney Tarrow. 1994. *Power in Movement: Social Movements and Contentious Politics*. Cambridge: Cambridge University Press; Kentworthy Lane. 1977. "Civil Engagement, Social Capital, and Economic Cooperation." *American Behavioral Scientist* 40(5):646-657.

development agenda is broad and covers production in agricultural and non-agricultural sectors as well as infrastructure, a very large number of stakeholders will be relevant. Even with a narrower focus on agricultural production, the number of stakeholders is large. They include:

- State and local level government;
- Agrarian producers, and their organizations;
- Laborers and their organizations;
- Owners of productive resources and their organizations;
- Consumers (buyers) of inputs and outputs and their organizations;
- Users of services and their organizations;
- Traders of inputs and outputs and their organizations;
- Suppliers of knowledge, technology, and know-how and their organizations; and
- Religious and other social groups.

The State sets up the formal rules and laws, including policies that determine such key issues as land reform and private sector development, as discussed in many of the SAs included in this volume. It also plays a crucial role in allocating the budget, including donor financing, to various regions and programs based on political pressure group dynamics. It defines private property rights and insures that they are protected. Formal institutional arrangements and the local social organization support or react to these. For instance, in Southeast Turkey the highly patrilineal norms often violate the State imposed gender equity built into the land consolidation process. The local governmental apparatus, influenced by the tribal groups, do not always have the power to enforce the formal rules.

Each group of stakeholders has its defining characteristics, and it takes specialized expertise to identify them. In the agrarian production sub-sector of sheep raising, the producers themselves belong to different social groups, and their interests may not coincide. Various ethnic groups compete for resources, and herders' interests are different from the interests of the crop producers. Subsistence producers are organized differently from producers of various cash crops; certain groups of producers, such as cotton growers, have formed associations that bring producers of different social origins together. Some producers have formed farmers' associations to improve access to inputs or markets while others come together in multi-purpose cooperatives. Some producers are landless while others have various assets with different ownership patterns shaping the stakes vis-à-vis other producer and stakeholder groups. Depending upon how producers are organized, they may have a strong influence on local or national politics.

Often, large landowners play key roles in local/regional politics and hold key political posts in political parties and even parliaments. Because of their potential for capital accumulation, they are often among the prominent merchants, traders or even industrial producers. Thus, their stakes as agrarian producers may overlap with their stakes as traders, buyers, and suppliers of inputs. Some members of a few large lineages monopolize key stakeholder positions in economic and political spheres and use their monopoly to block entry of other social groups. Sometimes there is conflict between stakeholder groups. For instance, traders of livestock who smuggle cattle across borders are able to supply animal products to markets at low prices, and local producers are not able to compete. Men of influence are able to exert pressure on the government to impose taxes on the import of certain crops to protect their interests against importers.

In many ECA countries, including Turkey, there are still large number of state farms that produce major agrarian products and buy inputs, services, and technology on a large-scale. Their laborers, wage and salary earners, professional staff, and managers all have stakes as does the State. With globalization, producers, traders, suppliers, and other sector stakeholders in other countries play an important role in local decisions. Indeed, governments and many farmers are aware of this; they take these stakeholders into account in their decisions on subsidy arrangements, crop diversification, irrigation expansion, inputs, and imports/exports. Sometimes they are forced to do so through international trade rules, international waters agreements, donors' constitutions, and many other institutions that once were far removed from the reality of the peasant life. Imperfections in the information/communications systems and opportunistic incentives of each set of actors will balance their potential to maximize net benefits.

The consumers or buyers, providers of services and technology, and input suppliers are likewise diverse. Foreign and local input or technology producers often have conflicting interests and different linkages between trade groups. Some are well organized and others are new to the local scene. Some have better social connections within a region than others. Some have large margin of flexibility through membership in national associations of people with similar interest and can exert collective pressure on both the national politics and associations of other stakeholders who are likewise brought together around common economic interests.

These illustrations are applicable to much of the rural landscape in Turkey. There numerous producers, buyers, traders, and sellers associations co-exist with local branches of national political parties and state organizations, strong tribes and lineages, ethnic groups, the research and academic community, and large numbers of NGOs. A thorough examination of how each of these plays a role in rural development is not an easy task. When the question is narrowed down to livestock, for instance, the institutional analysis would involve the identification of the stakeholder groups involved in commercial cattle breeding, free grazing, meat, milk and milk products, skin and skin products, wool, and bones. For each, there are producer, consumer, buyer, trader, importer, exporter groups and their interests; there are also the interests of laborers for each of the stakeholder groups.

An assessment of the political feasibility of the governments' ability to implement reforms is also important. Often other branches of government block policy changes or local governments have sufficient power to oppose reforms. The assessment of the feasibility of government support to solutions proposed by the poor and other vulnerable groups through major reforms ought to be complemented by a sustainability analysis so that the credibility of government promises and government stability can be taken into consideration. When government stability is low, ensuring the support of other key stakeholders would be especially important.

It is important to stress the membership overlap between stakeholders and the implications this has for the formation of powerful rent-seeking groups. Wedel points out that in each of the FSU countries "a small pool of people who are, simultaneously, businessmen, bureaucrats, members of parliament, and representatives of the press are active in policy, government, the economy, and the international arena...These powerful groups have every opportunity to pursue their private agenda...and undercut legitimate state institutions and governance, especially where the rule of law is weakly developed."[35]

A thorough stakeholder analysis for a specific sector requires time, resources, and a great deal of local knowledge and expertise. Because these are not easy to mobilize in the context of a project, much of stakeholder and institutional analyses is simplified to several categories: people, communities, formal public sector institutions, and NGOs. In more sophisticated analyses, "people" are divided into various ownership, gender, and ethnic categories; "communities" are defined as villages or neighborhoods; formal institutions are listed, and the names of a few advocacy or humanitarian groups are mentioned. There is a virtual absence of a complete analysis of the civil society, including NGOs, trade unions, professional associations, user groups, consumer groups, lobby groups, the media, private (or profit) sector institutions, their relationship with the governmental and administrative apparatus, and their role within the political system. Stakeholder diversity may not appear to be wide at the outset of project preparation, but it often broadens during project implementation due to the rapidity of change in transition economies.

### Formulating the Participation Strategy

Participation is perhaps the most difficult of all challenges in the SA process. A majority of the SAs now available in the Bank do not concern themselves with stakeholder participation in a broad sense. Rather, they focus primarily on the participation of the poor and, occasionally, of certain segments of civil society organizations. This is understandable in that the Bank as a development agency is particularly concerned with poverty alleviation and social inclusion. However, to extend concern for stakeholder participation to other critical groups is important for many reasons, including to build critical alliances to promote the participation of the poor and to reduce resistance of others.

SA should involve the consideration of several types of participation:

- Stakeholder participation in the design of a development initiative or the SA process;

---

[35]      Janine R. Wedel. 1994. "US Aid to Central and Eastern Europe, 1990-1994: An Analysis of Aid Models and Responses." In *East-Central European Economies in Transition. Study Papers Submitted to the Joint Economic Committee Congress of the Unites States.* US Government Printing Office. Washington D.C. P. 314

- Stakeholder participation in the benefits of development initiatives, with a focus on the poor and the excluded; and
- Participation in project implementation and maintenance and operation arrangements.

Participation is often sought for project design but the stakeholders are limited primarily to the central governmental institutions. Indeed, the "client orientation" that has found many advocates in the development institutions has in practice meant in the Bank the establishment of a dialogue with more than the primary borrower and the organization of workshops or other forum to receive the endorsement of several governmental ministries and select non-governmental organizations. The incorporation of the SA process to Bank-financed projects meant a broader and more systematic approach to stakeholder participation. Needless to say, the SA process itself can facilitate stakeholder participation but not ensure it. First, in some countries the principal "counterpart" agency or the "key borrower" may not support broader participation. Second, stakeholders whose active participation is sought may be uninterested or reluctant to dialogue; it may even be important for them to demonstrate uncooperative behavior for continued support of their constituency. Thirdly, direct participation in policy/project design may not be feasible. For instance, when the development initiative covers a large population, social and institutional analyses could provide important information based primarily on representative surveys about the views of the poor and vulnerable populations and the manner in which they are systematically excluded from benefits. In other instances, the policy/project context may preclude democratic representation in the determination of targeting arrangements. Nevertheless, there has been progress in designing methodologies to listen to the needs of the poor and vulnerable. Also, there is greater realization that consultations with NGOs and other organizations are not a substitute for direct involvement of the poor in the design of the type of assistance aiming to reduce their poverty.[36]

Often, it is easier to facilitate stakeholder participation at the outset of the development initiative; indeed, most of the SAs included in this volume were successful in conducting systematic consultations with a range of stakeholders during their preparatory phase. Few, however, propose concrete steps to sustain such participation with respect to the monitoring, evaluation and re-design of the projects. In the Turkish case, an evaluation carried out with stakeholders with communities that have already benefited from initiatives similar to those proposed for the sub-region where the Bank-financed project will be implemented will influence the design. Also, a participatory evaluation may be included in the proposed Project should an agreement is reached with the borrower. In the case of Azerbaijan, an explicit participatory monitoring component is incorporated in the Project. Elements of participation are also built into the monitoring of the Kazakhstan Project. In the case of Uzbekistan, however, such elements are largely missing.

Mechanisms for sustained participation in the SA process are not easily identifiable. In other words, how a broad range of stakeholders would be able to participate in the evaluation

---

[36]     Indeed, it was the President of the World Bank J. Wolfensohn who explicitly pointed this out in 1996.

of the implementation of a development initiative has not been systematically discussed. Beneficiary assessments provide one approach but focus on a partial set of stakeholders. Few projects succeeded in monitoring development efforts from the perspective of beneficiaries, adversely affected populations, local governments, civil society organizations such as the trade unions, but have not yet reflected on the implications of these specific cases for other projects.[37] The success of large-scale projects that involve reforms such as rural enterprise restructuring, land reform, privatization, etc., often depend on the support of the public at large as much as the stakeholders more directly involved. Therefore, each stakeholder group as well as the public at large requires adequate information. They also need mechanisms through which to express their perspectives. Thus, for projects that support important reform elements a carefully designed information/communication (I/C) strategy is needed to support the SA process in ensuring participatory stakeholder feedback.[38] Such a strategy usually has several elements:

- Mechanisms to share the information from the social and institutional analyses with the broader group of stakeholders and partners including national and international governments, civil society groups and communities;
- Mechanisms to ensure continuous provision of information to key stakeholder groups, including the public, throughout project implementation as an incentive to their participation; and
- Feedback mechanisms to ascertain stakeholder responses to the information provided.

Concerning the participation of stakeholders among beneficiaries of development initiatives much has already been written over the past four decades and these often concern issues of targeting. Targeting issues first center around institutional and organizational arrangements as discussed in the next section. They also concern issues of information/communication. It has also been shown that information is as crucial in agricultural production as is land, labor and capital particularly in areas of poverty reduction, increased participation, and improved governance.[39]

What happens, however, if rural people or certain segments of the rural people are excluded from development or reform benefits? The strategies consist of exit—leaving the rural landscape—or voice. While peasant revolutions, uprisings or other forms of voice have occasionally took place in history, peasant societies are more known for their tendency to opt for "exit" in the form of outmigration in response to the difficulties they encounter in the rural

---

[37]     Ayse Kudat. 1998. "Participatory Monitoring of the Russia's Coal Sector Restructuring: A Social Assessment Case Study for Learning Exercises." Presentation in the Social Assessment training courses. The World Bank.

[38]     Needless to say, such a strategy would not be needed and affordable for many development initiatives; however, those of a reform nature would substantially benefit from the support of an I/C component.

[39]     Willem Zijp. 1994. "Improving the Transfer and Use of Agricultural Information." The World Bank.

sector. "Past exclusion of the peasant from participation in the decision main beyond the bamboo hedge of his village deprives him all too often of the knowledge needed to articulate his interest with appropriate forms of action. Hence peasant are often merely passive spectators of political struggles ...and are slow to rise."[40] Voice can also be expressed on behalf of peasants or certain segments of the rural populations by political parties, civil society organizations and the voluntary sector and creating alliances to put pressure often on outside agencies. Increasingly, the advocacy role of intermediary institutions to seek alliances within the rural sector and to support the development initiatives' potential for reaching the beneficiaries is recognized.

Perhaps a more neglected dimension of participation and one that should constitute a major focus for the SA process concerns the mobilization of the rural people for self-help. Basically, development ought to be supporting the human potential for fulfillment, growth and enrichment. As such development projects should merely be catalytic in energizing the rural people to attain their potential for change along the direction that they themselves choose. Ideas, experience, knowledge, information, labor, capital and organization are among the key factors that people contribute to projects. Among these, participation in cost recovery (financial contribution), labor and its organization (community participation) are the more frequently explored issues. While self-help measures focus on creating values through the mobilization and utilization of a range of resources, tending to be pro-active and positive sum, many of the advocacy measures aim at redressing imbalances or inequities, seeking for instance different subsidy arrangements, law enforcement, etc., and tends to have zero-sum (redistributive) effects.

The potential for communities to pool their resources in order to achieve common objectives and to protect themselves against various types of shocks, including severe economic shortfalls is a known and insufficiently explored issue, especially in the specific context of projects. Also, participatory mechanisms that rural societies have traditionally had were also explored although inadequately in the context of the Former Soviet Union. For instance, communal systems that had been central to Russian peasant societies survived the days of serfdom from the seventeenth century through the first half of the nineteenth century, the pressures of the 1861 emancipation and the 1907 Stolypin reform's attempts to dismantle them. The Bolsheviks were forced to support the peasant communal system which resisted outside interference until collectivization and dekulakization.[41]

The ability of the Russian peasants for communal production and mutual support has not only helped policy makers to mobilize peasants throughout the period subsequent to 1929 but also during the transition. The participatory or communal traditions and the habits of collective work continue to be important in providing social safety nets to the rural people during transition and are likely to help the formation of community based organizations such

[40]     Eric Wolf. 1969. "On Peasant Rebellions." *International Social Science Journal* 21. P. 369.

[41]     Christine D. Worobec. 1995. *Peasant Russia: Family and Community in the Post-Emancipation Period.* Dekalb: Northern Illinois University Press.

as water user associations, production associations or credit and saving associations. Likewise, in Central Asia and Turkey, the indigenous systems of mutual help and cooperation for use of and production on communally held land and pastures were always important. These provide important social capital on which policy makers were able to build a land consolidation implementation agenda. They can also enhance cost recovery of rural infrastructure projects, such as roads, water supply, etc., by having the villagers pool their labor to reduce wage payments for implementation.

Many of the above mentioned communal and self-help traditions that mobilized the peasant societies and secured community participation while protecting the vulnerable co-existed with other traditions, namely patriarchy. This meant that cooperative norms would be enforced and well-defined leadership structures would constitute key social organizational elements to mobilize community contributions. However, this also meant the existence and continuity of exclusionary practices particularly with respect to women.[42]

Community based and self-help systems and other participatory arrangements were evident in many other spheres of live in socialist countries. Eastern European reform economies experienced a significant increase in self-help activities such as building of one's own house with the help of professionals or friends such that by 1988 two-thirds of residential property constructed in Hungary was organized by private owners who enjoyed rental incomes.[43] Self-help initiatives in agriculture and other sectors were important elements of the emergence of the private sector as well. Employee self-management systems developed in some of the socialist countries constituted "best practice" examples for participatory management for western countries.[44] In other words, the state imposed, top-down, anti-democratic, non-participatory images created for the socialist economies were neither universally valid in geographical scope nor were the dominant forms throughout the history of individual states; in each, a mixture of community, private and state sectors existed each displaying different modalities of participation.

The issues of participation in cost recovery have only recently been included in the social scientists' agenda but as mentioned above were embedded in different community based and private sector arrangements both in the socialist economies and Turkey.[45] The considerations of willingness to pay and affordability (or ability to pay) are common to both the economists and non-economic social scientists. The concern with affordability ultimately addresses the question of exclusions and how can goods and services be provided to those who cannot afford to pay for them. The concern with willingness to pay not only shows the

---

[42]    Ibid.

[43]    Janos Kornai. 1992. "Ownership and Coordination Mechanisms." in Poznanski, ed., ibid.

[44]    Bernard Wilpert and Ayse Kudat. 1979. *Participatory Worker Management*. Kent University Press.

[45]    Ayse Kudat. 1995. "Baku Urban Water Supply and Sanitation Project." The World Bank; Kudat 1996. "Uzbekistan Rural Water Supply and Sanitation Project." The World Bank; Kudat 1996. "Turkmenistan Urban Transport Project." The World Bank.

type and level of services people want but also the degree to which communities are prepared to contribute to receive these. There is yet another and perhaps more important dimension of inquiries on cost recovery and this has to do with revealed preferences.

Uphoff (1998) points out that opportunities for participation in decision-making, implementation and/or evaluation are shaped structurally by the kind (sector) of institutions involved.[46] These will affect people's participation in benefits. Rather than be satisfied with assessing simply whether or not people are participating, it is essential to ascertain exactly who participates, how, at whose initiative, with what incentives and regularity.

### *Third Pillar: Institutional and Social Organizational Issues*

An adequate stakeholder analysis is closely connected with institutional analysis; institutions are the rules of the game as they govern the relationship of stakeholders within and among themselves.[47] Stakeholder groups are formal and informal organizations that constitute the major elements of institutions. The poor and vulnerable groups, the Bank's development focus, belong to families, lineages, tribes, ethnic groups, village communities, farmers' cooperatives, water user associations, and many others. Each group has norms, the formal and informal rules that define membership and regulate relationships among groups. Social analysis of culture, norms and values, and social organization is a critical ingredient of institutional and stakeholder analysis. Because social systems are diverse and bring together many different nationalities, ethnic, and indigenous groups, the social or informal institutions are far more complex than is usually considered in the analyses of formal organizations.

One important element of the pillar is the analysis of the relative influence of formal institutions and informal and traditional ones. Experience has shown that while formal institutions establish the official rules of the game (laws, regulations, and so on), it is a serious mistake to assume that people behave according to the official rules. Rather, informal institutions may actually govern people's behavior. Failure to understand the nature of formal and informal institutions, and their respective incentive structures, can lead to project designs that are either unworkable or further exacerbate potential sources of conflict or corruption.

Uphoff usefully distinguished institutions from organizations while emphasizing their overlap. "Organizations, whether institutions or not, are structures of recognized and accepted roles, supported by rules, procedures and precedents. Institutions, whether or not

---

[46]     Norman Uphoff. 1998. "Local Institutions as Channels for Participation and Voice in Development." Mimeo. The World Bank.

[47]     North defines institutions as the humanly devised constraints that structure human interaction. They are made up of formal constraints (rules, laws, constitutions), and informal constraints (norms of behavior, conventions, self-imposed codes of conduct) and their enforcement characteristics. Together they define the incentive structure of societies. Douglas C. North. 1986. "The New Institutional Economics." *Journal of Institutional and Theoretical Economics* 142:230-237.

organizations, are complexes of norms and behaviors that persist over time by serving collectively valued purposes."[48]    Accordingly, marriage is an institution, family is an organization that is also an institution, but an individual family is only an organization. Another useful example is to think about "money" as an institution, "central bank" as an institution that is also an organization, but a local branch of the central bank is merely an organization.

The diverse effects of institutions on economic growth have been abundantly demonstrated in the New Institutional Economics literature.[49] The role that certain types of social norms can play in affecting growth-enhancing activities has also been shown. In contrast, the effects of economic development in undermining the efficiency of the family, clan and community as sources of trustworthy information are frequently observed as are the resistance of the traditional institutions to new laws, regulations and norms. Much of the development literature advocates the utility of building on existing institutions, instead of creating new ones, and historical evidence shows the resilience of societies' capacity to preserve its social capital.  Nevertheless, the efficiency of using the traditional institutions to support the creation of new institutions and/or in targeting project benefits is seldom considered, especially in the context of the transition economies. While institution building for equity enhancing reforms is a core goal for many, it meant different things to different people. Institutional reforms "require a more elusive motivation" and the "cooperation and participation of numerous agencies."[50]

The relative success of some societies to adapt to changing circumstances and the social losses that often accompany transition from one set of institutions to another have concerned students of rapid change. The examples chosen for this volume illustrate situations where the rapidity of induced change challenges the validity of old institutions and where new institutional arrangements are still too weak to guide the reforms. Indeed, in transition economies many of the institutions, namely collectives that governed rural social relations have become redundant. While others, including private ownership, co-existed with communal property regimes for many decades prior to their reintroduction as new institutions. At the same time, the communal traditions that once prevailed were undermined during the collectivization process and had lost their strength. In addition, the rapidity of change and the severity of the adverse impacts of transition hindered trust in new institutions and increased transaction costs.[51] Also, the institutional diversity that characterize FSU and their sub-regions

---

[48]    Norman Uphoff. ibid. P. 3

[49]    Mustapha Nabli and Jeffrey Nugent. 1989. "The New Institutional Economics and Its Applicability to Development." *World Development* 17.

[50]    Carol Graham and Moises Naim. 1998. "The Political Economy of Institutional Reform in Latin America." In Nancy Birdsall et al., eds., *Beyond Tradeoffs*. Brookings Institute Press. Washington D.C.

[51]    In transition economies, the creation of informal markets and new property rights elicit a variety of institutional and organization responses.  In the establishment of the new institutions, the transaction costs are often high and attempts to reduce them are visible in, for instance, the Bank-supported efforts of the Government of Azerbaijan in establishing a decentralized system of land registration and titling.

made it difficult to ensure the success of reforms based on universalistic policies with little flexibility for adjustment to sub-national characteristics. Therefore, the interrelationship of institutional and development has become a central concern in transition economies and in other situations of rapid change. Because stakeholder dialogue is important for the design of institutional arrangements and social sustainability of reforms, SAs have become a more central element of project preparation than would have been where policy makers expectations for change were more modest in terms of scale, scope and rapidity.

There are "historical, geographic, religious, legal, and ethnic factors that shape institutions, including the institutions of government. To understand the development of institutions, it is useful to know how much of their variation can be accounted for by these fundamental factors."[52] These factors not only define national institutions but also sub-national, regional and local ones as social heterogeneity characterizes systems at all levels. However, within the SA context the assessment of these factors is often difficult both because of the resource constraints and of the need to carry out this assessment separately for each issue identified. For instance, the institutions relevant for improved gender equity do not necessarily overlap with those dealing with intellectual property. Also, the organizations involved may be different, requiring the participation of separate sets of stakeholders. In this sense, the institutional analyses included within the SA process different from the broader policy context often used in Bank-financed projects.[53]

The institutional analysis carried out in the SA process complements the institutional analyses carried out in the context of the technical, economic, and financial assessments of the project. It focuses on the feasibility of proposed targeting measures, the sustainability of the proposed participation arrangements, the interface between beneficiaries and formal implementing institutions and the relationship between formal and informal institutions, particularly as they respond to new incentive structures. Equally importantly, it identifies social capital on which to build the development initiatives and points to local institutions that can help mobilize stakeholders to achieve development objectives and to preclude the creation of new institutions whose principal objective is to take advantage of new incentives. For instance, the SA of the Azerbaijan agricultural development project makes evident the widespread trust community members has for one another. People borrow from each other without contractual arrangements and families bring their land holdings together in order to enhance the efficiency of cultivation. The SA thus supports the establishment of saving associations to facilitate the development of credit systems when formal banking institutions are weak. In Kazakhstan, the SA has documented the extent to which informal and formal credit are used for different purposes, proposing modalities through which formal institutions must change to reach rural customers.

---

[52]      Rafael La Porta et al.1998. "Fundamental Determinants of Governmental Performance." Mimeo. The World Bank. P. 2.

[53]      Sue E. Berryman et al. 1998. "Guidelines for Assessing Institutional Capability." Mimeo. The World Bank.

## Types of Institutional Analysis

Institutional analyses that focus on policies/projects often ask several questions.

- Who are the actors that influence decisions/outcomes and what are their stakes?
- What formal and informal rule systems characterize the decisions?
- What incentives do the actors have?
- What is the information/communication structure? How can it be changed to support the desired outcomes?
- What changes in rules and incentive systems would be needed to affect desired outcomes?

Several steps are necessary to develop a good understanding of institutional and social organizational issues. The first step may consist of the analysis of local level institutions and the fit between informal institutions and the political and administrative apparatus.[54] The study of the formal and informal rules of the game provides important inputs to the design and implementation of a development initiative and the interaction of these.[55] However, a great many analyses of institutions focus primarily on organizations. Others focus on formal governmental institutions to the exclusion of informal, community, kin, or solidarity-based organizations. The roles of culture and tradition are also generally neglected. Clearly, solidarity networks based on ethnicity, or gender, race, poverty based networks can create economic and political opportunities for their members or can restrict access to opportunities. Some of the steps used in the SAs included in this compendium are summarized below; however, they are not detailed in the chapters presented or contain all the steps required for a comprehensive analysis of institutions. Nevertheless, they offer a useful guide for future SAs of agriculture projects in the ECA region.

### Macro Institutional Issues

In each country, the nature of property rights, subsidy arrangements, and regional investment policies have important implications for social equity and inclusion. The process of land reform in much of ECA has created formidable difficulties in the re-establishment of

---

[54] A. Lijphart. 1991. "Majority Rule in Theory and Practice: A Study of a Flawed Paradigm." *International Social Science Journal: Rethinking Democracy*; J. Manor. 1992. *Rethinking Third World Politics.* London: Longman; E. Ostrom, R. Gardner and J. Walker. 1994. *Rules, Games, and Common-pool Resources.* Ann Arbor: University of Michigan Press.

[55] E. Blackwood. 1997. "Women, Land and Labor: Negotiating Clientage and Kinship in a Minangkabau Peasant Community." *Ethnology*; C. Grootaert. 1998. "Social Capital: The Missing Link." Mimeo. The World Bank; M. O'Connor. 1990. "Women's Networks and the Social Needs of Mexican Immigrants." *Urban Anthropology*; R. Waldinger. 1995. "The Other Side of Embeddedness: A Case Study of the Interplay of Economy and Ethnicity." *Ethnic and Racial Studies.*

private property over collectively Soviet history.[56] Eastern Europe and the Soviet Union "have been forced to radically reform their economies because the traditional property regime-public ownership and state allocation- became completely inefficient given the drastic increase in scarcity of productive factors."[57] As a result, "if we look at the history of the socialist reform countries, we find that without exception, reform blueprints or programs were in circulation before the actual period of the reform."[58] The reform debates of mid-1950s resulted in the emergence of a significant private sector and the most important inroad of private activity in socialist economies occurred through private farming. China developed the family responsibility system, in Poland and Yugoslavia the private farming was never abolished, and increased role of private plots emerged in Hungary and many other countries.

At the outset of the transition period, in many countries there has been government reticence to distribute state land to families.[59] In some cases, former power holders resist change, either by making the distribution of land and assets slow and/or by delaying the process of land titling and land market development. In other cases, the lack of information and facilities related to farm inputs and marketing, coupled with the small size of landholdings distributed to individual families, has led to a slow growth in the spread of individual family farming. Some of these constraints are coped with through the re-emergence of production associations based on kinship, neighborhood organizations, or other types of social capital. In other cases, land consolidation is taking place through the ability of the better off segments of the population to buy or lease land from others and scale up their production.

The SA in Uzbekistan provides a good example of how changes in property rights are affected by other institutions. Particularly important is the interface of macro and micro institutions. Local level variations in the pace of land reform and the development of private property are also market. In Russia, for instance, private ownership of land is legally recognized in all but 10 ethnic republics. Although all countries in East and Central Europe recognize private ownership of land, the rights of the owners are circumscribed. The size of land ownership is subject to ceilings in Latvia, Bulgaria, Romania and Hungary and there are limitations to the leasing or the amount to be leased in a few.

In many ECA countries, land registration is a major challenge in the rural privatization process. Despite the fact that privatization and registration have been ongoing for some time,

---

[56]    This has been shown to be generally true for transition economies and is documented in an unpublished Retrospective Review (December 1998) prepared by Laura Tuck with inputs from Chaba Csaki, Michele de Nevers and Pascale Herengt Talai of the World Bank.

[57]    Kazimerz Z. Poznanski. 1992. "Property Rights Perspective· on Evolution of Communist-type Economies." in Poznanski, ed., ibid. P. 71

[58]    Janos Kornai. 1992. "Ownership and Coordination Mechanisms." in Poznanski, ed., ibid. P. 99

[59]    The de facto and de jure situation has been highly variable in 1995 with, for instance, Russia using some 87 percent of the land within the collective and state sector and Albania utilizing 90 percent of the land within the individual sector.

little knowledge of their social impacts is documented. SAs could help close this knowledge gap, especially if social impact assessment is incorporated into the monitoring and evaluation component of the projects. A new Bank-financed SA is underway in Kazakhstan that documents the role and effectiveness of a pilot land registration project. One preliminary finding of the SA is that the cost of registration in rural areas can be higher than the value of the land sale that is being registered. The SA process is currently ongoing to ascertain the relationship between formal and informal costs in the registration process to find ways to cost reduction mechanisms.

The issue of subsidies is likewise important but also inadequately analyzed with respect to social impacts. In many countries it appears that the balance of different interventions is negative for agriculture, and that urban populations benefit disproportionately from government resources. SAs also attempts to ensure that "subsidies for the rich and cost recovery for the poor" does not continue to be a practice for development initiatives that they attempt to support. Subsidies offered for different crops have social implications and these are investigated as well. For instance subsidized production of wheat on irrigated land in Turkey not only displaces landless agricultural laborers but also creates opportunities for unjustifiable profits for non-producers.[60] Land based agricultural projects tend to favor the interests of landowners to the exclusion of the landless and small holders who cope by engaging in multiple income generating opportunities.[61] Several of the SAs included in this volume address this issue and show that benefits will be enhanced for the landless and small holders if public policies have a broader focus on rural development. The SAs also provide important insights to help strengthen rural opportunities in non-agricultural activities.

## Local Level Institutions

An important challenge for and a comparative advantage of social scientists concerns the analyses of local level institutions. Uphoff distinguished between 10 levels of institutions ranging from the international to the individual level. "Local" levels include "locality" referring to a set of communities with economic and social ties, "community/village" level and the "group" level referring to a set of persons who share some common connection. The "local" level is distinguished from other levels in terms of potential for collective action. At this level there are public institutions that represent the state authority, the private sector, as

---

[60] Two recent events are particularly worth mentioning here. First, in the proposed project area, landowners have recently been organizing to exert collective pressure on the government to increase subsidies for cotton production. Second owing to large price differences between imported wheat and wheat purchased by the government, merchants have been selling the cheaply imported wheat to the government as if they produced them and getting a large amount over and above their own import price. After substantial sums of funds have been wasted in this manner, policies have been modified so that unless crop sellers can provide certification as farmers, payments would not be made to them.

[61] It has been shown that agricultural subsidies "benefit the middle class and the rich more than the poor." See "Turkey: Economic Reforms, Living Standards and Social Welfare Study." 1999. The World Bank. P. 51.

well as the traditional and informal institutions, including the "voluntary sector, the self-help sector, the participatory sector, or the collective action sector".

Pointing out that these distinctions are important in that "different kinds of claims can be made on the performance of a local institution depending upon its nature, Uphoff illustrates these by considering different kinds of hospitals as local institutions: government hospitals, local government operated hospitals, private hospitals, charitable hospitals, and cooperatives. A patient would have equal rights of access to a public hospital as other citizens and would have to bureaucratic channels or a parliamentarian to raise "voice. In the private case, on the other hand, the patient is buying services on the terms offered and has the right to go elsewhere in case of dissatisfaction (thus exercising an "exit" option). In the cooperative case, the management is accountable to all members of the cooperative and a patient/member has options to participate in determining the changes that can be achieved to meet member expectations. Because localities, communities and groups substantially differ from one another with respect to a range of characteristics, the study of local institutions within the SA context should aim at capturing the diversity of local institutions, and assessing the likelihood of success of development initiatives that aim to have wide geographical scope.

## Sociology of Institutions

Although the sociology of formal institutions has been a basic element of organizational theory or sociology of organizations, there are not many examples of this type of work within the Bank. As institutions, organizations also constrain and structure human interactions, but in addition they are action groups. As discussed under the stakeholder section, they are composed of individuals bound by a common purpose and include political parties, village councils, regulatory agencies, firms, cooperatives, and trade unions. In addition to including political and economic bodies, organizations also include social bodies such as churches, clubs and educational bodies such as training centers.[62] A recent SA of the civil service reform in Yemen showed the importance of tribal representation in public sector institutions and the care required to factor these linkages in attempts to downsize the sector.[63] A similar study is included among the recommendations of the SA carried out for agriculture in Turkey. It is obvious that a far better understanding of the operations of local institutions could be developed if similar research were carried out for local level institutions in other cases. In Uzbekistan, the SA finds the workings of the local government to be one of the crucial variables in providing social services to the rural population.

The SA for a real estate registration project in Kazakhstan, currently underway, uses participant observation to obtain detailed knowledge about the functioning of new offices created under a pilot land registration project from the view of clients. One objective of this

---

[62]     The interrelationship between institutions and organizations has been discussed in Alberto de Capitani and Douglas C. North. 1994. "Institutional Development in Third World Countries." The World Bank.

[63]     Gloria La Cava. 1998. "Draft Note." The World Bank.

exercise is to determine what changes need to be made to bring a client focus to the new offices. Another purpose is to identify ways to limit the rent-seeking opportunities of staff in the new offices.

### Blockages to Equitable Access

The poor and vulnerable groups (women, youth, and older people) who are intended beneficiaries of project initiatives may encounter difficulties in accessing project resources. The reasons are various: formal and informal institutions, local customs, inter-group relations, social institutions (family, kinship groups, tribal or ethnic affiliations), formal and customary laws, information, communications systems, and others. In some cases, formal and informal institutions mutually reinforce exclusionary practices and deny minority groups access to goods and services. In other cases, formal institutions counteract traditional exclusions although their social acceptability may meet resistance for a period of time. Therefore it is important to analyze the institutional basis of exclusions and evaluate the potential success of new institutional arrangements throughout the SA process.

In the case of Southeastern Turkey, the tribal traditions and patrilineages block the direct access of women to the ownership of productive assets. Although women are eligible for land ownership when the Government implements land consolidation, de facto changes in women's status will take longer, and they will be more easily adopted by the landless than by the prominent tribal families that adhere more strictly to traditions. Nevertheless, the legal changes brought about in the context of land consolidation carry an important promise for the improvement of the status of women; the institutional blockages imposed by social norms can be removed by formal legal rules. Once the Bank-financed Project is formulated, gender-specific impacts can be assessed and a determination of the relative success of institutional arrangements in bringing greater ownership rights to women can be evaluated.

Other SAs included in this volume also attempt to determine whether or not systematic, structural blockages exist and how they can be overcome. In Azerbaijan, the SA discusses gender issues and shows that latent social norms encourage or support the systematic early departure of women from state farms. The gender discrimination in formal employment brought about by the farm restructuring process not only works against women's ability to access pension income, but is likely to hinder women's equitable employment in new jobs that are created as rural development takes off. The SA also shows how the image of women as "wage earners" changed into "housewives" in the transition process despite the active role women play in agricultural production. Because of the danger that this new image of rural women as dependent wives may hinder the policy makers from investigating special institutional arrangements to target women in the context of credit and advisory services, the SA process incorporated extensive consultations to do so.[64]

---

[64] Recent research indicate that in some ECA countries women are primarily responsible for the cultivation of household plots but how the revenues generated from the sale of their produce, no matter how small, is divided between household members is unknown. Julia Eckert and Georg Elwert. 1996. "Land Tenure in Uzbekistan." GTZ.

The Kazakhstan SA highlights the need to offset the unemployment expected to result from farm restructuring. It also advocates an analysis of the economic structure of farm communities to identify opportunities for the Project to promote private entrepreneurship to reduce the economic stranglehold of the residual elements of the former state and collective farms. A detailed institutional analysis, examining the formal and informal rules of the game in determining access to new opportunities, shows that pre-transitional power structures block certain opportunities for farmers. It points to the need for impact monitoring, and the Project incorporates a monitoring component to ensure that existing blockages do not continue and that new ones do not emerge.

### Strengthening Institutions

Local-level and informal rules—norms, values, and belief systems that shape the attitudes and behavior of social groups—may affect implementation arrangements for development initiatives. The SA focus on institutional analysis therefore develops proposals for modification of existing arrangements, or even the creation of entirely new institutional structures to overcome them. The process of institutional development refers to the capacity of institutions to perform their functions and involves strengthening inter-institutional relationships. It also involves the assessment of cumulative and/or counter-productive effects of various institutions that characterize the policy or project setting.

The SAs included in this volume provide good examples of proposals for institutional strengthening. For instance, mechanisms to strengthen the democratization of water user associations are proposed for Turkey to facilitate access of the landless and the poor to services provided by the proposed Project. The SA shows that unless the associations of the poorer segments are strengthened, the traditional leaders of the area would continue to dominate the management of resources. However, strengthening one set of institutions requires reducing the dominance of others and this is not easy to do without meeting resistance or imposing external values to indigenous systems. Thus, in the case of the Turkish agriculture project the SA basically proposes to introduce new information through study tours and other mechanisms so that merits of a more democratic arrangement could be assessed by the people themselves.

In Azerbaijan, the use of local women's NGOs is proposed in order to target agricultural advice and credit to female partners. Based on the institutional analyses, it is shown that intermediary organizations would be necessary to ensure women's access to information and credit. The SA also proposes that although extension services may be gradually privatized, and the partnerships of the public and the private sector as well as NGOs would be needed for farmers to obtain relevant timely information. A number of legal reforms and changes in the regulatory framework are proposed for Uzbekistan in order to expedite the farm privatization process. In Kazakhstan, new institutional arrangements are proposed so that farm advisory centers financed by the Project assist in commercialization of restructured farms by mobilizing local expertise and re-directing it to a major new application.

It is important to stress that the SA cannot provide all of the institutional analysis required to design and implement a project. Rather, its task is to help design the social engineering required for poverty reduction and social inclusion. The SA aims to strengthen local level and community institutions, enhance self-help capacity, and to remove institutional constraints to inclusionary practices. The SA incorporates the institutional concerns that are directly pertinent to the achievement of the social development objectives, including those that deal with mitigation of adverse impacts of resettlement caused by a development initiative. For instance, a resettlement plan would deal with the adequacy of the legal and institutional framework to compensate affected populations, including coordination of governmental and non-governmental institutions. In many cases the institutional analysis also focus on budget processes to verify that the proposed arrangements are adequate to deliver the inputs intended under a resettlement action plan. Similar institutional analysis would also be required for environmental, financial, technical, and other aspects of projects; specialized teams normally address these other concerns.

### Defining Implementation Arrangements

A "good practice" SA defines the specific responsibilities and monitorable contributions of each stakeholder group—central ministries, local government, NGOs, citizen groups, the private sector, and donors to enhance ownership and commitment to inclusionary policies.[65] The stakeholder dialogue helps determine implementation options including institutional changes, capacity building, targeting, sequencing, subsidies, and incentives. The implementation plan also includes a joint evaluation of the social development benefits and risks including potential conflicts and costs. Three of the SAs included in this compendium make specific contributions in this regard addressing the issues of feasibility and desirability of various types of partnerships as well as performance of roles by different agents. For instance, the SAs for Azerbaijan and for Turkey concern with the appropriateness of private sector involvement in the delivery of extension services. In Kazakhstan, the SA emphasized the need to have an independent institution determine whether or not the holdings of farm managers were obtained through coercion.

As mentioned earlier, an important element of participation of stakeholders is the issue of cost sharing. The ability and willingness of different social groups to pay part of project costs as well as the behaviors they reveal regarding these burdens play a key role in the design and implementation of development initiatives. Often powerful social groups that are able to pay do not do so. Rather, they exert pressure on the governments to subsidize their operations while those who are in need of government assistance are left alone to cope with their limited means. Specific attention is paid to these issues in several of the SAs in this compendium

---

[65]    These are policies that promote equity and eliminate discrimination against individuals or groups on account of their gender, race, age, beliefs, etc.

which discuss the type and level of contribution that different groups ought to make towards cost recovery of various project components.

### *Fourth Pillar: Monitoring and Evaluation/ Social Impact Monitoring*

The inclusion of monitoring and evaluation (M&E) procedures is mandatory for Bank-financed projects as it should be for all development initiatives regardless of their source of financing. The purpose of the M&E program is to provide ongoing feedback to different stakeholders to enable them to modify operations to increase effectiveness and assure that project objectives are achieved. SA provides inputs to the M&E component by focusing on inputs, processes, outputs, and outcomes that pertain to the social development objectives of a project. Specifically, the SA identifies monitoring indicators for the participation of the poor and vulnerable groups or for other social objectives. These include input indicators (benchmarks); process indicators; output indicators; and procedures and impact measures to determine whether intended social development impacts actually materialize. In addition, the SA helps:

- Define transparent evaluation procedures, including participatory approaches;
- Ensure that monitoring and evaluation procedures are established for the mitigation plan; and
- Ensure that all M&E proposed is carefully scheduled, fully budgeted, and properly supervised.

Perhaps more importantly, social impact monitoring (SIM) of projects provides critical information for reality and theory construction for situations in transition and rapid social change. Because of the rapidity, scope and the scale of the changes involved, the social dynamics of transition economies are insufficiently documented. With respect to transition of the rural landscape, there are some general observations but the situational specificity in people's response to change is yet to be studied. Case studies and surveys undertaken during the past five or six years provide an overview but do not detail how different social groups within a country adapt to change. Similarly there is inadequate understanding of how a given ethnic group responds to change in different countries. The rapidity of change has made it difficult to establish adequate baseline knowledge in most transition economies especially with respect to scholars' ability to identify group differences in the use of social capital[66] in mobilizing resources. Therefore, SIM provides an important instrument to establish a baseline and document change. SIM also provides practical information on whether all social groups benefit equally from development initiatives.

In Kazakhstan, the need for transparency in the implementation of land privatization and farm restructuring is high. SIM for the Project could make an important contribution to

---

[66]    Deepa Narayan (1999) provides a useful overview of alternative definitions of social capital in "Complementarity and Substitution: The Role of Capital, Civic Engagement and the State in Poverty Reduction." Mimeo. The World Bank.

this end, provided that its results are shared. In Southeastern Turkey, rapid change is accompanied with growing exclusions from development benefits, thus requiring rapid corrective responses. Therefore, SIM of the Bank-financed project would help identify corrective action. In Kazakhstan, the initial differences in the status of sub-regions or social groups require customized responses that might not have been possible to formulate at the outset of the relevant Projects owing to inadequate knowledge of social capabilities.[67]

Several of the SAs in this volume have made specific contributions to the design of the M&E component of the Projects they have supported. The Azerbaijan SA paid particular attention to the design of a participatory M&E approach, focusing primarily on issues of small-scale credit and credit targeting to women. The M&E of these will be an important input to the definition of triggers for continuity of Bank support to the country's rural development efforts. The proposed Bank lending will be adaptable, with each step of the project defining the implementation arrangements for Adaptable Project Lending (APL).[68]

The SA in Kazakhstan included a component for the project to monitor project impact on farm members and managers. An important ingredient is monitoring the divestiture of social assets and changes in the economic structure of communities, both of which are expected to be affected by the commercialization of farms. The negative effects of these changes will be highlighted and provisions will be made to address them in the second phase of the Bank project. In this case, the new APL offers the opportunity to use SA findings and monitoring results to modify both policy requirements and implementation mechanisms as the project progresses from one stage to another.

Currently, a small proportion of the Bank-financed projects incorporate systematic social impact monitoring. This is likely to change as a result of Bank's renewed emphases on development effectiveness. To assess whether the intended results are achieved, a systematic monitoring and evaluation framework is required for projects. As this new effort gains momentum, the need for social, participation and institutional analysis will be better appreciated; consequently more projects will have an integrated SA. Both experimental and non-experimental methods of M&E are grounded in systematic quantitative approaches. Incorporating qualitative methods enriches the interpretation of quantitative results. Participatory quantitative and qualitative methods, on the other hand, involve the shareholders, particularly at the local level and substantially improve development effectiveness.[69]

---

[67] Sen (1997) defines social capabilities as the freedom to explore opportunities at an individual level and as the ability to participate in social activities. In his later work, he redefines it as a sense of belonging to a larger group in Amartya Sen 1997. "Social Exclusion: A Critical Assessment of the Concept and its Relevance." Mimeo. Asian Development Bank.

[68] This type of lending is introduced in the Bank in 1997 and allows the Bank and the borrower to agree on support for a long-term development initiative. Thereafter the first phase is jointly defined. The monitoring of this phase facilitates the identification of the type of support expected of the Bank in subsequent stages.

[69] Kene Ezemenari et al. 1999. "Impact Evaluation: A Note on Concepts of Methods." Mimeo. The World Bank.

A recent review of development effectiveness of Bank financed operations shows a steady improvement. Accordingly, more than three-quarters of development projects supported by the Bank have achieved satisfactory outcomes, including seven percent rated as outstanding, according to OED's 1998 Annual Review of Development Effectiveness, which presents the evaluation results for operations that exited the Bank's portfolio in FY97-98. At the same time, the review warns that weak financial and social sector institutions in developing countries undermine project success and render countries more vulnerable to external shocks. "Institutional development is slow and difficult to achieve. In addition to increased institution-building, it is essential that social development be placed at center stage, both in assessing development effectiveness and in financing country assistance programs."[70] Since impacts cannot be adequately assessed without a proper understanding of baseline conditions, including those relevant for the participation of the excluded groups and institutions, comprehensive SA based on all four pillars will likely become a more integral part of the Bank-financed development initiatives in the immediate future. Similarly, methods of establishing controls in impact evaluation, especially for those situations where time and resources do not allow a comprehensive social research at the outset of development initiatives.

## Defining Mitigation Plans

All Bank-financed projects must develop mitigation measures where adverse impacts are identified for certain social groups. Further, institutional mechanisms are needed to minimize or cushion the adverse impacts. These concerns would have been relevant for the projects included in this compendium if, for instance, major irrigation was involved, or land acquisition and resettlement became an issue. Although normally land consolidation and the construction of irrigation systems also result in land acquisition, in the context of projects discussed here no involuntary asset acquisition took place and no resettlement was induced by Bank projects. Therefore, social policies that deal with involuntary resettlement and/or with indigenous peoples were not incorporated into the SAs included in this volume. Otherwise, mitigation assessment and mitigation action plan preparation would have been an integral part of the SA process.

Although the SAs included in this volume do not involve such adverse impacts, they do describe the inequitable impacts of the farm restructuring, land privatization, land reform, land distribution, and land consolidation processes. Many of these adverse impacts relate to loss of income and/or employment but do not fall into the category of adverse impacts regulated by the Bank's social development operational policies. Therefore, the relevant SAs attempt to identify mechanisms to reduce these adverse impacts. For example, the SA for the Kazakhstan irrigation project introduced a decision-making mechanism to enable farm

---

[70]      OED. 1999. "Precis: Development Effectiveness, 1998: Opportunities in a Volatile Environment." The World Bank.

members to over-rule the investment decisions of farm managers. Similarly, it recommended that loan officers and business advisors carry out mandatory review the status of land share holdings and leases on restructured farms to ensure that exploitative managers do not obtain credit through the project.

## Methods of Information Gathering/Sharing for Social and Stakeholder Analyses

The SA process always falls within a pre-defined action framework. It is initiated within a specific context and often within a pre-defined geographical area based on the Country Assistance Strategy and the Bank/country dialogue. Often some of the key components of the projects under preparation are also pre-defined, although substantial modifications can be made should the SA process indicate inadequacies in the pre-defined action framework.

Given a broad sector and project framework, the SA defines an information strategy to identify the key social actors; their interactions; and the social provisions needed to achieve the project's specific economic, technical, and social goals. The information will cover socioeconomic characteristics of the key stakeholders; their problems, constraints, and needs; and their ideas for alternative solutions. These data will be used to define eligibility and targeting criteria, to confirm beneficiary identification, and to determine the appropriateness of alternative solutions for the targeted social groups.

An information/data gathering strategy may include both dissemination of information and more procedures for targeting different stakeholders. Based on the SA, an earlier Kazakhstan irrigation project included two major information elements: a program to disseminate information ranging from farming technology to legal arrangements; and a consultation and decision-making process that required engineers to meet periodically with farm members to inform them of the implications of different design choices and enable them to make decisions based on their own interests.[71]

The rigor of the SA data-gathering process is important. The arsenal of social science research techniques used for conducting a rigorous SA is broad. Often, a spectrum of social science research tools is brought together. Every program/project specific situation demands sociological imagination—a different combination of methods and procedures tailored to the issues and actors. Nevertheless, the initial data gathering must be systematic since the data provide an empirical basis for analysis and the baseline for monitoring and evaluation.

---

[71] Despite the presence of large ethnic groups in Kazakhstan and Uzbekistan, language issues are not systematically addressed in the recommendations of the SAs. Kazakh is the official language in Kazakhstan but Russian is also spoken in national and local administrations. With 51 percent of its population composed of Kazakhs, the country hosts Russians (32 percent), Ukrainians (4 percent), Germans and Tatars (2 percent each). In Uzbekistan, likewise, the majority are Uzbek, but Russians, Koreans, Turks, Pers, Tatars and Karakalpaks each speak their own language. Dinora Azimova. 1998. "Ethnicity and Ethnic Relations in Central Asia." Mimeo. The World Bank.

Large-scale representative surveys are instrumental in gaining the confidence of stakeholders in the diagnostic elements of SA. These provide non-anecdotal, systematic, and scientific knowledge on the characteristics, values, attitudes, and aspirations of diverse social groups that live in areas to be served by proposed development initiatives. It is usually easier to use qualitative information once a systematic base and the full extent of social diversity is captured through surveys. Based on the personal experience of the author, the policy makers in many ECA countries have greater familiarity with information obtained through scientific representative surveys, and they are more likely to trust the findings. This is especially so when quantitative baseline data are not available. The pre-SA perceptions of the borrowers may run counter to SA findings and stand in the way of the intended reforms. For instance, if the borrowers believe that people are unable or unwilling to pay for improved services, they may hesitate to raise tariffs.[72] If the SA shows that people pay a great deal for inadequate services and have a high willingness to pay once they improve, the reform has a chance of succeeding. Therefore social information based on surveys provides a useful first step in the SA process.

The World Bank has supported the capacity of the governments to gather systematic poverty data because it too has a tradition of relying heavily on systematic data to guide its policies. Different types of bias are introduced in development research;[73] therefore it is advisable to complement systematic data with alternative information gathering techniques which include focus groups, feedback systems, and others. Rapid rural appraisal and participatory rural appraisal are described by Mikkelson and Chambers;[74] some social scientists advocate the use of participatory methods on their own.[75] However, the assessments included in this volume rely on complementary utilization of qualitative and quantitative techniques of data gathering and analysis.

---

[72]     The SA carried out for the Baku urban water supply and sanitation project showed a high willingness to pay and thus substantially helped the water agency to get parliamentary support for tariff increases. Ayse Kudat and Ahmet Musayev. 1995. "Social Assessment of the Baku Urban Water Supply and Sanitation Project." The World Bank.

[73]     Chambers refers to these biases as "spatial, project, person, dry season, diplomatic, and professional biases." R. Chambers. 1993. *Rural Development: Putting the Last First.* London: Longman.

[74]     B. Mikkelson. 1995. *Methods for Development Work and Research: A Guide for Practitioners.* Delhi: Sage; R. Chambers. 1992. "Rural Appraisal: Rapid, Relaxed and Participatory." Mimeo; D. Morgan and R. Krueger. 1993. "When to Use Focus Groups and Why." In D. Morgan, ed., *Successful Focus Groups: Advancing the State of the Art.* Newbury Park: Sage; W. Desvouges and J. Frey. 1989. "Integrating Focus Groups and Surveys: Examples for Environmental Risk Studies." *Journal of Official Statistics.*

[75]     For purposes of institutional analysis, methodologies have been developed by such disciplines as management sciences, organizational theory, and sociology of organizations. Person perception as developed by social psychologists, power analysis carried out by political scientists, and participant observation as developed by anthropologists are also appropriate. Analysis based on trust and social capital, formal institutional analysis of administrative structures, etc. were used in the SAs. Although reference to these theories is not explicit, they are easy to deduce.

When social analysis is limited to a socioeconomic survey, some of the key stakeholder perspectives cannot be captured. An early stakeholder analysis or scoping would facilitate the design of an information (or data collection) strategy and ensure that the views of all strategic stakeholders are adequately reflected. For instance, in the Turkey project, the major stakeholder (borrower) for the proposed project was the Southeast Anatolia Project (GAP) administration. In addition, there were large numbers of local representatives of central government ministries; representatives of chambers of agriculture, commerce, and industry; other NGOs with important functions in rural areas; irrigation associations, village administrations, private sector organizations that provide agricultural goods and services; and many others. Semi-structured interviews and consultative sessions were held with each of the individual stakeholders. Because the poor people of the project area are the real client and they are socially diverse, it was important that their perspectives would first be covered through systematic surveys. In addition, focus group discussions were held with each of the key social groups so that the views of women, youth, and the landless were adequately captured as well. By holding additional participatory meetings with mixed groups of people from the project area some of the inter-group dynamics was also captured.

A broad range of social science information gathering methodologies has been used in the SAs included in this volume. Historical, cross-sectional, and statistical background information were gathered and analyzed. Representative socioeconomic surveys of farm households were conducted along with in-depth interviews, participatory observations, and focus group research. Many of the surveys followed rapid rural appraisals; in others rapid appraisals followed needs assessments surveys. In all instances, qualitative methodologies have been used to complement the quantitative work. In Azerbaijan qualitative and quantitative research methods were used interchangeably over 20 months to strengthen knowledge and stakeholder participation in farm credit, marketing, and extension services. In Kazakhstan, focus groups and in-depth interviews informed the household survey, which was followed by more focus groups to explore new issues and clarify survey results. In Uzbekistan a household survey was conducted, but data were also gathered from official sources, expert consultations, and a survey of local officials.

There has been less systematic focus on information sharing, feedback and dissemination issues. Although most SAs analyze how farmers and other stakeholders receive information, what their information needs are, and how best to make information available to them, they rarely consider the strategic information needed by the SA process itself. Data are gathered, but feedback is often not provided to those who supply the information. Although stakeholder seminars, media releases, and publications constitute alternative ways of providing feedback, a coherent strategy has not been an integral part of the case studies in this volume. There are, nevertheless, guidelines and established procedures for the incorporation of information and communication (I/C) elements in the SA process and in the projects they aim to support.[76]

---

[76] Ayse Kudat. 1994. "Information/Communication Guidelines for Environmental Activities." Mimeo. The World Bank.

## State-of-the-Art and Cost of Social Assessment

Until recently, very few SAs were built on all four pillars. Indeed, only one such case was documented in the previous collection.[77] In a sample of 100 projects recently completed for Bank financing, some had adequate social analysis but few incorporated social monitoring, a sustainable participation framework, and institutional restructuring. The quality of a social assessment depends on various factors including the availability of resources and time to carry out the required work. The ultimate test of a good SA is its impact on project design and implementation, and this depends in large measure on the specificity of recommendations that emerge from the analytical and consultative elements for achieving social development objectives.

SA is relevant to any development initiative that aims to reduce poverty, whether financed by the Bank, other donors, or the governments themselves. In the private sector, SAs are routinely carried out under different names including market research, consumer research, consumer relations, community relations, and client satisfaction. Many companies increasingly appreciate the importance of understanding the needs and expectations of their clients, the differences between social groups in responding to goods and services produced, monitoring the feedback from actual and potential consumers, relations with communities in which companies are located, labor relations, and understanding the cultural constraints of demand. The Bank's focus on the "client" (the poor and the excluded) and on a range of other stakeholders is an integral part of a global corporate focus on the consumers, producers and communities in which the productive activities take place. The case studies included in this volume integrate people-focused concerns of this nature. Within the Bank, SA is not mandatory, but there is a general understanding that projects require an SA where there are:

- Populations that are historically disadvantaged or excluded from development initiatives;
- Large social and economic inequalities and social exclusions;
- Post-conflict situations;[78]
- Acute social problems, institutional discontinuities, or turmoil;
- Large-scale enterprise restructuring including farm restructuring; and

---

[77] Even today, there is a general misconception that socioeconomic surveys and other social science observations constitute an SA. Often one hears task team leaders talking about commissioning SAs, referring basically to qualitative and/or quantitative surveys. Such surveys provide relevant and useful information for the SA process and can be considered as consultative tools to facilitate listening to some of the stakeholder groups. However, they are not synonymous with a comprehensive SA.

[78] Post-conflict and natural disaster emergency operations often require accelerated processing but can benefit substantially from SA if it is undertaken expeditiously. In these cases, SA guides project identification towards areas with significant numbers of affected people. In addition, social impact monitoring can build commitment and support for a process of re-engineering project components as experience warrants, an approach that is often necessary in the situations of uncertainty intrinsic to emergency lending.

- Potential adverse impacts such as involuntary resettlement, contact with indigenous peoples, and loss of cultural heritage.

Projects where an SA is advisable include operations where:

- Changes in existing patterns of behavior, norms, or values are required;
- Community participation is essential for sustainability and success;
- Insufficient knowledge exists of local needs, problems, constraints and solutions; or
- Beneficiary targeting mechanisms or eligibility criteria are unknown.

Even when no SA is required, it helps improve project relevance, build commitment, and measure project impact against development objectives. With the renewed emphasis on measuring outcomes and impacts, as distinct from inputs and outputs, the SA process and especially social impact monitoring become absolutely imperative. SA is also advisable for policy development with a focus on ensuring the quality and appropriateness of development initiatives. It greatly enhances understanding of the social context and the sustainability of Bank lending operations.

The new and comprehensive definition of SA provided in this chapter integrates social analysis, a participation framework, institutional analysis and structuring, and social impact monitoring into specific projects under one umbrella. It was formulated by the Social Assessment Thematic Team in late 1998 and disseminated to some of the Bank staff through learning exercises held in the headquarters and some regional offices. The Team intended to provide a roof under which social development concerns could be incorporated into Bank-financed initiatives. Although this integrative approach is very recent, many social scientists and project managers within the Bank have already adopted it. Nevertheless, there is a general tendency to associate social assessment with qualitative or quantitative social analysis carried out for project design and consider participation, institutional, and monitoring aspects as separate. The conceptual and project level difficulties that emerge from this fragmentation are many, and it is one of the many reasons for weak linkages between social assessments and projects.

The recent stock taking exercise based on the examination of the quality of 100 lending operations at entry mentioned above indicates progress in integrating participation into the Bank's lending operations. It also points to innovative approaches with regard to participation and social targeting. Highly satisfactory projects appear to have good institutional analyses. Where relevant, the compliance policies regarding resettlement and indigenous populations also appear to be integrated into projects. The stock taking exercise also shows that Bank Operational Directive (4.15) requiring structural adjustment loans to analyze adverse impacts on the poor has not been internalized.

Social analysis has been as important as economic and financial analysis since the 1980s as specified by Bank operational manuals. Over the years, there has been progress in the internalization of social development objectives and stakeholder views in Bank financed development initiatives. Some regions have had greater success than others. In South Asia, for example, where the social team has worked in the region a long time, there is more demand for and acceptance of social analyses. In other regions, such as ECA and MNA, the team is newly established. The ECA region itself was created after the collapse of the Soviet Union. Since it is a new region, there has not been great opportunity to develop a tradition of social science research. Success depends upon the availability of qualified social scientists and their experience of social science teamwork within the region. Several generalizations can be made:

- Specific projects have had greater success in mainstreaming SA than analytical sectoral work (ESW) and structural or sector adjustment loans. Also, rural development and infrastructure projects are more often based on SAs as compared to public sector management and private sector development loans;
- Socioeconomic surveys, rapid appraisals, stakeholder meetings, and/or consultations are included in projects;
- Comprehensive SAs, integrating social analytical, participatory, institutional, and impact monitoring elements, are found in only a small number of projects. The exceptions are projects that are regulated by compliance directives and thus require resettlement and/or indigenous action plans.[79]
- The integration of social impact assessment into projects is especially rare.
- Weak linkages between the design and implementation of projects and the SAs are frequent.

*I was invited to carry out the SA for a water supply project in two regions of a country. Some 2,000 communities were involved and the two regions were dramatically different in their social outlook. I explained to the task manager that SA was a process and with the $30,000 he was willing to make available, systematic fact finding, consultations, institutional analyses could not be carried out. He said he had an expert on the team to deal with institutional concerns and that I should confine myself to carrying out a survey of villages only, as he intended to confine the project to rural areas. However, the borrower was insisting on expanding and upgrading a main pipeline. Thus, I pointed out that focusing on rural areas alone would be misleading. He said he intended to discourage the client from a pipeline. In a country that no other Bank social scientist had worked until then, I identified a group of distinguished local scientists and we carried out surveys in a representative sample of villages and interviewing thousands of households. We also collaborated with local NGOs and scientists to test the quality of water. First, we established remarkably wide spread poverty and low affordability for water. Then we found that nearly all hand-pumps were polluted and ground water quality was extremely low in most sub-regions. We concluded that low cost community based solutions were not feasible, the hand-pump system could not be expanded, and the people could not participate in cost recovery in a foreseeable future. The task manager then said we should start surveys in the urban areas. We did that too and recommended that the water supply system not be expanded unless the waste water system is established. By then, far more than $30,000 was already spent, but the SA team was asked to specify the communities in which the hand-pumps could be installed, how incomes can be enhanced, and how the urban cost recovery systems could be improved. The answers required a great deal of stakeholder consultation and institutional analyses. It took additional time to convince the technical teams that this was necessary before answers could be provided. The first reaction pointed to the 'usual useless contributions of social scientists', but at the end of the SA process we would all see some light. A social scientist.*

Weak linkages occur for many reasons. The SA process may be ongoing at the time of its reporting and it may not be possible to identify the contribution of specific pillars to various components of the project except through SA recommendations. Also, the SA might

---

[79]     We have to stress that many projects that require compliance are primarily restricted to social mitigation issues and do not necessarily cover broader social development concerns.

have been prepared as an input to a sector review and the technical teams preparing this review may not attach as much importance to consultation with stakeholders as to technical, financial, and economic analyses. A more frequent reason for weak linkages between the SA and the project has to do with the timing of the SA. Often, the SA process starts relatively late in the project preparation cycle and after the major project components have been identified. When this happens, the SA is expected to endorse the social appropriateness of the project. It could also help modify the institutional arrangements for benefit targeting and provide inputs to measure development effectiveness, but the SA team may meet resistance if they propose major alternations in the project design.

Weak linkages also occur when the SA is limited to one or two of its pillars. For instance, the SA may be identified with a social survey and merely aim to inform the technical teams of the socioeconomic characteristics of the beneficiaries (e.g., the farmers).

Inadequate financing of the SA activities inevitably creates weak linkages. When the first SA volume was published in 1997, SA costs were estimated at $75,000-85,000. Today, this is no longer possible, especially if the SA is to provide a systematic assessment of social, participation and institutional issues, help design an institutional and participation framework, specify the required monitoring activities, and prepare a mitigation plan if needed. A project is normally designed after a Bank team visits a country three or four times, staying about two or three weeks each time. If a Bank social scientist participates in three missions for three weeks each, most of the resources allocated to the SA process would be spent for travel and gross salary, leaving insufficient funds for good quality empirical work and consultations. Whether or not the relevant fieldwork and consultations are carried out by local social scientists or international experts, when insufficient time and budget is allocated to the SA process its empirical content, data quality, and geographical coverage would be poor.

*"After long debates that SA should be integrated into the sector review and with insistence of a supportive donor, I was asked to prepare a terms of reference (TOR) for the SA. I was told that its empirical coverage should be representative of all regions of the country and some 9 million villagers. The work had to include economic assessment and develop pilot projects. We were also told to look at all the previous credit programs and develop new proposals that would work. Some 80 key stakeholders were involved in the sector. We had to organize a workshop to kick the process off. Traveling through a huge country, conducting systematic interviews in hundreds of villages, mobilizing many field teams, five senior scientists, statisticians, computer specialists, participation experts, going from one little local Bank to the other, writing and re-writing reports in two languages had to be done with $53,000. Because the money was barely enough for the field work, local scientists worked basically free and I spent every night and weekend for two months analyzing and interpreting results. The SA team was never invited to the missions of the technical team. Nor were they given an opportunity to present their results as a team. The first criticism for the draft SA report was that the results were not in accordance with the observations of the technical team that had visited a few villages! We were also told that no one could be bothered reading a long report and we should merely put forward ideas for pilot projects. Our claim that the problems could not be resolved through pilot projects but major institutional changes were required was dismissed. The report was cut from 140 to 10 pages. Then came a request that the social team carry out historical investigations, expand on their recommendations, add more text and delete tables. This too was done, and all without additional resources. By the time a new report of 40 pages was finalized the technical team changed. The new task leader did not want to read the SA report and declared it useless after it was subjected to many levels of review within the Bank. His manager also made statements that the SA was useless because his staff member said so and did not respond to calls for a discussion. The client team also changed and all those who were very supportive of the SA process left the Ministry. Now we have a report highly demanded by many outside readers and NGOs, but the chances that it will be integrated in the future planning of the sector is highly questionable." A social scientist*

SA may generate results that cannot be accommodated by the project for which it was intended. At one level all SA processes are somewhat limited in scope; an assessment intended for a road project would not help design an energy project even when local needs point in that direction. Many of the SAs included in this volume were launched for agricultural projects and address findings that are directly relevant to agrarian production. Several SAs note that rural infrastructure inadequacies are large but the relevant projects could not incorporate infrastructure components. The Azerbaijan case lists issues "falling outside the scope of the project", including the problems of internally displaced populations. Because the proposed Project could not address their needs the Bank is responding to them through another project. No single project can respond to all findings of social assessment, but SA observations may be useful in guiding the Bank's subsequent initiatives. Also, within the APL instruments mentioned above, subsequent phases of a project may be able to integrate additional SA recommendations. Other reasons for weak linkages between the SA and the relevant development initiative may rest with the SA itself:

- The SA may be of bad quality;
- The SA team or the staff in charge of SA may not be persuasive;
- The SA specialist may not have time to work as an integral part of the project technical team;
- The Bank SA staff may recruit local social scientists and not be able to spend sufficient time with them in the field in order to articulate SA observations to the Bank technical team;
- The technical team may not attach value to potential SA contributions and may consider SA as a product, a background paper; or
- The SA may be assigned lowest priority and given little or no budget.

These are all high probability scenarios and it is not rare to encounter several of them in a single project.

The relevance and utility of SA for the design and implementation of development projects and policies are more widely recognized now than before. As their quality improves, SAs will be an integral part of most development initiatives. The quality of the SAs presented in the volume is somewhat mixed, and not all are built on four pillars equally strongly. They are also being report at different stages of project cycle for instance, SA of Turkey is recent and the relevant Project is at the outset of its preparation. The Azerbaijan case is reported as the Project's design is being completed. The next chapter provides a brief summary of each SA included in this volume.

# Chapter Three

# Highlights of the Case Studies

## Ayse Kudat

## Introduction

In the previous chapter we have defined social assessment (SA) as process through which social development and stakeholder concerns are incorporated into development initiatives. A "good" SA is said to contain four elements: identification of social development issues, analysis of institutions relevant to those issues, definition of key stakeholders and a framework for their participation, and social impact monitoring to evaluate development effectiveness of policies and projects. We have emphasized that this holistic and comprehensive conceptualization of SA is new and reflects an "ideal" situation rather than providing a description of the actual state of the art of SAs available to date.

In what follows, we summarize the SA cases included in this volume using the "four-pillars" as a framework. In general, these cases:

- Provided voice to some of the key stakeholders, particularly to the poor and vulnerable groups. They facilitated the dialogue of the governments and donors on issues of inequity, social diversity, vulnerability, transparency, social capital and trust;

- Analyzed social issues not merely from the perspective of individuals and households, but in terms of social groups and their interactions;

- Showed the wide range of local responses to development initiatives and drew attention to the importance of formulating community specificity and targeting of policy/project interventions;

- Demonstrated that the process of social change or transition is dynamic, taking different paths in different countries, regions and sub-regions, indicating that a great deal of more change is yet to come;

- Focused on several local level institutions and the interaction of "traditional" and formal organizations, but pointed to the need to further strengthen institutional analyses;

- Concerned the sustained participation of the low-income, the landless, and the other vulnerable groups with respect to the design, implementation and monitoring of the projects.

As with institutional analyses, SAs were less instrumental in helping design monitoring and evaluation arrangements for the projects for which they were launched. Nonetheless, they provide useful insights which can guide assessments of the social impacts of the relevant initiatives.

## Social Assessment of the Azerbaijan Agricultural Development and Credit Project

### *Background*

A sharp deterioration of agricultural output and the large problem of refugees and internally displaced populations have led to decline in living standards in rural Azerbaijan since 1988. The situation was exacerbated by the Caspian Sea environmental crisis. Currently, up to 60 percent of the rural population lives below the poverty line and access to social services, previously guaranteed, continues to erode. "With the distribution of land, there is danger of a rapid increase in rural unemployment and poverty, as the new small holdings may often be too small to maintain a family and land concentration is likely to occur. Independently of oil developments, the prospects for a substantial proportion of Azerbaijan's rural population depend on the recovery of agricultural output and the re-establishment of growth in agriculture and agro-industry" (FAO/CP 1998: vi).[1] The Government of Azerbaijan (GOA) and the Bank agree that the agricultural sector faces two major challenges. The first is to complete the transition to a privatized system and the second to minimize the risk of further loses in agricultural output that may be caused by the anticipated oil and gas revenues. The GOA has invited the Bank to help support its efforts to meet these challenges through a new project on Agricultural Development and Credit.

### *The Social Assessment*

The SA aimed to identify the social development concerns of relevance to the project, evaluate institutional and social organizational issues, define a participation framework for stakeholders with a focus on small holders, and establish mechanisms for monitoring and evaluation to ensure that development effectiveness objectives of the project are met. To carry out the SA, several basic steps were followed:

- The relevant global, regional, and country specific experience was reviewed and lessons learned were incorporated in the SA;

- A rapid qualitative assessment of the results of the Bank-financed Pilot Agricultural Privatization Project was carried out with a team of Bank and local social scientists;

---

[1] The March 15, 1999 issue of the Economist raises serious questions concerning the accuracy of expectations of national revenue from oil. This suggests that a development focus on agriculture is even more important than anticipated at the outset of project preparation.

- A survey of 800 households was conducted in five regions where the Bank-financed Pilot Project is being implemented;[2]

- Focus group meetings were held in communities in eight regions and consultations were carried out with a range of stakeholders, including the private sector;

- A stakeholder seminar was held sharing analytical results of the SA with NGOs, the academic community, the relevant government ministries, and key donors;

- Several thematic assessments were initiated with feedback received from stakeholders, the borrower, and the Bank in order to understand the special needs of female farmers, the elderly, the displaced populations, social acceptability of alternative institutional arrangements for credit, and the targeting issues with respect to credit and advisory services; and

- The results of the analytical and consultative processes were integrated into the draft Project Appraisal Document for discussions within the Bank and consultations with the client during the Appraisal.[3]

### *The Four Pillars*

#### Key Social Development Concerns

Although several social assessments had been carried out in Azerbaijan, there was very little knowledge about rural conditions and the social dynamics of change. A rapid shift from a command to a market economy and expeditious government action to restructure state farms left individual families to make the adjustment from being wage earners with state guaranteed employment security and income to becoming independent farmers, relying primarily on social capital in reorganizing their productive activities. Despite ethnic and other types of heterogeneity and partly because of the threat of external violence, family, neighborhood, and community (as defined by the settlements supporting state farms) bonds were mobilized to facilitate private farming. The SA had several concerns:

- Does social capital help ensure the viability of the newly emerging peasant sector?
- Will exclusionary practices emerge?
- Are these changes indicative of more changes to come?

Some two years ago, at the outset of the SA, the team felt that pre-transition structures would be replaced by peasant farms. It now appears that change permeates the rural landscape, and the capabilities of different social groups to adjust to change determine their relative success.

---

[2] The SA team was unable to get permission to work within the pilot communities; rather it reviewed issues in other communities in regions where the pilot project is under implementation.

[3] As this draft was prepared, the Bank project team, including a member of the SA team, was in the country appraising the project.

The SA attempted to find mechanisms to enhance the equitable integration of small holders, including women, in the agricultural reform process. New institutional arrangements appeared to be necessary to ensure that women are better informed of legal and procedural changes and have access to know-how and skills required to make a transition from a wage based rural economy to a peasant economy. They are also crucial in guaranteeing that men and women receive equitable access to farm inputs, technology, and marketing facilities. This involved an understanding of the current problems and constraints farmers face and the institutional mechanisms that need to be modified to enhance the benefits of reform for rural people. It also required an in-depth understanding of the potential solutions farmers felt to be appropriate. The SA raised the question of how best to sustain participation of small holders in the restructuring of the agriculture sector in Azerbaijan.

The well-being of the internally dispersed populations (IDP) dispersed throughout the country as a result of armed conflict was another major concern. IDPs constitute a large percentage of the population of Azerbaijan; their problems have been addressed by other Bank-financed projects. The SA included in this volume incorporated a focus on issues of post-conflict reconstruction based on the results of sequential SAs carried out in the context of the Azerbaijan Reconstruction Project, the Azerbaijan Poverty Assessment, and the Caspian Sea environmental crises. Although the proposed project will not deal specifically with the needs of IDPs, the capacity created within the agricultural sector will no doubt be of direct support to those who are being re-integrated in liberated areas. The GOA previously limited its assistance to re-integrating the IDPs in their original settlements and thus did not include them among beneficiaries of land privatization in the host communities. Recently, it has modified its policies and sought the Bank's assistance in preparing a new project to assist their integration in the host communities. The particular focus is on IDPs who are living in tents and other temporary shelter in rural areas. Thus the SA carried out for the agricultural development project will provide valuable information for the preparation and implementation of the IDP project.

### Institutional Issues

At the micro level, the SA noted the newly strengthened role of the family and the household in the restructuring of rural Azerbaijan. It showed that as transitional vulnerabilities increased and the wage economy lost its predominance, the social capital based on family, neighborhood, and community ties started to cushion adverse impacts of farm restructuring. Also, the household-based economic interdependencies augmented and home plots gained increased importance in the survival of the family. At the community level, voluntary associations of farmers emerged as a new phenomenon, building on pre-transition social capital based on pre-reform work relationships and leadership structures. At the regional and national levels, new institutions are also expected to emerge and/or be strengthened: rural credit unions, irrigation associations, extension services, land titling/registration systems, and others. There is trust in these emerging institutions because they are based on existing social and community bonds. Institutions that are viewed as a continuation of Soviet period administrative and political structures are somewhat less welcome. The extent to which these new institutional requirements will be met through private sector provisioning or will have to be provided by the public sector is being debated, but the need to ensure transparency in the delivery of these services is clearly indicated by the SA. On the whole, there appears to be greater social capital in rural Azerbaijan. This may be a key factor in explaining the rapid and positive progress of the newly initiated farm restructuring and farm privatization process.

### The Participation Framework

Defining the participation framework for the proposed project requires a double focus: first, for the preparation of the SA and, secondly, for project design and implementation. The participation framework for the SA consists of systematic and qualitative consultations with a range of stakeholders, particularly with the poor rural population and the small holders. As for the project, several partnerships have already been defined with respect to different components, and others are in the process of being defined.[4] In the proposed project, the participation of small holders is particularly relevant in the delivery of advisory services and credit and less important for policy development and land registration. Transparency in land registration and ensuring individuals' rights to land, and in particular those of women and young people, are of critical importance. Other stakeholders, however, also require consideration. This can be best achieved by incorporating an information/communication (I/C) component and specifying I/C requirements for each project component.

### Monitoring and Evaluation

Helping define a monitoring and evaluation framework for the project involved the identification of input, process, output, and impact indicators for the project to ensure that its development objectives are achieved. A preliminary list of M&E indicators is provided in the chapter on Azerbaijan. A strong participatory element is built into this component so that stakeholders have an opportunity to provide feedback on project implementation arrangements and development effectiveness. Since the proposed project is intended to provide long-term continuous support to agricultural development through Adaptable Project Lending (APL), impact monitoring is central to defining the next steps in lending.

### *Social Assessment Impacts on Project Design*

The SA process reported here covered the identification and pre-appraisal stages of project preparation. Both active research and consultation elements of SA are continuing to guide the appraisal; therefore, it is early to list the impacts of the SA on project design conclusively. Nevertheless, based on the various stages of the SA process completed to date as well as preliminary technical, economic, and financial assessments, implications for project design are listed with respect to four basic elements: policy design, land registration, advisory services, and credit. The SA confirms that these elements correspond to the priority needs of the people. However, the need for policy development is expressed primarily by policy makers and donors rather than by farmers. In addition, the SA identifies a more detailed specification of these elements and outlines additional issues which fall outside of the scope of the proposed project but will nonetheless have to be addressed in the future. Continuous SA activities, as formulated by the

---

[4]     As this section was being written, the project was not yet appraised. During appraisal all relevant findings of the SA will be incorporated.

M&E arrangements, will provide further inputs for implementation. The major components of the proposed Project are summarized below.

### Involving Stakeholders in Development of Agricultural Policies

Azerbaijan offers high potential for agricultural production. The transitional problems caused by agricultural restructuring and the expectation that the oil potential of the country will provide sufficient income for everyone decreases the motivation of the youth to engage in agriculture and pushes them out of the rural areas in search for jobs. This undermines the productive capacity of the country and poses a threat to the wellbeing of large numbers of people. An equitable integration of small holders into the agricultural sector is not feasible without a solid policy base that specifically addresses the question of what motivates able-bodied men and women to engage in agricultural production. It is therefore essential to develop strategic options for agricultural developments through broad-based stakeholder participation.

The SA not only supports a policy focus but also proposes participatory arrangements for its development. It suggests that an information/communication strategy be built into the project to support the policy development objective of the project and other components. Since there is a general tendency for policy makers to base their planning efforts primarily on technical inputs, there is a need to create mechanisms of consultation as alternative policy scenarios are being prepared. In addition, there is a need for a specific focus on key stakeholders in the implementation of this component so that the alternative policies developed consider the future roles of the public, private, and civil society sectors, as well as those of the small holders, women, and other social groupings. The initial stakeholder seminar held to share SA results brought the NGOs, the academic community, and policy makers together around some of these key policy issues; however, further stakeholder feedback mechanisms should be integrated into the terms of reference required for the implementation of this component.

### Supporting the Creation of a Land Registration System

The need to increase the transparency of the privatization process and ensure that land rights are actually transferred to people emerges strongly from the SA; it justifies a focus on land registration to enhance the security of private land ownership. The SA shows that prolonged implementation of privatization and land and asset distribution reduce people's trust in reform. Therefore, communities that complete the privatization process enjoy greater public support for reforms. Also, land ownership without receipt of title and the ability to trade assets hinders the reform process. Therefore it would be important for the Bank to support the development of a land registration system. The SA also demonstrates that people's ability to pool their resources together to form production associations and to have access to credit will be enhanced once security of ownership is ensured. If land markets are to develop and land is to serve as collateral to activate the credit systems, it is particularly important that gender and regional equity is ensured. Project monitoring will therefore address issues of social inclusion and equity.

The participation framework for this component would attempt to ensure that the right to land ownership granted by the GOA to all its citizens, including women and children, is not eroded in the process of registration. In addition, the implementing agencies need to make sure that small holders and less influential members of local communities in rural Azerbaijan are not subjected to corrupt practices in their attempts to obtain titles to their land; this has been observed elsewhere in ECA. A well-designed information and communications (I/C) activity will support the land registration component by avoiding the erosion of entitlements and reducing the risk of corrupt practices. Although the GOA disseminates new laws and regulations through newspapers, it is difficult for ordinary citizens, to understand their content or implications despite their relatively good educational backgrounds. Therefore, it is important for the proposed project that farmer friendly I/C strategies are developed to inform farmers of their entitlements, including those relating to land registration, credit, and advisory services.

## Providing Appropriately Targeted Advisory Services

The transition and privatization processes have created substantial knowledge and skill gaps in rural areas. Many people who received land do not have a farming background. Others who were once responsible for specialized tasks within the collective state farm system are unfamiliar with diverse aspects of farming. Farm inputs are obtained in different ways than they were before. Now each farmer, rather than a specialist, is expected to make a judgment about the quality of these inputs. Likewise, individual farmers or associations, rather than state farms, market farm outputs. The knowledge base required to support a major shift from a rural society consisting primarily of wage earners to a largely peasant society is weak. Likewise, the management and financial knowledge needed to support the household and the new contractual relationships between private farmers are lacking.

These observations, based on the SA, suggest that the proposed Project support the provision of improved advisory services that are prioritized and targeted. This will require the establishment of new institutional arrangements and partnerships including those that would help women gain access to appropriate information and farm advice. Equally important is to prepare individual farmers for the next transition, one from a largely subsistence-based peasant society to one based on commercial agriculture. Currently, the shortage of land and of farming skills encourage farmers to establish voluntary associations, referred to as firms, either within the family or among neighbors; formalization of some of these firms and the creation of new business entities will constitute the next phase of the rural transition.

The existing institutional framework to deliver advisory services to individual farmers is inadequate in terms of know-how and organizational structure. The pre-transition institutional arrangements allowed each state farm to work with its own staff of experts and specialists. The transition to the peasant mode of production requires, however, that all farmers have broader know-how and skills. It also requires that the capacity for extension services be developed outside of the state farm system to respond to the needs of peasant farmers and their associations.

In the long term, advisory services could be made available by the private sector; thus, it is desirable to provide early support to the private sector to build the relevant capacity. However, given multiple constraints in getting access to critical farm inputs—seed, fertilizer, pesticide, machinery, irrigation water, credit, and know-how—farmers will have to establish priorities for their purchases and investments. Although the ability of newly independent farmers to farm more productively depends critically on know-how, advisory services are a low priority for the majority of small farmers given their acute cash shortages. Thus, without subsidies, advisory services provided by the public or private sector will be inaccessible to small holders.

Throughout ECA, people increasingly rely on garden plots to meet many food requirements; this is also the case in Azerbaijan. However there is little knowledge concerning the constraints households face in making effective use of their garden plots. It is appropriate to expect advisory services to stimulate increased specialization and intensification of cultivation and provide know-how for home based processing opportunities. The project's advisory services and credits components should pay specific attention to increasing the efficiency of garden plots.

### Establishing Institutions to Facilitate Access to Credit and Other Farm Inputs

The SA shows that credit is needed to alleviate major input and marketing constraints of farmers. Farmers associations that pool the assets and labor of rural families also need credit. Farmers borrow primarily using social networks; small amounts are lent and repaid in short periods of time. When their needs are solicited, however, preference is given to longer term borrowing for larger sums to facilitate purchasing of land and equipment or to open a new business. Small holders need credit primarily to purchase livestock, as they are less certain of their ability to pay back the loans. Given that different stakeholders need credit for different purposes, it would be important to design a credit component that has sufficient flexibility to meet different needs. Providing seed capital to saving associations and allowing them to establish the implementation arrangements does this. It is also important that affordable interest rates are set so as to allow small holders to gain access to credit; financial analysis shows that the sustainability of credit institutions would depend on interest rates higher than farmers appear to be willing to borrow. On the whole, the project's response to SA findings related to credit is mixed; the intention is to provide short-term credit for high interest but leave flexibility in implementation.

Lack of credit is just one of many problems farmers face. The lack of good quality seed, affordable pesticides and fertilizer, and opportunities to market produce are among the most pressing problems for small holders. Larger land owners and those who have been able to assume the management of newly established collectives experience these constraints less and can make more effective use of credit. The capacity of the rural society to ease input and output constraints depends upon the availability of capital to re-build agribusinesses or create new ones—including facilities to produce, import, and/or distribute farm inputs; and production or distribution units that purchase farmers' produce.

### Monitoring Social Impacts to Define Next Stages of the Project

The first stages of the SA, including rapid rural assessments and systematic surveys, pointed to the need to continue the SA process. Conditions in the project areas were rapidly changing, resulting in corresponding changes in actual living conditions and people's perceptions. The next phases of the SA identified new institutional arrangements to provide targeted advice, including to women. The consultations were carried out jointly with local social scientists and women's NGOs to ensure that women gain equitable access to advisory services and credit and that their needs are taken into consideration in the design of agricultural strategies and land registration processes. These consultations resulted in decisions to involve NGOs in credit and extension work to be financed under the project.

Another important follow-up to the initial phases of the SA was the need to understand the formation and functioning of voluntary associations of farmers for production purposes and to assess the feasibility of introducing water user associations. A separate assessment was needed to show how people utilize their garden plots and what they need to maximize the returns to their investments—including labor. In addition, an assessment of the institutional arrangements required targeting credit to individual small holders or to their production associations was needed. Again, these issues were uncovered through support of local social scientists and NGOs, and relevant recommendations were incorporated in the project.

Social assessment can provide a better understanding of the farmers' needs beyond the five regions which were the focus of the recent qualitative and quantitative data gathering activities. Resource constraints made it difficult to expand the coverage of the SA. Nevertheless, qualitative research has been carried out in different rural areas to broaden the understanding of cross regional issues. The issues relevant for local corruption were also studied and their impacts on the poor and on independent farmers were noted.

The dynamic nature of the rural transition in Azerbaijan as demonstrated by the sequential qualitative and quantitative studies and consultations convinced the Project team that a flexible project design and implementation arrangements and an intensive monitoring process were necessary. Therefore, the SA team helped identify elements and procedures for impact monitoring to ensure that the advice provided reaches the intended beneficiaries and the expected behavioral changes take place.

### *Issues That Fall Outside the Scope of the Project*

There are many useful findings of the SA that could not be accommodated within the proposed Project. First, the SA points to the rapid evolution of a peasant society in rural Azerbaijan. The transformation of the rural society into a peasant society requires the establishment of a diverse economic base, including but not restricted to agriculture. The development of the non-agricultural base of the rural society thus falls on other sector projects, but the credit component could help support agri-industries and create industrial employment. Second, a number of issues emerge that point to current and future vulnerability of the rural populations outside the productive systems. For instance, the restructuring of the state farms and collectives and the growing number of unemployed in rural communities signal an increased number of people, mostly women, who will not have pension incomes in the future. Private pension insurance systems do not exist and are unlikely to be attracted to rural areas in the near future. However, the need to respond to rural social protection systems also falls outside the framework of the proposed project.

Rural people face problems of social assets and infrastructure. Rural education and health systems have been adversely affected and much of the infrastructure is broken down. Shortages of electricity directly affect farm production and irrigation. Deteriorated transport infrastructure likewise has direct adverse impacts. The rapid loss of social assets constitutes a major push factor and induces out-migration. There is little that the proposed project can do about these issues. However, the project's policy development component will have to take these issues into consideration as the availability of labor in rural areas will depend in part upon the availability of social services and social insurance systems.

## Social Assessment of the Kazakhstan Agricultural Post-Privatization Assistance Project

### Background

The Republic of Kazakhstan initiated the Agricultural Post-Privatization Assistance Project which seeks to support the development and commercialization of privatized farms and agribusinesses in two agricultural regions of the country and to improve rural productivity and incomes. " Kazakhstan's agricultural sector, including irrigation, merits particular priority for Bank lending because of its potential impact on rural employment and poverty alleviation and also because it is unlikely to attract large external resources on rural employment and poverty alleviation" (The World Bank 1977:16).[5]

The changes required at the enterprise level to increase productivity and to achieve and sustain commercial viability involve: splitting-up existing enterprises; changes in ownership, management, corporate status, and accounting practices; new investment; and changes in product mix, input usage, and production practices. Some of these changes have already been in effect but their social impacts were not adequately assessed. The social appropriateness of other changes has not been evaluated. Therefore, a social assessment was carried out in order to understand the nature and impact of privatization on rural families, to identify needs and opportunities for intervention, and to formulate mitigation measures to eliminate or minimize hardships on the people.

The SA was carried out in Akmola and former Taldy-Korgan Oblasts in early 1997. In April 1997, oblast boundaries were changed. Subsequently, the national capital was shifted from Almaty City to the City of Akmola, renamed Astana, and Akmola Oblast was expanded to include most of Kokshetau Oblast. Akmola Oblast is located just west of the center of Kazakhstan in the northern third of the country on the open steppes. Most parts of Taldy-Korgan are now included in Almaty Oblast, located in the southeast corner of the country north of the Zailiyskii Alatau in the Tien-Shen Mountains which form the border between Kazakhstan and the Kyrgyz Republic. Akmola Oblast has an estimated population of 880,500 and 1,632,900 people live in the Almaty Oblast. Almaty Oblast is predominately rural and Akmola Oblast is predominantly urban. Despite

---

[5]        Memorandum of the President of the International Bank for Reconstruction and Development and the International Finance Corporation to the Executive Directors on a Country Assistance Strategy of The World Bank Group for the Republic of Kazakhstan. The World Bank, Washington, D.C., 1977. This document commits the Bank to provide advice and financing for comprehensive farm restructuring and rural financial services as well as support for redefining the role of the government in providing agricultural services.

considerable emigration of Germans, Russians, and Ukrainians since independence, Kazakhs are still a minority in Akmola; in Taldy-Korgan, Kazakhs are the majority of the population.[6]

## *The Social Assessment*

### Objectives

The purpose of the SA was to ascertain the extent to which the rural peoples of Akmola and Taldy-Korgan Oblasts understand and are actively involved in the privatization and restructuring processes; to evaluate the reaction of independent farmers and farm workers in agricultural enterprises to economic changes underway; to identify the needs and aspirations of independent farmers and farm workers; and examine the social structure that has emerged as the result of agricultural reforms. An attempt was made to study emerging social development, participation, and institutional issues and to recommend appropriate remedial action and a monitoring framework.

### Methodology

Three social research methods were used in the assessment in addition to consultations with the borrower and the Bank team:

- A systematic survey of 600 households—farm workers and independent farmers;
- In-depth interviews of 100 managers or heads of base farms—the large remnants of former collective and state farms; managers of procurement, input, supply, marketing, and consulting enterprises; managers of agricultural processing and storing enterprises; and officials of district and oblast agriculture departments; and
- Fourteen focus group discussions with farm workers, heads of base farms, and independent farmers.

## *The Four Pillars*

### Key Social Development Issues

Privatization of the agriculture sector in Kazakhstan was carried out during the economic crisis following independence from the former Soviet Union and the transition from a centrally planned to a market economy. The SA revealed growing impoverishment and a high level of dissatisfaction and discontent among people in the two regions. With the exception of farm managers and independent farmers, most farm workers were pessimistic about the future.

---

[6]    Despite the presence of large ethnic groups in Kazakhstan and Uzbekistan, language issues are not systematically addressed in the recommendations of the SAs. Kazakh is the official language in Kazakhstan but Russian is also spoken in national and local administrations. With 51 percent of its population composed of Kazakhs, the country hosts Russians (32 percent), Ukrainians (4 percent), Germans and Tatars (2 percent each). In Uzbekistan, likewise, the majority are Uzbek, but Russians, Koreans, Turks, Pers, Tatars and Karakalpaks each speak their own language. Dinora Azimova. 1998. "Ethnicity and Ethnic Relations in Central Asia." Mimeo. The World Bank.

Privatization was seen as an imposition, a government policy that ultimately benefited farm managers and officials alone.

Privatization and restructuring trends in the two regions had both similarities and differences, and thus regional equity was a concern for the SA process. The privatization process started in 1992 with no significant impact on day-to-day life. Beginning in 1994, privatization and restructuring were still merely formal in nature; farm members were officially allocated land and property shares related to their length of service, but farm structures remained essentially intact. A third phase started in 1996 after the adoption of the new Civil Code of the Republic of Kazakhstan. The Civil Code required all economic enterprises to reregister in a recognized form and, if needed, change their management structures to conform to the Code.

Land and property shares were divided according to standard methods common to the whole Republic. However, certificates for conventional land shares were not given to the majority of the rural people in either region. Rather, they were kept in the offices of the base farms. Property (capital stock) was not divided physically anywhere. Two-thirds of the farmers who had separated from the base farms did not get their property shares due to various excuses. Because farm members generally received merely promises of ownership, privatization did not significantly change the status of the farms or their workers. Consequently, a major part of the rural population still considered privatization to be merely a formal act. They did not understand the potential value of their land and property shares and did not really consider themselves to be joint owners of former state property.

The SA indicated that more than 20 percent of the farm workers who knew that they had land and property shares had transferred their land shares, and sometimes their property shares, to someone else. In 75 percent of the cases, the transfer occurred free of charge. Officially, such transfers were considered to be voluntary, but compulsion, blackmail, and fraud were quite often involved in the process. The most common ploy was for managers to promise to pay wages once farm members transferred their shares to them and threaten to fire farm members who did not comply.

Rural living standards declined following independence and has not recovered. Family income decreased for more than 75 percent of those surveyed and 45 percent stated that their income significantly decreased. Chronic non-payment of wages during many months and years was still common even for more than half of the separated farmers, and it was not compensated by any alternative source of income. Wages paid in kind and various forms of rationing of vitally important products (special bread coupons in Akmola Oblast, for instance) were common in both oblasts.

Throughout the last decade, the availability and quality of social services had declined in the two oblasts. Kindergartens were closed, as were some schools; other schools were closed part of the year to reduce heating costs; and the hospitals and polyclinics on farms were closed in most places, as well. Meanwhile, informal costs of education and medical care increased greatly, further burdening rural families and compounding the effects of non-payment of wages and disguised unemployment in the countryside. The imminent collapse of many farms, as well as restructuring, threatened to exacerbate the situation, resulting in increased hardship and the prospect of a lost generation of rural youth.

### Stakeholders and Institutions

The principal stakeholders in the project are owners and managers of base farms, farm workers, independent small farmers, commercial banks, input and service providers, the district and oblast agricultural administrations and social service providers. They were all consulted during the project preparation process, either directly or through the SA. The SA concentrated primarily on farm managers, farm workers and independent farmers to understand the dynamics of the privatization process, the status of various stakeholders, and the needs and aspirations of rural residents; and to give voice to those who are most affected by privatization but least heard by decision-makers.

A majority of new peasant farmers faced major difficulties in managing their farms for various reasons, including lack of machinery and credit. Most independent farms were not specialized; they grew a mixture of crops, with field crops predominating. Thirty-nine percent of the farms also raised livestock. Practically no farms were engaged in processing agricultural products or providing services. The technology used on most farms was simple, because it was virtually impossible to acquire machinery, and the price of spare parts, fuel, and fertilizers was prohibitively high. Indeed, independent farmers reported that the lack of equipment and spare parts was their main difficulty in operating their farms, followed by working capital for fuel, lubricants, and inputs. By the time of the SA, one of the biggest motivations for people to separate from the base farm was desperation—the desire to escape the collapse of the base farm. This contrasted markedly with the earliest privatization period in which the main stimulus to separate was the aspiration to be independent.

About a fifth of the independent farmers were able to make a profit during the 1996 agricultural season, and many broke even. Some of the new institutional arrangements appeared to affect the farmers adversely. For example, many found taxes to be a problem. Despite the serious challenges they faced, very few of the independent farmers had tried to return to the base farm or wanted to sell their land. Many wanted to expand their farms but were prevented from doing so by a lack of credit. Many reported that the base farm administrations complicated their work, primarily by arbitrarily limiting access to land, machinery, and property. The dominance of pre-transition institutions reduced people's trust in reforms.

Despite the efforts of farm managers to consolidate land and property shares, it was obvious that the number of independent farms would continue to increase over time and that a shift from state to private ownership of land rights was inevitable. Twenty-one percent of the workers on the base farms wanted to separate and start their own farms and 53 percent said they would like to become independent if they had access to capital and credit. Lack of money and lack of machinery were cited as principal reasons for remaining on the base farm. The farm workers who did not want to be independent said they had no funds or no experience that would enable them to separate. They were also discouraged from separating by the administration of the base farms. The heads of the base farms, on the other hand, unanimously stated that they did not receive orders and no one dictated from the district and oblast administration (although there were statements to the contrary in focus groups). A majority of the independent farmers said that the base farms did not restrict their actions and they got no real assistance from the State.

On most base farms, economic indicators had sharply decreased and farm debt had increased steadily. The increase in farm debt was due to the imbalance between farmgate prices for agricultural products and the prices farms for the industrial products they needed to operate. The price differential deprived most farms of the sources of self-development. Any profit they obtained was immediately used to repay debts, particularly for utilities. Barter was the dominant form of exchange everywhere, to the detriment of primary producers. On most farms practically all products produced were exchanged for spare parts and fuel. Livestock had been the main currency for barter, which resulted in a sharp decline in their numbers. In addition, barter had greatly increased transaction costs and slowed the development of market relations in the countryside.

In most cases, both the base and independent farms were unable to obtain credit. State and semi-state banks did not give credit to unprofitable farms. Private banks were not familiar with lending to agriculture and viewed it as risky and high cost. In addition, farm managers and independent farmers considered the interest rates of the private banks to be very high. Farm workers and independent farmers sought loans for different purposes. Workers gave the highest priority to buying and repairing machinery, while independent farmers said they needed working capital to buy inputs and spare parts. The owners of independent farms would have liked to join or create associations to solve different production, financial, and social problems. However, they did not have the financial resources needed to organize such associations nor did they know of groups that could help them organize.

### Participation

In the past, the participation of rural people in transition decisions was weak. The consultation and information exchange between local and national institutions and between the rural people and local managers was largely unstructured and very limited. As the privatization process unfolded, strong vested interests limited the flow of information and manipulated contacts and institutions to strengthen their hold on rural assets. At the same time, the adoption of new legal structures and the emergence of new players provided countervailing influence, opening new opportunities for rural residents to pursue their individual and collective interests despite the opposition of various elements of the establishment. As the project is to be demand driven, information flow and knowledge dissemination will be critical factors in generating the demand for project initiatives.

The lack of credit in the countryside, except on a barter basis, stymies growth and depresses entrepreneurial tendencies. The SA revealed that farm workers and independent farmers have little knowledge about credit and even more limited experience with credit, although farm managers have some experience. Although most people agree that credit is needed, the potential use of credit varied among different groups. In aggregate, credit was desired mainly to buy machinery, raw materials, seeds, fuel and spare parts.

Similarly, lack of knowledge about where to get advice and assistance, as well as a general mistrust of both private and public sector institutions, prevented rural residents from taking actions that might improve their condition. This gap in information and knowledge also applied to such issues as creating farmers associations. Many people would like to combine their resources and energies with others, but they did not know how to create institutional mechanisms that would respond to this need. Access to information was limited on all fronts, but people nonetheless looked for help and advice when they needed it.

Friends, neighbors, and acquaintances were the most common source of information regarding farming issues, followed by radio and television, then newspapers. People argued that their choice of information sources was very limited and irregular. Media exposure was limited by electricity shortages, particularly during the winter, so radio and television were not dependable information sources and they devoted too little time to rural problems. Standard newspapers did not devote much attention to rural issues either. Further, they were considered to be expensive and one paper was often passed to several families. In short, it was clear that the hierarchy of information sources indicated above reflected chance rather than choice. Nevertheless, the provision of information to those whose lives are affected was clearly the first step towards a more participatory implementation of reforms.

### Monitoring and Evaluation

The first phase of the Project will essentially be a pilot program. Systematic monitoring of demand, supply, impacts, and opportunities to ensure development effectiveness will be required in order to maximize learning in the pilot. The project is expected to benefit rural people in four ways. First, the technical and business information and capital it provides will enable independent farmers and managers of farms and agribusinesses to make their enterprises commercially viable. Second, in doing so, it will significantly redress the existing gross inequity in access to information in the rural areas. Third, it will introduce new, client-centered institutions, and reorient the approach of existing institutions, notably banks. Fourth, by delivering information to rural residents about their rights and responsibilities, as well as alternatives and opportunities, it will encourage nascent entrepreneurs to break away from existing relationships; and by providing capital, it will empower them establish new enterprises that produce the goods and services needed for a vibrant rural economy

Given this ambitious set of anticipated outcomes, the monitoring program is designed to assess:

- Changes in the composition, magnitude, and content of demand for advisory services and credit;
- The adequacy, responsiveness, accessibility, and effectiveness of the advisory centers and participating financial institutions;
- Economic and social impacts of public information campaigns, advisory centers and networks, restructuring plans, investments, and the divestiture of social infrastructure at the farm and community levels; and
- The impact of small-scale investments in local commercial and service enterprises, and emergence of new opportunities.

### *Implications of the Social Assessment for the Project*

The SA had a number of implications for the design of the project intended to help farmers in the aftermath of privatization and restructuring and to promote further restructuring. They are summarized below.

#### Use Information and Communication to Enhance Social Capital

Many rural people are poorly informed of the privatization process, their rights and options, and the implications for their future. Without adequate information, they remain unable to protect their own interests and they mistrust the intent of the reforms and those who support them. The SA thus contributed to the identification of key problems that can be resolved through information and communication, and provided guidance to the specification of the media, medium, messages, and target populations.

#### Regularize Land Share Transfers and Lease Arrangements to Ensure Equity and Transparency

Many farm members have transferred, sold, or leased their land shares under arrangements that are open-ended, without specifying the reciprocal rights and responsibilities of the farms and farmer managers, and without bringing real benefits to the people who relinquished their shares, either temporarily or permanently. Therefore, the SA recommends that before project funds are made available to any corporate farm, farm members are informed of their rights and options, and an assessment of the status of shareholdings is made.

#### Promote Flexible Arrangements among Shareholders to Broaden Opportunities.

An uncertain number of farm members may eventually wish to exercise their legal option to become independent farmers. Consequently, the project promotes restructuring and financing arrangements that preserve the option of farm members to separate from the base farms and pursue other options.

#### Create Equity in Capital Flow

Farmers and farm managers understand that their opportunities are limited due to the lack of capital for operations and investment, respectively. Capital is needed to induce restructuring; to establish new farms, large and small; and to increase, production, productivity, and profits. A portion of the credit incorporated in the Project is reserved for small independent farms to ensure that the credit demands of large units do not overshadow the needs of smaller units.

#### Make Business and Technical Advice Accessible to Enhance Inclusionary Practices

Advisory services will need to prove their value to farmers who are justifiably skeptical of initiatives promoted by the public sector and unfamiliar with private sector operations. Responding to the findings of the SA, priority will be given to the provision of financial and legal advice and market information to all categories of farmers, regardless of gender, ethnicity, and other group membership.

### Mitigate Adverse Social Impacts

To become profitable, many farms will have to divest social infrastructure and reduce their labor force. Although direct mitigation is beyond the scope of the Ministry of Agriculture, efforts will be made to help local administrations assume responsibility for social infrastructure and to provide start-up capital for local commercial and service enterprises, both to reduce unemployment and to serve local needs through the private sector.

## *Issues That Fall Outside the Scope of the Project*

Social infrastructure in the countryside deteriorated sharply. Cultural-educational institutions vanished and kindergartens were closed everywhere. The schools in Taldy-Korgan Oblast did not function for a long time due to lack of heating. In addition, medical services began to require payment. In the districts surveyed in Akmola Oblast there were no special consulting services. In Taldy-Korgan Oblast, the agricultural administration established a marketing-consulting center but had insufficient time to demonstrate results. Nonetheless, practically all the heads of base farms and about half of the owners of the peasant farms sought the advice of specialists. Most farmers said that they would welcome the creation of special consulting services. However, they also expressed the wish that initially such services should be available free of charge or at low prices.

# Social Assessment for the Uzbekistan Agriculture Enterprise Restructuring and Development Project

## *Background*

The Government of Uzbekistan (GOU) has committed to transfer management of farms and the rural economy from the public sector to private hands. Starting in 1994, GOU approved and began to implement a number of reforms, although it made little progress. To support this effort, GOU formally requested the Bank to help design and support specific reform programs for its rural sector. The Bank responded with the Agriculture Enterprise Restructuring and Development Project, the first Bank project targeting Uzbekistan's rural sector. During the early stages of project preparation, the Bank, working with local groups, conducted a social assessment.[7] This summary presents the findings of the SA's first phase and recommendations in designing and implementing the Uzbekistan Agriculture Enterprise Restructuring and Development Project.

The Republic of Uzbekistan comprises the Republic of Karakalpakstan and 12 provinces (oblasts), 124 cities and 157 counties. This semi-arid country covers 444,000 square kilometers.

---

[7]      The social assessment team comprised sociologists from SIAR-EXPERT in Tashkent, led by Alisher Ilkhamov and joined by Igor Pogreboc (research design and data analysis), Arustan Zholdasov (survey execution) and a consultant from Harvard University, John Schoerberlein-Engel.

Gross Domestic Product (GDP) is about $15 billion. Negative growth and hyperinflation have plagued Uzbekistan for most of the past decade, but inflation is now near zero. Sixty-one percent of Uzbekistan's rapidly growing population of 23 million live in rural areas. Agriculture comprises 30 percent of GDP and has grown faster than the economy generally. Only 10 percent of land is cultivated and 95 percent of that is irrigated. Cotton and grain are the most important crops, while meats, fruits, vegetables, milk and silk are also produced. Cotton is the largest export while food is the largest import.

GOU has launched a number of economic reforms but they have not been widely implemented. Methods of payment, land reform, price reform and privatization are all needed but have yet to be genuinely implemented. Three major stakeholders—GOU, regional and district governments and the management of state/collective farms—are responsible for lack of progress. As a result, they and the other major stakeholders—independent dehqan farmers, lease-hold dehqan farmers, household farmers and non-agricultural labor—remain impoverished.

### *The Social Assessment*

The SA of Uzbekistan focused on the four pillars—key social development issues, the institutional and organizational capacity for addressing these issues, the mechanisms for participation by various stakeholders in developing and implementing policies and programs, and the creation of a system to monitor and evaluate progress. The SA identified key social development issues, targeted project benefits to the appropriate stakeholders, helped design the institutional and participation mechanisms to encourage popular participation in programs, and ensured that the intended beneficiaries actually received the benefits of the project.

#### **Objectives and Methodology**

The SA first addressed the needs of the rural population, the strategies it adopted in coping with the farm restructuring process and potential mechanisms through which stakeholders can participate effectively. These objectives were achieved with several tools: 1) a survey of 981 households across 65 communities randomly selected from six provinces in Uzbekistan; 2) a survey of 51 local, regional and national government leaders; 3) interviews and consultations of households, community leaders and government officials; and 4) a review of literature, official documents and data.

#### **Key Stakeholders**

Key stakeholders in the farm restructuring process are state/collective farms, cooperatives (shirkats), independent dehqans (peasant farms), leasehold dehqans (unregistered farms), smallholders (subsidiary plots), private livestock farms, mahalla/qishlaq councils and district administrations. The district administrations control the allocation of land, irrigation and the purchase and distribution of agricultural products, making them very powerful. The mahalla/qishlaq councils fundamentally are extension of the district administrations. State/collective farms and cooperatives account for most of cultivated land, despite their relative small number. Cotton and grain predominant their production. The large number of smallholder and dehqan farms account for a much smaller proportion of cultivated land. These different types of farms share a large proportion of their labor with each other. About 37 percent of state/collective farm members also cultivate private plots as smallholder farmers. A significant portion of these workers also engages in non-agricultural service work. Many dehqan farmers also work on state/collective farms. The most independent farms in Uzbekistan are the 1,600 private livestock farms, which possess on average 400 head.

## *The Four Pillars*

### Key Social Development Issues

Rural Uzbekistan suffers from a number of problems that cause income and employment to decrease. Some existed during Soviet rule—rapid population growth and shortages in water, gas and credit. Others are problems resulting from the restructuring process—dilapidated social welfare and public goods such as education, extension/technical assistance, water and gas. Another set of problems results from incomplete implementation of sound reforms—small land plots and land shortages, shortages in agricultural processing, poor communication of government reform efforts and little or no popular participation in developing and implementing reforms. These problems of incomplete reform are most critical.

**Decline in Income and Opportunities**. Average annual per capita income is between $76 and $125 and has been falling over the past 10 years. Using the black market exchange rate lowers these figures by 25 percent. Most or all of this income is spent on basic necessities for most families, precluding expenditure on education and healthcare as well as savings and investment.

**Lack of Confidence and Initiative.** Most citizens lack confidence in their ability to address their declining incomes by working in the private sector. Many doubt opportunities exist outside of the state/collective system and they count on government, not themselves, to fix their problems. Lack of experience, training, information, opportunities and start-up capital cause this pessimism.

**Decline in Public Goods and Social Safety Net**. The difficulties in making a living are compounded by a parallel decline in the social safety net and provision of public goods such as education and utilities. Twenty-four percent of the population age 16 and over has no source of income whatsoever, belying government statistics indicating low unemployment. Eighty-four percent of the unemployed does not receive any welfare. Doctors now charge patients for care and medicines. Ancillary educational costs such as books, school supplies and transportation now preclude many from attending school. Other services such as drinking water and pre-school care are no longer provided by the government. The unemployed, the sick, some women and a small number of pensioners are most vulnerable to these conditions.

**Ethnic Tension**. Seventy percent of the population is Uzbek and the remainder are Russian, Tajik, Kazakh, Tatar, Karakalpaks, Korean, Turk and Pers. Violent conflict emerged in the last decade due to political borders dividing cultures rather than uniting them, GOU's fear of growing Islamic fundamentalism, GOU aggrandizement that agitates Uzbeks in neighboring countries and threatens their governments, GOU discrimination against minorities, and too many diverse people competing for limited resources.

### Stakeholders and Institutions

Widespread and inefficient state/collective farms, overly powerful district administrations, weak local governments, weak but improving regulation of financial institutions, lack of rural financial institutions, lack of private industry and civic associations and lack of marketing facilities mark the need for further institutional and organization development. District administrations implement and enforce the settlement accounts system and state control of prices, quotas, water, inputs and processing. State/collective enterprises utilize their close relationships to district administrations to control these assets and prevent more efficient private farmers from competing freely. Weak mahalla/qishlaq committees locally do not have the power or resources to provide adequate welfare, education and healthcare. Institutions for providing credit, training, technical assistance and marketing facilities do not exist in rural areas. Industry associations that could assist in all of these areas, such as water user associations, do not exist. There also is no intent or institutional capability to involve the rural population to identify and develop solutions to these and other problems. The only area of significant progress has been GOU's effort to strengthen legal, regulatory, accounting and bank supervision capabilities. Such improvements are strengthening lending capabilities and opportunities to attract foreign investment.

**Management and Workers on State, Collective and Cooperative Farms**. Farm members are organized in a lease system which are unfavorable to members. Although these farms are favored by the administration, they are so tightly controlled that they are little changed since Soviet times. State and collective farms are failing, thus there is a move to turn them into joint stock companies. Farm members constitute one of the largest labor groups in the country, and one of the most poorly paid. They principally depend on off-farm income and production to survive.

**Independent Peasant Farmers**. These units are theoretically independent, but they rely heavily on the larger state and collective farms for irrigation, inputs and marketing. Consequently, their fate is tied to the larger units, although they have started to expand and develop some independent traits, especially in the areas near major cities.

**Smallholders**. Households among the smallholder group are part-time farmers, at most. They provide for family needs on their small household plots and generally have a number of household income sources from agriculture to transport, petty commerce and the provision of services. Currently they are prevented from expanding their units, but their entrepreneurial interests would undoubtedly bring some dynamism to the sector if the ban on expansion were lifted.

**Other Farms and Farmers**. Other specialized farms and farmers and farm laborers have varying interests and needs that can be addressed in the project. Some units are dynamic; others will undoubtedly disappear as the privatization process unfolds.

**Administrators**. District and oblast administrators have active roles in the sector, much to the detriment of agriculturalists. By regulation, coercion and monopolistic and monopsonistic practices, they strangle farmers and farms of all types, exacerbating major weaknesses in the sector.

Local administrations, on the other hand, are powerless and without resources, but they have the potential to articulate the interests of rural people and advocating policies that direct resources to revitalize the rural economy.

**Currency**. In lieu of money, many transactions are handled by debiting the buyer's "account" and crediting the seller's "account." Transactions are time-consuming and/or difficult to accomplish. Additionally, the government uses these accounts to administer the economy and control behavior.

**Input Supply**. A small number of state-controlled enterprises produce and distribute inputs and services such as fertilizers, pesticides, seeds and machinery. These enterprises act as monopolies, dictating the terms and conditions of supplies. Because of the traditional ties between these state enterprises and the state/collective farms, state/collective farms receive inputs first at the expense of private dehqans and household farms. Dehqans and private household farms try to get the inputs they need from the state/collective farms but only sometimes are successful.

**Irrigation and Gas Supply**. Distribution of water without charge and technical shortcomings in the irrigation system are the primary contributors to widespread water shortages and increasingly salty soils. Economic, political and ethnic disputes as well as declining crop yields have resulted. Gas supplies are hampered by state-controlled monopoly distribution and lack of GOU attention. Forty-one percent of local officials state gas shortages are the most critical problem they face.

**Processing**. Eighty-four percent of local officials surveyed stated dissatisfaction with agricultural processing. Seventy-three percent believe that agricultural processing should be the government's highest investment priority. Although GOU says it has "privatized" the agro-processing industry, it retains control through numerous measures on most aspects of the industry. These measures take the form of:

- monopoly and monopsony in most commodity sectors;
- state ownership/management control of processors;
- administrative controls on most aspects of business; and
- a ban on exports.

These measures depress demand for agricultural products and the prices paid for them. However, a small number of genuinely private agro-processors have begun to emerge which over time could impose genuine competition throughout the industry

**Credit**. Thirty-one percent of local officials surveyed stated credit is the most urgent problem in rural Uzbekistan. Financing cotton and wheat production accounts for 90 percent of all agricultural lending. A system of "centralized credits" is used where government controlled banks lend to input suppliers and trade associations which advance their physical inputs to the producers. Other loans are made with opaque eligibility criteria and a demand that loans are only available when the borrower guarantees a specific quantity of product at a fixed price to a predetermined buyer. Interest rates are normally around 100 percent annually. Collateral is not well understood or readily available. As a result, only 15 percent to 20 percent of private farmers have obtained loans and about half of state/collective farms have difficulty receiving them. However, GOU is strengthening the sector's legal and regulatory framework, adopting internationally accepted accounting systems, strengthening bank supervision and developing capital markets.

**Incomplete Implementation of Reforms**. The reforms have focused primarily on four areas: divest GOU assets from agricultural production; transfer land-use rights to those that utilize land; deregulate prices and eliminate monopsonistic government procurement; and reduce taxes and forgive debt incurred by state/collective enterprises. These reforms are insufficient, have significant shortcomings and have not been fully implemented. They maintain most production under the control of state/collective farms, which are far less productive than private dehqan and household farms. Survey respondents feel reforms implemented to date benefit only the privileged elite and fail to give the general population economic and political freedom from elite control. Most citizens want more reform, particularly regarding land, which has been difficult to obtain and use effectively.

### Participation

Participation and communication is poor in the reform process. There is a disconnect between the government and the population—government officials favor state/collective farms whereas the general population sees them as the problem; and government officials are far more optimistic than ordinary citizens about conditions in Uzbekistan. Further, 82 percent of the citizens do not fully understand the reforms. Inaccurate government perceptions and poor understanding of reforms by the general population result from poor participation and communication. Government does not disseminate much information about the reforms. When it does, it is through media that are the most reliable but least used—newspapers and specialized literature. Radio and television are also used to disseminate technical and legal information, but the treatment is more superficial and thus less useful to rural residents. No effort is made to hold interactive meetings and the provision of extension/advisory services is insignificant. Information and communication will be critical to the success of the project.

### Monitoring and Evaluation

The Bank intends that the proposed project provide long-term, continuous and adaptable support to agricultural development in Uzbekistan. Therefore, the results of initial phases of the project must be recorded, assessed, and used to develop additional solutions to problems. Specific output and impact indicators of the project's progress will be monitored and assessed. Bank staff, GOU officials and local populations will work together to assess and develop these solutions regularly. The M&E program should monitor: 1) the abolition of the settlement accounts system; 2) the development of private distributors in fuel, inputs, cotton, cereals, water; 3) the development of private processors; 4) improved policy formulation capacity with adequate popular participation and strengthened technical assistance; and 5) the decontrol of prices for cotton, cereals, water and inputs. It should also attempt to measure: 6) the rate and equity of land distribution; 7) the rate and equity of collective share distribution; 8) reductions or elimination of subsidies for cotton, inputs, water; and 9) increased institutional lending to rural clients by commercial banks.

## *Implications for Project Design*

Privatizing government functions, building government and private institutions, and involving citizens directly in economic reform and rural development will address the major weaknesses described in the preceding sections. Resolving problems with national and local governments, agricultural input suppliers and producers, processors and other agribusinesses, irrigation and gas providers, and financial institutions will significantly improve incomes and employment in Uzbekistan, as well as reduce ethnic tensions and help the social conditions for women.

### Privatization and Institution Building

**Genuinely Privatize Collectives**. Shares in state/collective enterprises must be distributed in a fair, transparent process to all entitled farmers, dehqans and collectives. Laws governing this process must be passed by GOU and address the distribution of land and agricultural equipment as well as basic procedures for enterprise governance.

**Improve Water Management and Increase Gas Supplies**. To use water more efficiently and improve agricultural production, the SA recommends that GOU fix technical problems with the existing irrigation system, implement full-cost recovery and establish water-user associations comprising private farm representatives and others. GOU should facilitate entry of private fuel supply and distribution companies that can compete effectively with the existing state monopoly.

**Liberalize Input Markets**. GOU should facilitate establishment and financing of private input suppliers in all input industries, decontrol prices and phase out existing subsidies.

**Assist Farms in Marketing**. GOU should eliminate state prices and quotas on cotton and cereals and end the purchase/distribution of cotton fiber, cereals, water and land by district administrations. In lieu of its current role, GOU should establish private service associations to facilitate marketing activities, including exports, and construct physical market facilities. It should no longer retain hard currency earnings on exports.

**Privatize and Liberalize Processing**. The SA recommends that state orders and the settlement accounts system be abolished, permitting prices and quantities produced to be determined by the market and permitting farmers to be paid in cash. To instill competition, GOU should encourage additional private sector participation in processing. Private processors should receive financing, technical assistance and other business advisory services and help in developing storage, grading and packing facilities.

**Build Credit Institutions and Markets**. The SA recommends that the World Bank provide finance to suitable banks in Uzbekistan to extend loans to restructured farms and agribusinesses, and that GOU abolish the "settlement account" system. Lenders should finance activities that guarantee an early increase in productivity. New financial instruments such as mortgage financing, leasing, use of warehouse receipts and crop insurance should be offered. Financing should be extended at interest rates to permit banks to earn a reasonable profit. Training to facilitate the loan application process should be provided to both lenders and applicants. To help force responsible borrowing, borrowers should be required to risk some of their own capital as part of the loan agreement.

**Strengthen Mahalla/Qishlaq Councils**. GOU should empower these local village committees with tax collection authority and budgets to provide social welfare and services programs and implement the state/collective share distribution process.

### Public Mobilization and Technical Assistance

**Promote Popular Participation**. All levels of government need to establish public forums for discussion of policies and programs with ordinary citizens. Minorities need to be recruited heavily for these events. As part of this effort, GOU needs to improve its gathering, analyzing and dissemination of objective information. GOU and local community organizations may want to consider celebrating days of cultural importance to minorities.

**Provide Technical Assistance**. GOU needs to facilitate provision of technical assistance and handbooks to all stakeholders, including women and minorities. Such assistance should address management, technical, finance and business topics.

## Social Assessment of the Sanliurfa-Harran Plains Agricultural Development Project in Turkey

### *Background*

The Southeastern Anatolia Project (GAP) is a regional development project comprising several large-scale irrigation and energy investments. One of the first projects within GAP is the Sanliurfa-Harran Plains Irrigation Project which aims to irrigate an area of 150,000 hectares, equivalent to 10 percent of all the cultivable land in Turkey, thus adding more than half as much to the total irrigated land in the country. The completed project will cover an area of 400 settlement units and more than 200,000 persons. More than half of the Project is concluded and at the present time irrigation has already reached around 82,000 hectares.

The Agricultural Development Project proposed by the World Bank targets rural development efforts including on-farm improvement within the newly irrigated area of 178 village settlements. Land improvement refers to consolidation (agglomeration) of plots, land use planning, leveling for efficient irrigation, pavement of new in-farm roads, cadastral management, and agricultural extension activities. The social assessment is part of the Project preparation process and aims at identifying an appropriate mix of Project inputs to reduce poverty, help target benefits to the landless and small holders, and support Project implementation so as to ensure equity in the distribution of the benefits of Bank financed activities.

## *The Social Assessment*

### Objectives

Specific objectives of the Bank's social assessment completed in October 1998[8] were to:

- Understand the impact of the rapid agricultural modernization instigated by irrigation on the social structure of the area on various social groups according to ownership criteria and gender;
- Uncover the perceptions of the population benefiting from irrigation, their assessment of the material and technical services they are receiving, and their expectations concerning needed institutional changes;
- Identify stakeholders in order to enhance the possibility of participation of the population in sustained development based on project goals; and
- Propose monitoring and evaluation indicators for project management and implementation.

### Methodology

The SA's empirical focus was on 35 villages chosen for their variability with respect to the extent of technological change in agriculture.

- Village background information was gathered through questionnaires and discussions with headmen and village elders;
- Participatory stakeholder consultations were conducted with groups including women, landowners, landless peasants, and focus groups in these 35 communities and elsewhere in the region;
- Consultations with other stakeholders, especially agency representatives, the private sector, and non-government organizations were likewise carried out both at the regional and the national level; and
- Results were shared with the borrower, the Bank team, as well as with broader stakeholders in a symposium

## *The Four Pillars*

### Key Social Development Concerns

The rapidity of changes in production relations, income, and equity were significant over the past decade and substantially reduced poverty among the landless. Social inequities are nevertheless marked, with the landless, women, and younger generations having lower income and playing more marginal roles in the society. Large-scale investments in irrigation have substantially

---

[8]    The field work for the SA was conducted in November 1998 through a partnership of the Turkish Rural and Urban Development Foundation, Oklahoma State University, and Water/Construction Engineering and Consulting Group.

increased agricultural productivity and household incomes in the region. Household incomes from cultivation are approximately double in areas where irrigation canals exist. In absolute terms, incomes and living conditions have improved for all segments of the population including landowners, tenants, sharecroppers, and workers; relative to their situation prior to these investments, low-income groups have made the highest relative gains. For instance, while 71 percent of the landless were migrating to other regions as seasonal laborers prior to these investments, currently only 12 percent of the landless do so. The region now receives migrant laborers from other regions—some permanently and most on a seasonal basis. Further expansion of irrigation systems to other areas will have similar impacts provided that labor intensity in agriculture can be maintained.

As a result of transition to irrigated agriculture there has been a rapid shift in cropping patterns toward cotton; 93 percent of the households planted cotton. The labor intensity in cotton production has been instrumental in enhancing the employment of the landless, and irrigated cotton growth has increased incomes for sharecroppers although 70 percent of the harvest revenues go to the landowners. Besides the economic risks involved in concentration in a mono-crop, land productivity is not high in cotton production. There are many reasons why a shift away from cotton would be highly desirable, but several constraints need to be removed. At the policy level, subsidized government purchase of cotton discourages the cultivation of other cash crops. Urban markets, on the other hand, are not easily accessible and farmers are reluctant to shift to perishable crops. Lack of marketing cooperatives adds to the problems. Road infrastructure also hinders access to fields, but improving on-farm road infrastructure will also have a disproportionate return for landowners. If cost recovery could have been assured directly from the landowners without having them share the costs with the sharecroppers there would have been greater social justice in these investments. However, there is little incentive for cost recovery and the regional/local power relationships are likely to encourage continued heavy reliance on the State for major investments as those in power derive the largest benefit from these investments.

Another important outcome of irrigation is the increase in farm equipment and tractor ownership; one third of the families own tractors which, were bought within the last three years. Landlords provide tractors to their sharecroppers and lease them out to other cultivators. The concentration of land and technology in few hands enables the sharecroppers only one-third share of the harvest. It is also clear that further investments in land productivity that are financed by the State would bring larger benefits for the landowners rather than for the landless low-income households. Lacking collateral in the form of land, the latter cannot individually access credit for technology. This could have been overcome through support of associations or cooperatives of landless farmers. However, civic engagement or voluntary associations are relatively less developed in this region as compared to many other regions in the country, possibly as a result of strong loyalty to primary social groups and feudal traditions.

Income and asset distribution is highly skewed. Sixty percent of the households are landless; they work as sharecroppers or as wage workers. Large landlords lease some land out, but predominantly engage sharecroppers. The major source of income is farming; the size of land ownership determines family income levels. Sharecroppers cultivate an average of 5 hectares while those who own land cultivate an average of 18 hectares. Livestock ownership declined when former grazing lands were reclaimed for irrigated farming. Although only 18 percent of the households own cattle and 40 percent own sheep/goats, livestock is an important determinant of household income. Investments in agriculture whether they are to improve irrigation, land consolidation, land leveling, or farm roads are bound to disproportionately benefit the landowners and those who are already wealthy. A clear social development challenge of any Bank-financed investment in this region where traditional, feudal relationships dominate is the identification of mechanisms through which the landless and the poor directly benefit from public investments. This can be achieved through support of labor intensive modes of crop production, promotion of livestock management through cooperatives of the poor, promotion of non-farm employment, and through the formation of democratically organized irrigation associations and other civic establishments such as cooperatives.

The society is highly stratified and diverse. Income, gender, and inter-generational inequities are substantial, and the distribution of social capital is highly unequal. Some significant differences between the three main ethnic groups, Turkish, Kurdish and Arabic speaking populations, also exist but little is known about the Project specific implications of this. The heads of more prominent tribes not only have better economic status but the tribal affiliation provides solidarity, information, connections, and a social safety net for its members. The prominent families belong to tribes or large lineages, but many of the landless and the migrants have fewer family connections, and those they have are of little importance. Although most households have a nuclear family composition, extended family bonds prevail. Those with large families within the region and good connections in the towns and cities command respect in their communities.

The tribal traditions require utmost respect for the elderly and subordinate the young and the women. The position of women is especially low for women who have married into local families but come from outside the region or the village. Women's subordination takes many forms and excludes them from public life. They do not inherit land or other productive family assets. The tradition of bride price continues along with polygamy despite the existence of the Civil Code that has long abolished it. Most adult women are illiterate and uneducated but the gender gap is decreasing among school age children. An unexpected benefit from the consolidation of plots has been the ratification of women's property rights in land. Women inherit land but the male head of the household often exercises de facto tenure. Sisters and mothers are forced to transfer their titles to male members of the family. Under the existing legislation, property is registered equally during land consolidation and women thus gain the right of ownership. While it will take time for the feudalistic traditions to change, the Bank's support for the land consolidation process will have some positive impact on women's status in the region.

Improvements in village infrastructure have not kept pace with agricultural technology and worked to the disadvantage of women. In half of the households drinking water is the most important problem. Fetching water from distant wells is predominantly a woman's task; the solution of this problem would increase the welfare of women directly. It will also improve the health of children since the deficient quality of drinking water is the most important source of childhood disease and mortality.

Poverty and landlessness are associated with difficulties in obtaining education. Sixty eight percent of the villages claim that primary educational and health facilities are not sufficient. More than half of the school-age children does not attend school; poverty and the use of child labor in agriculture are the primary reasons for this. Children in poor households participate in seasonal labor migration with the entire family , which prevents them from attending school. The children in high-income households have significantly better educational attainment regardless of gender. While the Bank provides high levels of support to the Basic Education Program of Turkey, part of the solution lies in strengthening incomes of poor families. Gender is the single most important determinant of educational attainment among the adults, with far fewer women primary school graduates. Indeed, in household visits one can also easily see that many adult women do not even speak Turkish, especially if they are young brides from outside the country. The fact that Arabic is spoken in about 90 percent of the households is extremely significant for adult women's ability to benefit from the proposed Project's advisory services.

### Participation Concerns

Inequitable participation in benefits of land-based public investments is of critical concern. For instance, the leveling of fields, already implemented in some parts of the region, increases efficiency in water use. Average income from arable land is 38 percent higher in households where fields have been leveled. Investments in land improvement thus have a direct impact on incomes. However, investments in land consolidation and land leveling primarily benefit the wealthy. Those that are already better off enjoy a disproportionately high share of highly subsidized state investments in what is in essence private property. It is therefore important for the Bank to stress the importance of a socially equitable system of cost recovery for all investments in individually owned property.

Drainage is insufficient and its improvement is crucial for both economic and environmental reasons. Without proper drainage there are risks of salinity, soil erosion, and contamination of the drinking water. Therefore, these investments can be more easily justified in terms of their public health aspects; they are socially more just than other investments in private land. Also, there is some willingness on the part of landowners to participate in the costs of leveling and drainage although at low levels. The higher income groups participated in cost recovery through land contributions when they voluntarily gave up a certain portion of their land in the process of land consolidation and the construction of the irrigation system. In all other types of public investment, the landowners indirectly pass over their cost recovery contributions to the landless sharecroppers.

### Institutions

The institutional issues are complex and require far greater focus. The interface between central and local level institutions and between informal and formal organizations requires further analyses. Also, the influence of kinship-based local level civil society organizations need s to be examined in order to identify institutional mechanisms through which firmer associations of small holder and the landless can be encouraged. At the time of the reporting of the SA process, these central institutional issues were not adequately addressed; thus, the most important challenge is the completion of the analytical and empirical work to better understand local institutions.

Several observations can be made about organizational and agency related issues. There is insufficient coordination among various agencies responsible for rural development. Generally, knowledge among the public about the institutions responsible for development projects is insufficient and trust in public institutions is low. Civil society organizations, including economic organization of farmers, other producers, and marketers, are weak. Strengthening them so as to facilitate the participation of the landless would significantly enhance the economic status of those with fewer assets.

The eleven water user associations in the irrigated area are not working efficiently and democratically. Each village is represented by a headman and two representatives; the associations are managed by seven-member committees, which often consist of the leaders of lineages and tend to reflect the traditional social order of the region. Complaints about water user associations are widespread: 38 percent of the heads of household and 57 percent of the village headmen feel that the allocation of water is not fair. These associations will become more important as irrigation has to be managed daily. Enhancing participation of sharecroppers in water user associations is thus of particular importance.

Extension services are deficient. Information about optimum irrigation techniques, agricultural innovations, modes of crop diversification, and non-agricultural income generation possibilities is not available. Trust in government agencies providing advice is low and their areas of expertise is largely unknown: only 18 percent of the households obtain information from one of the extension agencies—the Provincial Agricultural Directorate, the Village Services Directorate, and the Land Reform Directorate.

Irrigation water is insufficient due to the lack of knowledge of irrigation techniques and nighttime irrigation that would use the water more efficiently. There are also clear inequities in getting access to information both through the formal and informal channels. The powerful landowners are also prominent figures in the tribal organization, and their families have many more sources of information. The landless, the migrant seasonal workers, and the agricultural workers whose families are small lack such opportunities and thus are less able to pursue innovative economic activities. Therefore, the design of extension services needs to pay special attention to formal and informal information networks and pilot new methods of service delivery prior to the establishment of firm institutional arrangements.

### Social Impact Monitoring

Social inequity in the distribution of the Project benefits and in participation require close monitoring during implementation. Of particular importance are community, ethnic and gender specific impacts of Project components as well as the impacts on the landless and small holders and younger generations. The institutional changes that take place in the process of rapid economic and social change also require close monitoring so that implementation arrangements can be modified as needed.

## *Implications of the Social Assessment for the Proposed Project*

### Supporting Agricultural Development

Irrigation increases incomes and benefits the entire population; its expansion should be expedited and management improved with explicit consideration of direct benefits for the landless and small holders. It should be accompanied by effective extension services that offer information about optimum irrigation techniques, about intensification, and about the cultivation of crops other than cotton. The cultivation of such labor-intensive crops as vegetables and fruits that do not require much land could increase benefits to those with little or no land and would enhance the potential for non-agricultural activities in the region such as in packing, cleaning, drying, canning, and freezing. A similar argument holds for keeping of livestock, especially cows, where feed can be grown intensively. These will also create employment for the landless and increase their income-earning potential. It is desirable to devise credit schemes that target this population in order to provide the required initial input of capital.

Irrigation, land consolidation, land leveling, drainage, and improvement of in-farm roads should be seen as integrated components of a unified development package. Consolidation of plots has been a success and is essential to justify long decades of investments made in the region. To enhance equity in the distribution of State and Bank-financed investments, participation of landowners in cost recovery should be secured and labor intensive modes of production should be emphasized. The willingness to participate in the costs and labor requirements of land-based development projects is far short of the full costs of improvements. Land consolidation could substantially contribute to gender and inter-generational equity if implemented fully and if, through information and communication (I/C) activities, the re-registration of female land holdings to men can be reduced.

### Improving Village Infrastructure

Drinking water is the most crucial input for the welfare of the population. Its ready availability will improve health conditions and contribute to the eradication of diseases, lowering infant mortality and increasing women's quality of life. The Bank should support drinking water projects and should encourage the participation of users in costs and labor. Schooling for children 6 to 14 is a priority. It may also be possible to devise a subsidy scheme for the poorer segments of the population whereby a small long-term loan, repayable after the children finish school, is given to families to compensate them for the labor they lose by sending children to school. This would also help generate additional work opportunities for adult laborers within the region and its surrounding areas.

### Institutional Development

The inequity in economic power and the dominance of tribal roles permeate other institutional arrangements and hinder the development and democratic functioning of the civil society organizations, including water user associations. A better understanding of the linkages of formal and informal institutions is required before concrete recommendations can be made. This is recommended as a next step of the SA. Improved institutional coordination is another need, and there is a lack of information at the level of farming technology and cropping decisions. While the GAP project foresees a development of labor-intensive farming that would improve the status of small holders, the farmers need better information about crop diversification and a wide range of other advisory services. The institutions currently delivering these services are not trusted and relying primarily on private sector provisioning will have inequitable impacts. Thus, there is need to build trust for public institutions. At the same time, there is need to better understand the current operations of civil society organizations and pursue partnerships with them.

Water user associations are potentially the most important participatory structures constituting the interface between the irrigation technology and farmers. Ideally these associations should be encouraged to take on responsibility for a wider scope of activities involving not only the distribution of water but also the technology of its use, field management, and cropping decisions. Eventually they may evolve as purchasing and marketing cooperatives. At the moment, however, water user associations are believed to work unfairly. Through exchange programs and other capacity building efforts, these associations could be encouraged to work in a more democratic and participatory manner in order for there to be trust and predictability in their operations.

# Chapter Four

# Azerbaijan Agricultural Development and Credit Project

## Ayse Kudat and Bulent Ozbilgin[1]

## Background

### Geography, Population and History

Azerbaijan covers an area of 86,600 square kilometers. It shares borders with the Russian Federation and Georgia, Turkey and the Islamic Republic of Iran, the Caspian Sea and Armenia. Azerbaijan's population of 7.2 million is relatively homogeneous, with Azeris (mainly Shiite Muslims) accounting for about 83 percent of the total and the remainder made up of Russians and other minorities. Azerbaijan includes the Nakhchevan Autonomous Republic, an enclave between Iran, Turkey and Armenia that has no border with Azerbaijan proper, and Nagorno-Karabakh, a landlocked mountainous area of 4,400 square kilometers.[2]

Azerbaijan was an independent state in the fourth century B.C. By the third century A.D. it was part of the Sassanid Empire, and in 642 Muslim Arabs conquered the region. Azerbaijan prospered under the khanates of Shirvan and Mughan. As the ruling Abbasid dynasty weakened in the eleventh century, the Seljuk Turks united the former Abbasid lands and brought a measure of stability to the region. Local khanates had a measure of autonomy. In the sixteenth century, the territory of Azerbaijan became a battleground for control between the Persian Safavid dynasty and the Ottoman Empire. During the following century, it faced a new incursion from the expanding Russian Empire in the west. In 1920, The Red Army invaded Azerbaijan and declared the Soviet Republic of Azerbaijan.

Azerbaijan was among the first states of the former Soviet Union to adopt a resolution of sovereignty which proclaimed its exclusive control over its land, water, and natural resources, along with the right to establish and maintain diplomatic relations with other countries. Its economy underwent fundamental changes in 1992 with the breakdown of the central command system and the weakening of traditional links with the states of the former Soviet Union.

[1]     Special thanks are due to Van Roy Southworth, the task team leader for the proposed Project, for supporting the SA process. The authors acknowledge substantive and editorial support of Jonathan Brown and Peter Gordon. Large numbers of social scientists in Azerbaijan and the staff of the Azerbaijan Women's Development Center have gathered a great deal of the survey and qualitative data and participated in the field work carried out by the authors.

[2]     The Nagorno-Karabakh region has been a growing source of friction between Azerbaijan and Armenia. Sporadic outbreaks of violence since 1988 and the ensuing heavy casualties have triggered massive movements of refugees in both countries and resulted in the closure of the main corridors for Armenian trade through Azerbaijan.

### *The Farm Privatization Project*

The World Bank has been assisting the rural reform efforts of the Government of Azerbaijan's (GOA) through the Farm Privatization Project, which is designed to be a pilot project. In the context of the Project, GOA has privatized what appeared to be typical or "representative" farms covering about 23,000 people (6,645 families) in the Barda, Udjar, Lenkeran, Salyan, Khachmaz and Sharur rayons. The main goals were to:

- Develop essential support services such as land registration, farm information, credit provision and irrigation to facilitate privatization;
- Build linkages between relevant government institutions and banks; and
- Develop community-based social support services.

The experience on these farms has been used to develop solutions for privatization on a larger scale. Some of the emerging needs and those that could not be responded to within the framework of a pilot exercise will be met through the proposed Agricultural Development and Credit Project.

### *The Agricultural Development and Credit Project Adaptable Program Loan—Phase One*

The proposed Project is being processed as an Adaptable Program Loan (APL). The APL is one of two instruments that World Bank created recently to make its lending more flexible and appropriate, the other instrument being the Learning and Innovation Loan (LIL). The Agricultural Development and Credit Project APL for Azerbaijan permits Government and the Bank to agree on a three-phase program of assistance over ten years, with performance benchmarks and triggers to move from phase to phase. The Bank's actual funding commitment is only for phase one, but multiple phases engage the Government, the Bank, and the various actors in the rural sector in a long-term partnership of cooperation and consultation. The APL approach is particularly dependent on a permanent process of integrated social assessment and stakeholder consultation as defined in Chapter 2 of this volume since they often provide the critical information on which performance benchmarks are established and assessed.

The objective of the Adaptable Program is to restore Azerbaijan's farming areas to former levels of productivity under a new system of agriculture characterized by individual families and group farms operating in private markets. The Program will support the development nationwide of:

- An accessible, secure and unified system for registering agricultural property (land, buildings and other property of the state farms);

- A distribution network for financial services in rural areas consisting of local financial intermediaries such as credit cooperatives and borrowers' groups to provide the newly established private farms and other rural enterprises with market based credit and deposit services;

- An appropriate mix of public and private sector rural advisory services for rural entrepreneurs; and

- A capacity to formulate appropriate policy and institutional responses to the likely impact on the competitiveness of the rural economy in the light of the anticipated build-up in oil revenues.

The first phase of the Program, the Agricultural Development and Credit Project, would last four years and focus on registering private rights to farmland and on testing mechanisms for a unified real estate registration system encompassing urban and rural land and buildings. It would also support the creation of a self-sustainable rural financial system based on the development of local savings groups and financial intermediaries. In addition, it would establish rural information and advisory services and strengthen research and veterinary services. The two subsequent phases of the Program, each lasting three years, would focus on expanding the geographic coverage of the rural advisory services and the number of local financial intermediaries. In addition, initiatives to encourage commercial banks to establish branches in rural areas would be added in the subsequent phases. For the real estate registration system, the focus in later phases would be on expanding the unified real estate registration system nationwide so that urban and rural lands could be registered into a common system.

Based on extensive stakeholder consultations, several systematic needs assessments carried out with farmers as the privatization process developed, and lessons learned from the pilot experience, the first phase of the Program would consist of two basic elements. First is the provision of security of private ownership and the development of land markets. The second is the provision of credit and information to a larger group of rural women and men, including the newly emerging private entrepreneurs. A brief description of the elements proposed for inclusion in the first phase follows.

### Real Estate Registration

The focus of Government to date has been on privatizing farms, creating transparency in the land allocation process, and the issuance of titles. A great deal of progress has been achieved in title distributions and systematic consultations with communities suggest that an expeditious completion of this process is necessary. Little has been done to establish and maintain a registry of these titles that is secure, up-to-date and capable of recording rights to land and land transactions (mortgages, liens, leases, sales, inheritance, gifts, etc.). Without immediate action to develop a simple registry system, transactions in land may not be uniformly recorded and the titles that have been issued may become obsolete. This is of particular importance for women, the elderly and the youth who have so far all received equal access to land. The social equity in the distribution of land titles has been one of the most significant achievements of the rural privatization process and needs to be safeguarded.

To enable the people to register their lands and buy and sell rural property freely, it is important to increase accessibility of registration offices. Currently, the allocation of land and issuance of titles are being handled by the Baku office of the State Land Committee. The Government is committed to decentralizing the process to regional offices so that once the process of land allocation is completed, recording land transactions can be conducted in convenient and easily accessible locations, especially considering the difficulty and high cost of travel to Baku from remote regions. Among many advantages of decentralization will be the greater ease with which the vulnerable groups—the elderly, women, disabled or the very poor—would conduct land transactions.

The land titling process has so far only focused on the land allocated to individuals. The management and control of municipal and state lands that constitute large areas is not being properly recorded and there is a need to incorporate these lands into a land registration system. The rural people do not understand the reasons for these allocations when their own land plots are excessively small. Available information for the communities included in the Bank financed pilot project show that about a third of the lands under the state farms were allocated as municipal and state share (Characteristics of Farm Privatization Project Areas). Consultations with ex-managers of these farms suggest that these include common grazing lands and cultivable land managed by the public sector. It is said that the revenues from the lands cultivated would be used for the operation and management of social services. It could not be confirmed, however, that this actually takes place. Thus, the decentralized management of land allocations would bring enhanced transparency to municipal and state land ownership and enhance the transparency behind the operation of newly created municipalities.

> **Characteristics of Farm Privatization Project Areas**
>
> **Barda rayon:** 325 families and 1,252 people in project areas. Agricultural land area is 2,954 ha of which 1,531 is distributed to people at 1.1 ha/person.
>
> **Lenkoran rayon:** 1,256 families and 4,333 people in project areas. 64 percent of 766 ha is distributed at 0.8 ha/person. Rayon suffers from Caspian environmental crisis and high population density.
>
> **Salyan rayon:** 3,154 people in 694 families in project areas. Agricultural land area is 3,888.5 ha. of which 2,317 ha is distributed at 0.66 ha/person.
>
> **Udjar rayon:** 2,098 people in 488 families in project areas. Of the available 4,153 ha., 1,628 ha is distributed at 0.77 ha/person.
>
> **Khachmaz rayon:** 9,057 people in 2,636 families. Of the available 5,254.5 ha., 3,170 ha is distributed at 0.3 ha/person. Region has many fruit gardens and the land received per person appears less because of that.

Based on the stakeholder consultations in the SA process and focusing on the needs of the low-income rural populations and their expectations of the public sector, the Bank would fund two aspects of the activities to be initiated with regard to real estate registration:

- Provide support to the Central office and ten regional offices of the State Land Committee for issuance of those remaining land titles and for the development of a decentralized land registration system and land ownership database; and

- Finance the implementation of a pilot unified real estate registration system and cadastral system in the Nakhchevan Autonomous Republic, covering both urban and rural land, as a model which could later serve as a basis for wider geographic replicability throughout Azerbaijan.

### Rural Finance

The objective of this component is to increase the general availability in rural areas of market based credit and deposit services for rural households, enterprises, and farms. Specifically, the component will provide:

- Funds from the World Bank to establish a credit line to be allocated through the local financial institutions (LFIs) created under the project to finance eligible sub-projects that would be co-financed up to 20 percent by World Vision; and

- Resources and agreement from an international cooperative banking institution to establish two types of local financial institutions: credit cooperatives; and groups of jointly liable borrowers.

Given the rural conditions described above, the proposed credit cooperatives will have to be formal, sound and competitive financial institutions with a wide ownership and client base. Consultations with rural people and farmers and the already-existing small-scale local initiatives such as farming groups or small savings groups point to the willingness of rural Azeris to take part in such community based initiatives. These cooperatives will provide comparatively simple savings and credit services to their owner-clients. Their credit services will be funded with equity contributions and deposits from their members and from the partnership of the World Bank and a reputable international cooperative banking institution. Groups of jointly liable borrowers with 10 to 20 individual farmers and rural micro-entrepreneurs—essentially village banks—will be used as transitional channels for World Bank and the implementing institution funds to formulate flexible responses to the diversity in credit demand in rural areas. These groups will be particularly important for non-farm micro-entrepreneurs in the small rural localities where there may not be a large enough population to support the fixed-cost structure required for the proper funding of credit cooperatives.

The close-knit social ties existing in rural areas and low-income communities make them ideal candidates for assistance through group-based financial institutions like village banks. All productive economic activities will be eligible for financing, including those carried out predominantly by women, such as handicrafts and petty trading in roadside markets. The Project will help organize the groups and help individuals within these groups obtain sub-loans on the basis of their joint liability. Social capital in the form of existing character information, peer monitoring, and local enforcement will help to guarantee the group's repayment. Groups of borrowers will not need legal incorporation or a banking license as their members will only borrow from project funds and will not be allowed to collect deposits. They will be provided with simple portfolio monitoring tools. Under these principles and partnership with NGOs, it is expected that disadvantaged groups, including rural women, would benefit from the operational principles to be followed by the local financial intermediaries.

### Information and Advisory Services

The new private farmers need managerial, technical and legal advice in mastering farming practices as they change from a largely commodity oriented, cost and market insensitive agriculture to a more specialized and sophisticated system. At present, there is no delivery system for support services, including agricultural research and extension. The current slow emergence of private sector agricultural consultants and pre and post-service assistance from agricultural suppliers and output purchasers is a good sign for the future but less than what is required by the farming community. In addition, few farmers can presently afford to pay for these services, thus making their private provisioning non-viable unless heavily subsidized.

The Project will fund the establishment of advisory services through the Ministry of Agriculture in some regions and contract out to a private company in other regions. The approach to the rural advisory function proposed in the Project recognizes that private advisory services are already beginning to operate and that these offer the best prospect of providing support for the wide-ranging technical needs of farmers in the longer term. The Project would encourage the private sector for advisory services. However, public sector capacity for planning and delivering public good extension will be retained. The design emphasizes decentralized planning and management of public good services, in view of ensuring responsiveness and accountability to the farming community. Using regional advisory centers to oversee the provision of training and extension activities at the local level is designed to ensure responsiveness and accountability to the farming community. At the mid-term review of the Project, a decision will be made on which mechanism to use in the remaining two areas during phase one and in subsequent phases.

The advisory services will consist of mass information campaigns, publications, price information systems, farmer training, farm management advice, technical advice, technology demonstration and private veterinary services. This component will also promote agricultural research through on-farm experimentation and funding research initiatives in conservation tillage, seeding practices, forage development, and experimenting with wheat varieties. In order to promote cost recovery in the long run, subsidies for activities that are private in nature will be progressively phased out. To ensure effective implementation of this component, monitoring and evaluation activities are crucial, especially in recognizing the diversity of needs of different categories of rural entrepreneurs and the need for different channels and mechanisms to serve their distinct requirements. Also needed is constant monitoring of changes in practices and technology, in markets, and in the economic environment, both within and outside the country.

The information needs of different types of groups and individuals as well as residents of various regions differ. Inexperienced rural people need information assistance on basic production methods, crop cycles, yield increasing methods, and overall farm management. Rural women need information in marketing if they are to engage in small borrowing and sell home-based crafts. A number of women, who had been managers in former state and collective farms, have been engaged in farming after privatization, but they need business and management assistance in continuing these roles in largely male-dominated rural enterprises.

**Agricultural Strategy Development**

Government of Azerbaijan aims to establish a sound analytical basis to define and implement a public policy for the agricultural sector in the context of the expected increase in oil-related revenues over the period to 2010. In particular, it will be important to define policies and interventions that will reduce potential adverse effects of the surge in oil revenues on key agricultural industries and for rural livelihoods. A secondary objective would be to increase the capacity of both the government and research institutions for sound economic policy analysis relating to the management of oil revenues. The need for policy development is felt primarily by the public sector and the academic institutions that would participate in its formulation.

Usually, information is conceived as an input to project implementation, flowing from policy makers and implementers to beneficiaries and other stakeholders. Two other critical types of information which flow from the people are almost always neglected: one helps policy makers design policies and projects, and the second, feedback from the people, makes it possible to re-adjust policies and take corrective action. As rural private sector and civil society institutions—including farmers' associations and chambers of agriculture—develop, it would also be important to seek their feedback on policy issues. The monitoring and evaluation component of the Project is expected to serve as a means to provide better facilitation of the flow of information and feedback from the beneficiaries to policy makers.

## *Rural Sector*

Azerbaijan is endowed with fertile agricultural land and ample mineral resources, including considerable oil reserves. It has a relatively developed industrial base, characterized by a high degree of specialization in production and traditionally strong trade links with the states of the former Soviet Union. Next to industry—which is dominated by the food, oil, textile, and machinery industries—agriculture ranks as the second major sector.

Reforming the agricultural sector in Azerbaijan has at its heart the improvement and transformation of peoples' lives and behavior. Public sector investment to improve the agricultural sector is important in view of the present and future impact of the nexus of three events on peoples' lives and thus on the success of rural reform:

- Transformation from a command to a market based system,
- Results of ethnic strife in the region, and
- Impact of energy sector development on the competitiveness of Azeri agriculture.

**Transition from a Command to a Market Based System**

Insufficient knowledge of pre-transition forms of farm management makes the design of reform implementation difficult. Rural change was evident prior to transition in many ECA countries, generating important lessons that could have helped shape new reform initiatives.[3] The

---

[3]    There were important internal and intra-country differences in the pre-transition period. For instance, a collective farm in the remote Siberian countryside was different from one close to a city because of the availability of markets. This, in turn, determined the relative importance of household plots as a key source of income. In remote

collapse of the command system of the former Soviet Union undermined one set of social and economic relations in which the pattern of people's lives was well established without clearly identifying paths for the future. Their current status, mores and aspirations are not well known. While the transition to a more market based economy is accepted by many as the vision for the future, there is still no clear understanding of what this requires, particularly with regard to an enabling framework of private and public sector support in the agricultural and rural sectors. The typical farmer in the former Soviet Union, a woman almost as likely as a man, was a salaried employee on a large collective farm with a specialized job, which may or may not have been directly related to growing crops or tending for livestock. Social services were provided by the collective farm or by government to all, including the large number of retirees who remained as part of the collective farm community. Decision-making centered around the farm manager and the local Communist Party apparatus. The most entrepreneurial aspect of this command system was the private garden plot which belonged to the farmers but which might use various inputs taken from the collective farm without payment.

The typical farmer in the more market oriented agricultural sector today is self-employed, often male, required to do a full range of farm jobs supported by his wife, who now holds the title of housewife. He is responsible for multiple aspects of family life, including a larger share of caring for elderly and the unemployed, and is required to make many more individual decisions than previously about economic and social aspects of life. For the new farmer, dealing with people outside the extended family is more uncertain than under the more regulated framework of the Communist system.

### Internally Displaced Persons

Civil strife between Azeri and Armenian ethnic groups in and around Nagorno-Karabakh in the late 1980s and early 1990s led to the displacement of almost 900,000 persons—equal to 15 percent of the population of Azerbaijan—and the occupation of nearly 20 percent of the territory of Azerbaijan, which still remains outside the control of Government. Of the nearly 600,000 people who are still considered internally displaced persons (IDPs), 53 percent live in urban areas. The rest are in rural areas putting further strain on land, rural infrastructure, and society. It now appears that the return of IDPs to their original homes will be substantially delayed and they will have to be integrated into the host communities. The Government has already adopted an action plan to support this process and a new Bank project is under preparation. If they gradually move out of the "temporary" camps where they have survived on humanitarian assistance without returning to their original homes and are assimilated in other rural areas, further hardship will be created in rural areas unless productivity is enhanced and employment is created.

---

areas household plots tended to serve subsistence needs, while in areas closer to markets they were used primarily to grow commercial crops destined for farmers' markets. Such differences were important in determining the degree of dependence of the farm workers on the managers. Where managers were the only group who had access to trading institutions and networks, workers felt dependent and unable to take the risk of independent farming; managers remained in a powerful position and were able to dissuade members from leaving the state farms.

### Impact of Energy Sector Development

The reality thus far of substantial oil and gas development for Azerbaijan from the Caspian Sea has led to high levels of foreign investment in the country and the prospect of large revenue flows from bonuses, royalties and energy taxes. These flows have begun already but are expected to accelerate and increase rapidly in the first decade of the next century. In addition, unlike many of the surrounding countries, Azerbaijan has conducted macro-economic policy prudently. This has added to the influence of the energy sector so that the exchange rate has appreciated. The appreciating exchange rate has made food imports progressively more competitive with domestic production. There is the very real possibility of a situation in which the terms of trade in an energy exporting country discourage unsubsidized domestic agricultural production.

### Agricultural Sector Conditions

Azerbaijan's agricultural sector accounts for 36 percent of total labor, between 26 and 30 percent of total GDP and 30 percent of total exports. Rural areas comprise about 46 percent of total population. In 1995, 60 percent of the country's population lived in poverty, and most of these people resided in rural areas. Half of Azerbaijan's agricultural land (2.2 million ha of a total of 4.4 ha) is cultivated and the rest is grazed.[4] Azerbaijan has a highly diverse crop mix that depends on irrigation and imports of key inputs such as seeds, fertilizers, pesticides, and farm machinery. Prior to transition and the subsequent occupation of large territories, the diversity was far greater as was the demand from other Soviet territories for the raw and processed agricultural, horticultural and fisheries products

Agricultural output has fallen due mainly to difficulties in transiting to a more market-based economy. This fall was compounded with the problems of armed conflict with Armenia and the consequent large-scale internal displacement of people. From 1991 through 1994 the overall level of agricultural production declined between 12 percent and 25 percent annually. Plots of land that are available are often too small to sustain a family, so most rural families generate income from multiple sources. Lands adjacent to the Caspian Sea are part of an environmental disaster—the rise of sea level has damaged infrastructure, increased salinity of agricultural land, and destroyed fisheries.[5] Irrigation systems and farm machinery have deteriorated and most farmers cannot obtain credit. The large number of small plots of land is vulnerable to consolidation by a few more enterprising landholders once the land markets develop; the predominant majority of the rural households received land, but their holdings are far too small to be economically viable. Also, development of large oil reserves and industry may draw resources out of the agricultural sector, thereby depressing its development.[6] However, indications

---

[4]      Staff Appraisal Report. Azerbaijan Republic. Farm Privatization Project. The World Bank. 1996.

[5]      The World Bank is also providing assistance to mitigate some of the environmental damage. A social assessment of the Caspian Sea environmental disaster has recently been published (Ayse Kudat. People's Response to Environmental Disaster: A Case of Azerbaijan. The World Bank. 1999).

[6]      Until recently, it was anticipated that the oil boom would occur rapidly. However, it now appears that the developments in the energy sector may be slower. Thus, there is a greater need to focus on rural development issues and enhance agricultural productivity.

of improvement appeared by 1998, particularly in the cotton and livestock sub-sectors. The rural economy may have reached its nadir and may be improving.

GOA intends to privatize about 70 percent of the 1.46 million hectares of land under the control of about 1,800 collective and state farms. It has passed relevant laws to reform the agricultural sector. The objective of these laws is to transfer lands permanently from government entities to entitled private groups and individuals. On November 12, 1995 Azeris adopted by referendum a new constitution that permits private land ownership. On July 16, 1996 Parliament passed the Land Reform Law that:

- Gives authority to regional and local agricultural reform commissions to implement provisions of the land law;

- Defines land which can be privatized and that which remains under government control; and

- Sets forth procedures for privatizing and transferring land between owners.[7]

Within this legal framework, local administrations are dividing the agricultural lands of state and collective farms and distributing them at the village level. The process has been completed in many communities and there has been increasing private ownership of land. Information about the regional scope of the privatization process is inadequate, however, as much of the empirical information presented here is based on consultations carried out in about 15 rayons (administrative districts). The future of the rayons under occupation is uncertain.

**The Social Assessment**

These three sets of influences—the transition, IDPs, and the energy sector—make the future of the rural sector even more uncertain than the present and place a special importance on the process of social assessment through the Project interventions. Fact finding efforts and stakeholder consultations are important devices to determine the current situation in the agricultural sector and the responses of different rural groups to reform initiatives.

Azerbaijan faces a fundamental challenge in its rural regions: to alleviate rural poverty and increase rural incomes. To accomplish this goal, Azerbaijan started to privatize the land and assets belonging to former state and collective farms. It is also searching for mechanisms to minimize the possibility of continued reduction of agricultural output during the transition to private farming. The Government of Azerbaijan (GOA) has invited the Bank to support a new project on Agricultural Development and Credit to meet these challenges. This report summarizes the findings of a comprehensive Social Assessment (SA) that began in 1997 and will continue for the duration of the project.

---

[7]     Other laws include the Land Code of Azerbaijan (1991), Reforms of Kolkhozes and Sovkhozes (February 1995), Bill on Improvement of Irrigation (October 1996), Rules of Allocating Land Plots (October 1996), Improving the State of Use of Saline Land (October 1996) and Definition and Use of Pastures (October 1996).

## *SA Objectives*

The purpose of the SA is to ensure that the proposed Agricultural Development and Credit Project is based on stakeholder ownership, and that the anticipated benefits are adequately targeted and socially acceptable. It aims to do so by identifying the project's major stakeholders and their needs and interests so that they can participate in project design and implementation. It also attempts to establish a good understanding of current variability in agrarian social organizations in order to design well-targeted and socially sustainable rural policies and projects. Identification of the social development issues, the institutional dimensions of project preparation and implementation, and how stakeholders, especially small holders, might participate in project implementation are also important components of the SA process. Additionally, the SA guides the design of the participatory monitoring and evaluation process to assess results and develop appropriate follow-up measures. This latter objective is particularly important since the Project is the first phase of an Adaptable Program and guiding the latter through social impact monitoring (SIM) will enhance its effectiveness. The SA is a continuous process of analyses and consultation and has so far guided various stages of Project's preparation.

During the design stages of the proposed Project, priority areas had to be identified to devise interventions to overcome the most critical bottlenecks for development in the rural areas. As a result, the Bank's ongoing assistance to rural development in the context of the ongoing project had focused primarily on land privatization and credit issues; thus, it had a comparative advantage among other donors to provide further support to these key areas. The consultations, fact finding efforts and project supervision activities carried out for the ongoing project provided important inputs to the SA process and further strengthened the rationality of a narrower focus than that justified on the basis of the needs assessment. Nevertheless, these broader findings are shared with the stakeholders to allow additional responses to support Azerbaijan's rural development needs, especially those that are most acutely felt by the people.

## *SA Methodology*

The first phase of the SA covered the rayons in which the ongoing Farm Privatization Project is being implemented, namely Barda, Udjar, Lenkoran, Salyan and Khachmaz (Figure 1). It focused on a number of specific issues, including main sources of livelihood, decision-making within households, constraints and opportunities associated with the privatization process, access to agricultural assets, perceived need for agricultural restructuring and credit, experience with entrepreneurship and marketing, and attitudes towards institutions and change.

The first phase of the SA comprised:

- Review of existing knowledge based on secondary data from the Farm Privatization Project experience, earlier SAs supported by the Bank, the 1995 Poverty Assessment, and various SAs of Internally Displaced Populations;
- Surveys of households and women following a qualitative rapid assessment;
- Semi-structured interviews of individuals;
- In-situ observation by Bank staff;

- Consultations with policy makers in Baku and local rayon administrations;
- Consultations with local and international NGOs;
- Discussions with ex-managers of state farms and community leaders, and
- A stakeholder seminar.

Figure 4.1 Scope of social assessment

The household survey involved interviews with 900 randomly selected households in Farm Privatization Project rayons (but not in the Farm Privatization Project farms) during January/February, 1998. In addition, a survey of 210 women in three of these six rayons was conducted to obtain information on the general impact of reforms on women and the constraints they face. Semi-structured interviews included in-depth interviews with farmers, farm managers, unemployed workers, community leaders, women's groups, local associations, technicians and government officials. A member of the Social Development Team for the ECA region lived with a farming family in the Khachmaz rayon to conduct an in-situ observation of the impact of farm privatization. In November/December, 1997 five focus groups comprising relatively homogeneous groups of stakeholders (farm managers, women and unemployed workers) were conducted to examine inter-group dynamics, common problems and opportunities and ways to gain the confidence and commitment of stakeholders. In the summer of 1998, a stakeholder seminar on the preliminary results of the SA included representatives from central and local governments, local social scientists, women's organizations and donors.

The second phase of the SA continued to guide the pre-appraisal stage of the Project preparation. This phase built on the results and issues identified during the first phase and enlarged the geographical scope of the SA to 15 additional regions not covered by the first phase. Further extensive interviews were conducted in a new set of rayons on specific issues relating to credit and extension needs and targeting requirements, with a focus on rural women's needs. This work also focused on learning more about the creation of voluntary association of farmers. An extensive attempt was made to prepare farm models for different types of producers. A focused attempt was also made to understand more about marketing and sales channels, local level and road-site corruption, and its impact on farm incomes.

## Four Pillars of Social Assessment

### *Key Social Development Issues*

The key social development concerns in rural regions focus on poverty, gender and social exclusion. The pre-transitional patterns of inequality and social exclusion can also be expected to produce different types and degrees of support to rural reforms. In connection with this, regional factors and region-specific issues play a crucial role in social and economic development in rural Azerbaijan. The transformation from a salary-based rural economy to a peasant society is creating enormous stresses in families. The pre-transition practice of garden farming provides a major source of subsistence to many; it is also the source of skills which can extend to the cultivation of newly acquired private plots. Yet, there are many families that played specialized or administrative/professional tasks in the state farm context and therefore lack farming skills. Other families lack able-bodied people because of out-migration of workers or disability and old age. It is therefore important to provide the skills required to make a transition from a wage based rural economy of large state farms to one in which peasantry and voluntary associations will play a larger role.

Equally important is to ensure that people are better informed of legal and procedural changes and have access to know-how. Also important is that male and female farmers receive equitable access to farm inputs, technology and marketing facilities. This involves an understanding of farmers' constraints and the establishment of mechanisms to enhance reform benefits to rural people. The success of rural reform efforts depends, to a large extent, on increasing the participation of rural stakeholders in developing the economy and providing them with incentives and opportunities to stay in rural areas.

As a result of the conflict with Armenia, the Azeri population of Nagorno-Karabakh and surrounding rayons, totaling about 650,000 people, fled to other parts of the country. About half of these IDPs live in rural areas, with the other half dispersed in urban congregations, especially in Baku. The occupied rayons were predominantly rural and agricultural, with an average rural population share of around 75 percent (against a national average of 54 percent). In terms of social infrastructure, the occupied areas accounted for around 18 percent of the schools and 13 percent of the pre-schools, though they had very few higher educational facilities. IDPs are among the worst-off in the country, but not all the IDPs are poor. They share some common characteristics. Wages are significantly less important for them, government transfers are significantly more important, in-kind benefits from NGOs result in a much larger share of their total income, and the income share from sale and consumption of own-produced food reflects the difficulties they have in gaining access to land. [8] IDPs in rural areas live in a wider range of shelter types than those in urban areas, of whom the majority live in either hostels or apartments. There is also anecdotal evidence that IDPs sometimes work as manual labor in agricultural land of their host rayons and travel, especially to Russia, to act as traders. In Baku and surrounding rayons, some IDPs work for summer-home owners in the summer homes during the winter, taking care of gardens and the household. These families move out during the summer.

Although the proposed Project will not specifically deal with the needs of IDPs, the capacity created within the agricultural sector will no doubt be of direct support to those who are being re-integrated in liberated areas. As land registration systems and advisory and credit services become more widely available, returning IDPs will also benefit. The SA recommends that the Policy Development component of the proposed farm privatization project focus on the viability of addressing the needs of the IDPs primarily through the reconstruction effort.

### Rural Poverty

**Demographics:** Families in rural Azerbaijan are generally crowded; the SA survey found average family size in project rayons is 5.3 (Figure 2), which is slightly higher than the official figures ranging from 3.4 to 4.5 people per household.[9] A significantly high 14 percent of the households have more than 7 members. The average age of male household heads is 46.4 while that of female household heads is slightly higher at 52.6. As is the case in most countries of the former Soviet Union, the education levels of the household heads in rural Azerbaijan is high. Most have at least a high school education and illiteracy is almost non-existent. However, household heads who are older than 60 usually have lower education levels compared to younger household heads. Unemployment is at high levels in all rayons; more than half of the household heads (with the exception of Barda) are officially unemployed. It is most visible in the 26-35 and 51-60 age groups. In Lenkeran, however, a number of household heads are reported to be working in Russia and sending remittances to their households.

### Figure 2: Household Size

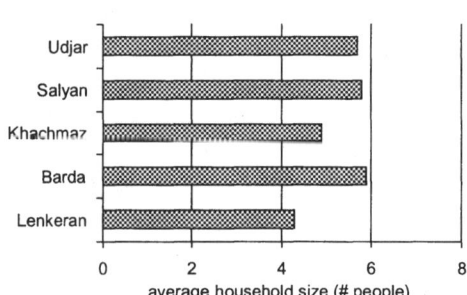

Source: Social Assessment Household Survey, 1998.

---

[8]       Azerbaijan Poverty Assessment. The World Bank. 1995.

[9]       Obtained from Farm Privatization Project, Project Management Unit, February 1998.

There is a significant outmigration from the rural areas; especially young people are leaving their homes in search of better opportunities and incomes. Usually, as these younger people leave villages, their older relatives stay behind to take care of their families or to be taken care of by these families. As a result, there is a significantly high level of multigenerational households (those containing people older than 60) in the rural areas while the incidence of households containing only older people is lower (Figure 3).[10]

## Figure 3: Age Structure of Rural Families

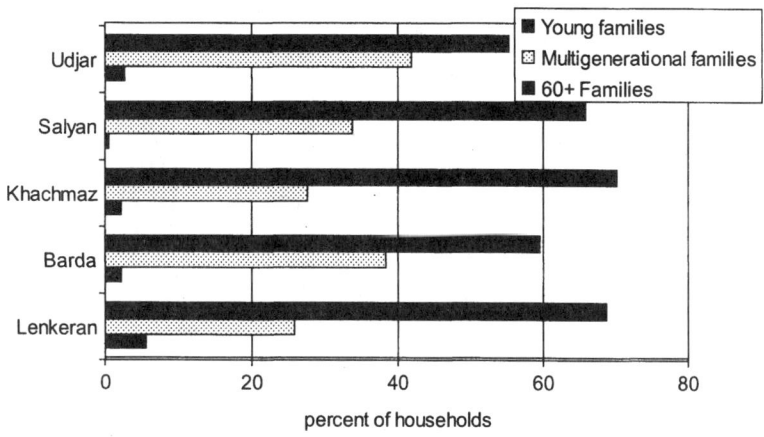

Source: Social Assessment Household Survey, 1998.

**Incomes, Expenses and Agricultural Production:** The poverty data gathered in May 1999 and compared with the poverty data from 1995 indicate that over the past years rural poverty has increased, although the trends are highly region-specific. In 1995, poverty levels were similar in Baku and in rural areas and about 20 percent of the population was very poor. By 1999, the proportion of very poor increased throughout the country but especially in rural areas and cities outside Baku. About 31 percent of villages were very poor and another 52 percent were poor.

Given the high levels of regional differences observed in poverty, it is not surprising that in the areas covered by the SA, there was an improvement in poverty. However, these results are merely indicative as the local level indicators for both 1995 and the 1999 poverty surveys are not representative at the regional level. Nevertheless, in the regions covered by the SA, the proportion of income spent on food has marginally decreased, indicating availability of disposable income over and above the amount required for the most basic needs.[11] Nonetheless, an important portion of the agricultural production is still used for subsistence purposes and smaller portions are sold (Figure 4). There are many constraints to enhance agricultural production and marketing and

---

[10] Young families are those whose members are all younger than 60. Multigenerational families include individuals both older and younger than 60 while 60+ families have only members older than 60.

[11] The 1995 Poverty Assessment found that the majority of the poor live in rural areas. Farm restructuring, access to credit, and price and marketing liberalization will help increase the incomes of the rural poor. As productivity gains increase, a large number of poor rural families are likely to benefit.

these cannot be removed merely by market mechanisms when they do not properly function. The support of government, based on substantially reduced cost recovery expectations, will be important in stimulating agriculture and contributing to the ability of small farmers to gradually sustain their operations.

**Figure 4: Use of Agricultural Production for Subsistence**

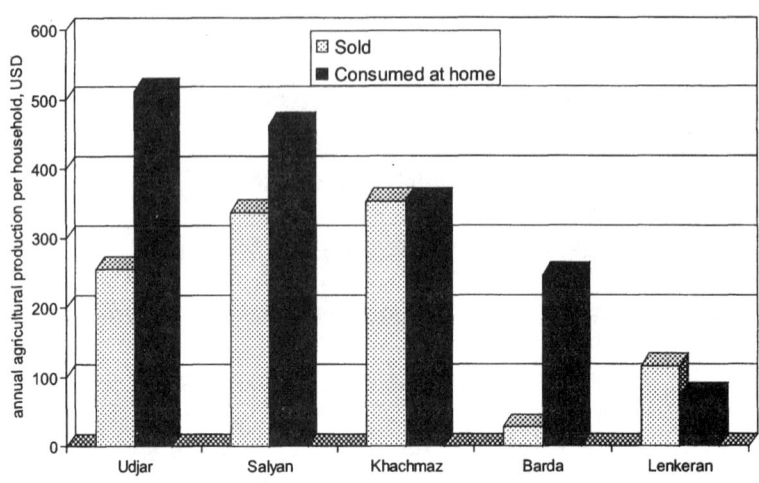

Source: Social Assessment Household Survey, 1998.

Further, most households have accumulated a range of durable goods such as refrigerators, radios, televisions, stereos, sewing machines and jewelry (Table 1). Finally, many households have kept their ownership vouchers, which act as a key component of savings and investment capital (Figure 5) [12]. This too represents an asset that they can still trade in the market. The household survey also found that unlike in Russia, multigenerational households with older heads have more livestock and property than younger households. These findings are consistent with the Azerbaijan Survey of Living Conditions (1995).

Table 1. Ownership of Household Durables ( percent of households)

|  | Barda | Lenkeran | Salyan | Udjar | Khachmaz |
|---|---|---|---|---|---|
| Refrigerator | 82 | 75 | 93 | 70 | 80 |
| Sewing machine | 58 | 48 | 59 | 33 | 45 |
| Radio | 45 | 52 | 60 | 31 | 36 |
| TV set | 83 | 86 | 96 | 82 | 88 |
| VCR | 11 | 21 | 15 | 6 | 9 |
| Stereo | 46 | 55 | 45 | 32 | 38 |
| Jewelry | 59 | 74 | 76 | 49 | 79 |
| Telephone | 15 | 46 | 46 | 18 | 10 |
| Automobile | 28 | 18 | 30 | 20 | 17 |
| N | 180 | 178 | 180 | 179 | 181 |

Source: Social Assessment Household Survey, 1998.

---

[12]     The difference in the value of vouchers is due to different family sizes and the fact that some family members sold their vouchers while others kept them.

**Figure 5**
**Voucher Ownership and Values**

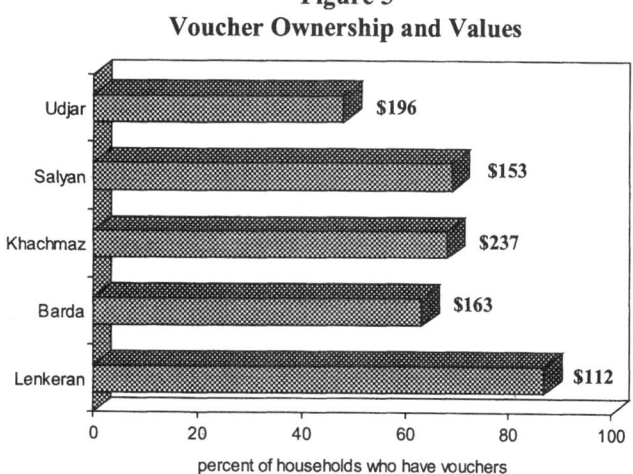

Source: Social Assessment Household Survey, 1998.

**Poverty Levels and Characteristics of the Poor:** [13]  In Azerbaijan, policy development and new investments for rural areas lag behind those in urban areas. The situation is particularly acute for people living in remote areas with inadequate access to markets, productive assets and social services.  Agricultural production and regional location are key determinants of rural incomes and poverty—where agriculture improves, incomes improve; where it stagnates, incomes stagnate.  Similarly, in areas close to major trading routes and/ or Baku, incomes are higher. However, strategies that would broaden the potential for agriculture are as important as those that would create a potential for supplementary income (Figure 6).  Only 14 percent of families do not receive income from agriculture and most of these are pensioners. It is also evident that a peasant economy has emerged with multiple income sources and activities, and a large majority of families rely on more than one source of income; indeed, very few are able to survive on agriculture alone.[14]

According to the definition of the poor at purchasing power parity of US$1/day/capita, 59 percent of the households in the rural areas can be characterized as living below the poverty line.[15] Poverty levels are strongly correlated to the agricultural production that the households undertook; while 58 percent of those engaged in cultivation are under the poverty line, a significantly larger 70 percent of the households who did not cultivate are poor.  The relationship between the level of

---

[13]     For the analysis, the absolute poverty line was set at purchasing power parity rate of US$1/day/capita, estimated at 116,700 manats per capita per month.  The exchange rate at the time of the survey was 3,890 manats to $1.

[14]     "Given the importance of different income sources, comparisons of rural and urban living standards must look at both cash and in-kind income.  In rural areas, almost 50 percent of total resources come from private plots.  For urban households, this share falls to about 7 percent.  The share of income spent on food has risen for all household types.  This increase reflects falling real incomes and indicates a clear decline in living standards for the majority of the population..." (World Bank-A 1997:48).

[15]     Social Assessment Household Survey, 1998.

cultivation that the household undertook and the monthly per capita income (expenditures) is also positive; as agricultural activity increases, incomes increase.[16]

**Figure 6**
**Household Reliance on Multiple Income Sources**

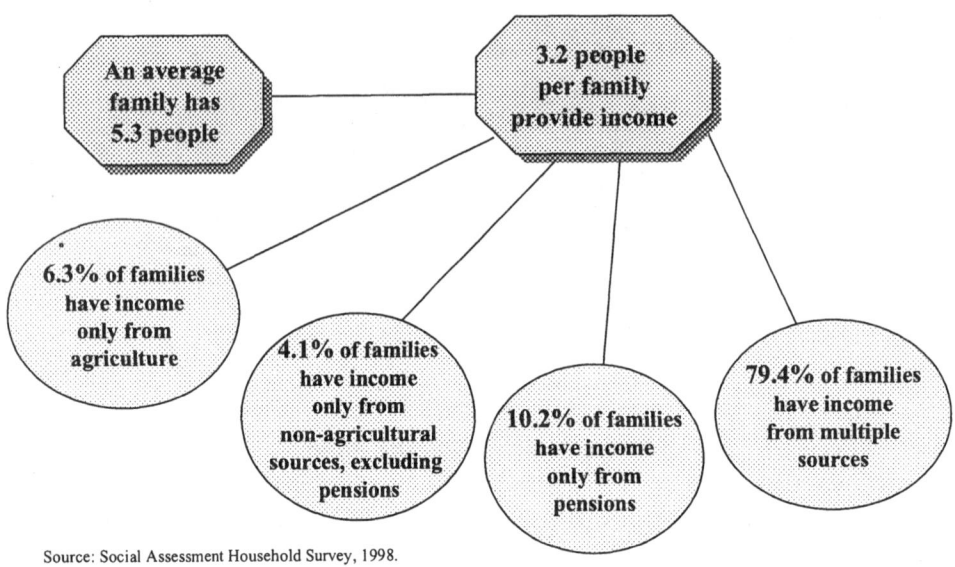

Source: Social Assessment Household Survey, 1998.

Related to crop cultivation, access to agricultural inputs and irrigation is also a determinant of household poverty; while 56 percent of the families with access to irrigation are poor, a significantly higher 75 percent of those without irrigation are under the poverty line. Access to seeds also affects household poverty, however, the relationship is not as definitive as in irrigation. Poor households also have fewer household assets (both productive and non-productive) as compared to non-poor households. It is a curious phenomenon that level of livestock ownership is not directly correlated with household poverty levels, however, the number of livestock owned is higher in non-poor households (Table 2).

Table 2: Livestock Ownership and Poverty

|  | Cattle ownership (%) | Sheep ownership (%) | Poultry ownership (%) | # Cattle owned | # Sheep owned | # Poultry owned |
|---|---|---|---|---|---|---|
| Poor | 55 | 48 | 58 | 2.85 | 14.55 | 18.37 |
| Non-poor | 45 | 52 | 42 | 3.20 | 17.38 | 21.73 |
| N | 646 | 297 | 804 | 646 | 297 | 804 |

Source: Social Assessment Household Survey, 1998.

---

[16] The analysis attempted to identify the relationship between per capita expenditure levels and other variables such as age of the household head, number of people earning an income in the household, annual crop production levels and household size. There is a significant linear association between the annual crop production levels and the household size (the significance level of t statistics is less than 0.05), with the annual crop production having a positive relationship and household size having a negative linear relationship with per capita expenditures. Other results explained in this section are cross-tabulations of poverty levels with a number of variables, with chi-square significance levels of 95 percent or more.

The second most important determinant of rural poverty is its region-specificity[17] (Table 3). By 1998, even though the conditions had been improving, still more than half of the rural population lived in poverty. There are other common traits that poor families share. Poor rural families are more crowded than non-poor families (household size is 5.68 for poor families as opposed to 4.79 for the non-poor). There is a significant relationship between the age and gender of the household head and poverty level of the household; households headed by older women are more vulnerable. The average age of poor female household heads is 53.8 while it is 46.4 for poor male household heads. Similar results are obtained in the 1999 Poverty survey for the country at large and are reported separately.

Table 3: Regional Dimensions of Poverty (percent)

|  | Barda | Lenkoran | Salyan | Udjar | Khachmaz |
|---|---|---|---|---|---|
| Poor | 73 | 72 | 33 | 63 | 54 |
| Non-poor | 27 | 28 | 67 | 37 | 46 |
| N | 180 | 179 | 181 | 179 | 181 |

Source: Social Assessment Household Survey, 1998.

The age of the household head or the presence of elderly members within the household is not a significant determinant of poverty in rural Azerbaijan. About two thirds of the multigenerational families and 56 percent of the families consisting of only younger people are under the poverty line. Also, the employment of the household head is not a determinant of household poverty. Many household heads had been unemployed for a long time by the time the survey was carried out. Thus, these people were generating income from informal sources, even though officially they were still unemployed. There is, nonetheless, a significant difference between poor households and non-poor in terms of their expenditures on food. Poor households spend about 48 percent of their total expenditures on food items while non-poor households spend 38 percent on food.

**Poverty Reduction:** Farmers in the regions included in the SA identify six key factors that would help reduce poverty in rural areas: 1) access to land, farming equipment; 2) access to agricultural inputs such as seeds, fertilizers, feed; 3) access to financing for inputs and land preparation; 4) access to information and advice; 5) access to markets; and 6) access to infrastructure, in particular to irrigation water.

In the short period of time since empirical data became available in 1994,[18] farmers' access to each of these factors has changed, with significant regional differences. Initially, access to land and farming equipment was limited and the major determinants of rural incomes and influence were positions held within the state farm system, other rural enterprises and

---

[17]      Data were broken down both by communities and by regions (rayons). The analyses revealed that community differences were much smaller than regional differences. In other words, communities within each region shared common characteristics.

[18]      This corresponds to 1994 when several SAs were carried out for Bank-financed projects. These did not directly address rural development issues, but they provided valuable insights on rural problems. The 1995 Poverty Assessment, on the other hand, included a representative sample of rural households.

administrative posts. These conditions changed with the distribution of agricultural land in the 1997-98 period. During this period, inputs such as seed, feed, fertilizers, and pesticides became hard to obtain. However, by 1998, the availability of these inputs become a secondary concern as their supply somewhat eased but farmers could not obtain financing to purchase them. During the same time, the lack of reliable information in rural areas has impeded the transition to a peasant economy.

Lack of access to markets in many cases still impedes the sale of farm produce and limits income earning opportunities; that's probably one reason why rural rayons closer to Baku tend to have higher income levels. Poor transportation facilities and high transportation costs limit the marketing of agricultural products. Many farmers now sell their produce along roads near their farms. There they receive exposure to only a few buyers and they often must make informal payments to local police and other authorities. Additionally, this dispersed form of marketing inhibits the broad dissemination of information—on prices, inputs, credit, irrigation and other resources—that could help farmers increase production and efficiency. Improving access to rural infrastructure such as irrigation water, electricity and transportation would also help improve incomes, although these are outside of the scope of the Agricultural Development and Credit Project.

However, a technical definition of these factors cannot account for the human behavior that defines responses to existing constraints and devises coping methods. The existing social capital in rural areas has been providing a cushioning mechanism for the population during the transition period and continues to do so as more people engage in agricultural activity in association with the people that they know and trust. It is with social capital that people have been able to cope with hardship, gain access to information, pool their resources together and join their land plots to increase efficiency of production. Based on mutual trust between family members, neighbors and ex-coworkers of state farms, they have also been lending to one another. In all likelihood, the future development of agriculture will require some sort of land consolidation. Part of this will be achieved when those who can afford to buy the plots of those less able to cultivate. Partly, however, land consolidation will be achieved through the voluntary associations that people will continue to form. A similar, parallel institutional development is expected in terms of availability of financial services. On one hand, formal rural finance institutions will develop over time. In the meantime, informal structures in which people exchange financing and information will continue to co-exist with the formal channels.

### Gender Issues

Most women favor the privatization process, although most feel its adverse affects more than men. Before privatization, women participated in most sectors of the economy. During the transition, many had to move from farming, education and medicine to housekeeping (Figure 8). Forty-two percent of women surveyed say the division of work in their families did not change, while 40 percent see positive changes. Where privatization has been complete, women's family relations tended to improve. Where reform has lagged, family relations have deteriorated as more burdens were put on women.

**Figure 7**
**Female Employment Patterns Before and After**
**Privatization**

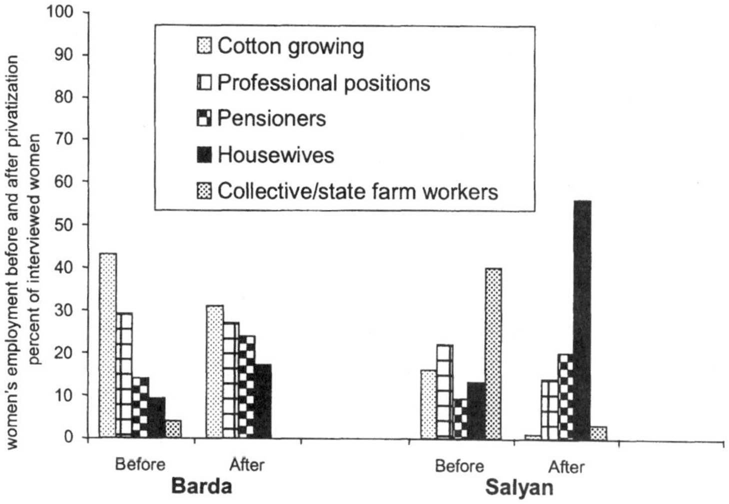

Source: Social Assessment Household Survey, 1998. Survey with Rural Women, Azerbaijan, 1998.

Figure 7 illustrates that women had a high labor force participation rate in rural Azerbaijan before the break-up of the Soviet Union. However, a close look at the graph shows that a significantly high number of women held part-time positions because they were mobilized for the cotton harvest; only about one-third of the women interviewed had full-time, professional positions.

In rural Azerbaijan, a number of factors constrain women's equitable participation in rural reform. Traditions of the male-dominated society are still alive in rural Azeri families in accordance with Soviet economic policies which determined a distinct, gender-based division of labor where women found themselves trapped in household economy and household labor. Demographic behavior orientated towards large families and high fertility rates on the one hand and inadequate social services for women and children (day care, kindergarten and schools) on the other hand cause women to be preoccupied with childcare.

Poor rural infrastructure (irrigation systems, pumps, community access to main roads and markets, water, gas, and electricity supply); limited availability and high costs of consumer commodities (clothes and food products which cannot be produced at a household level); and lack of services (laundry, dry-cleaning and barber shops) keep rural women engaged in basic tasks such as wheat and rice processing, bread baking and dairy processing (cheese, yogurt and butter). Women of all ages are also typically engaged in yard farming, including manual ploughing of land, irrigation, and collecting produce with significant participation of child labor.

A high number of small roadside vendors are older women (50-60 years old) who sell fruits, nuts, and dairy products at bus stops and crossroads. Lack of opportunities for women are due in part to the general high unemployment rate along with women's preoccupation with time-consuming household labor and domestic care. These chores leave little room for rural women's education or career development; women rarely read, watch TV or listen to radio, especially with electricity available only 2-5 hours per day.

Living conditions declined dramatically in the last two years; lack of trust in the positive impact of rural reforms has an adverse psychological effect on women, especially those of young and middle age; many women have expressed their apathy and hopelessness about the future. Women are concerned about their children and are not sure how reforms would improve their future. Mothers who consider their sons as future providers tend to favor them the most. Young males are encouraged to study and build their careers if possible, while girls are raised and treated as future brides and housewives. Thus, parents, and mothers in particular, contribute to gender bias.

Young women in rural areas have limited opportunities to develop their professional skills. The dramatic decline of the education system in rural Azerbaijan and lack of employment both at the village and rayon level lead to high unemployment among women as a whole. Early marriages for rural women have increased steadily in the last few years, and arranged marriages among girls of high school age (15-17 years) are common. Economic hardships and uncertainty about the future are among the main reasons why parents try to marry their children off as soon as possible.

When women register for land they may do so as part of a household rather than as an individual. This practice puts additional limits on their opportunity and flexibility. If care is not given during registration, there is danger that women's right to land would be diluted during the legalization process.

> *"... I used to work in the silk factory in town before it closed down. Now I have to work at the garden in my house to feed myself. I have two small children, and my husband is working in Russia, so I have to provide for my children, too. My husband sometimes sends money but it is not enough. Sometimes I graze animals for wealthier people and get paid, but I don't know what will happen to me or to my children in the future."*
>
> **Source:** A woman from the focus group in Sheki, October 1998.

[19]Women appear to have far less access to farm equipment and other assets. The criteria of property share attribution discriminate implicitly against women. These shares are calculated according to workers' length of service and wage levels. Since women retire five years earlier than men and their average wages are lower than men's, the value of their shares is consequently lower. The fact that women in the rural sector left state farms earlier caused the gender gap in unemployment to increase; this also affected women's access to property shares.

Historically, women have played an important role in livestock management, both within state farms and in home-based agriculture. Restructuring within the livestock sector has had

---

[19]     Russian law provides that individuals, not households, receive land shares, and that land share certificates are registered individually. Such a legal provision provides formal protection of individual, female land shareholders. However, Russian law provides that peasant farms are held in joint ownership and registered under the head of household. Kyrgyz law provides that only one land share certificate is issued per family. It is unclear under Kyrgyz law whether individual members have a right to withdraw their land share, or sell or lease their land share. In Kazakhstan, the household is the legal unit, with the head of the household controlling land transactions. In several countries, legal provisions provide that members of peasant farms may not withdraw land and non-land property in kind when they withdraw from a peasant farm. Both Russia and the Kyrgyz Republic have such a provision in the Civil Code. (Social Dimensions of Agricultural Development in ECA: Knowledge Vacuum. Ayse Kudat. 1998).

mixed impacts on women. Because women outnumbered men in the sector, the restructuring and privatization of livestock farms meant large losses of female employment, and fewer women search for employment as compared to men (Figure 8). However, families expanded their private ownership of livestock when they were distributed as an aspect of restructuring. Women continue to be the principal managers of livestock and, perhaps more than before, are able to generate cash by selling their products in nearby markets.

### Figure 8: Gender Specifity of Labor Force Participation

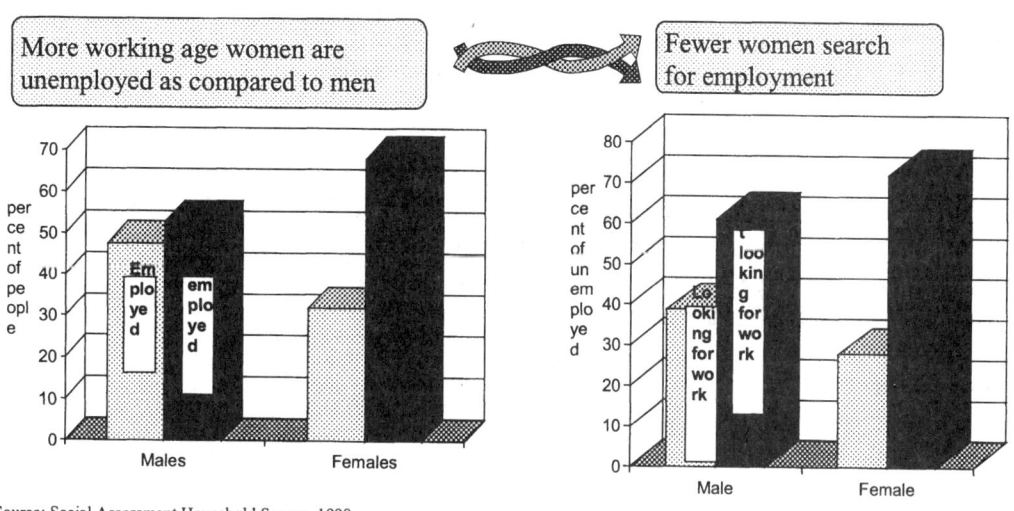

Source: Social Assessment Household Survey, 1998.

Privatization and the economic transition appears to have adversely affected the material living standards of most women; 61 percent said they did not buy anything specific for their households after privatization. [20] Of the rest, the majority said that their living standards, if anything, got worse after privatization. Most women do not have any spare time anymore because of household chores and the responsibilities in the household land plot. About 60 percent of the women nonetheless admit that their lives have changed due to the effects of privatization and about one-third of those think they now have the chance to earn their own living.

Women in rural areas have little information and widespread concerns about credit payback. Women also complain about not being able to have adequate access to farm machinery and equipment due to the high cost of the equipment. In the three rayons surveyed, the majority of women's families have sold agricultural products. In about 30 percent of the families the women themselves sold these products while in half of them the spouse was responsible for the sales. Main difficulties encountered in these sales are similar to those faced by all rural families such as poor transportation, lack of storage and processing facilities and unfavorable prices. For women, the most important source of information on the market is their neighbors, relatives and associates; more than 90 percent of the respondents indicate that they obtain their information from informal sources.

---

[20]     Survey with Rural Women, Azerbaijan, 1998.

Rural women are currently involved in such diverse activities as small household chores, roadside sales, household plot gardening, and handicrafts. These activities need to be supported so that women will benefit from rural reform. Currently, most women working in those areas are educated, but not experienced, and most of them do not have the know-how and marketing information even if they wanted to be engaged more in entrepreneurial activities. Therefore, any assistance that the proposed Project provides has to be flexible enough to meet these diverse needs. The Information and Advisory Services component of the project has a focus on group training with rural women to ensure that they are aware of their rights and responsibilities and they have access to know-how and marketing information on feasible activities. The borrowers' groups portion of the Rural Finance component is aimed at small-scale group lending which is suitable for rural women and will enable them to purchase inputs and engage in small-scale activities.

There is also a chance that women will not have equitable access to benefits of the rural reform program such as land registration. Therefore, within the scope of the Project, it is important to take precautions to ensure that women receive equitable access to the services of the Project. For instance, women should be included in registering land titles and equipment which are distributed to heads of households (predominantly men).

## *Institutions*

### Stakeholders

The power base of the rural society is changing. It is important to focus on the dynamics of this change and do so with local specificity. The key stakeholders in Azerbaijan's rural development efforts are not easy to identify for several reasons:

- Inadequate knowledge of the Soviet era, the institutional and social fabric, and the implications of interactions with the broader Soviet system as felt by local communities;

- Lack of research on the social organization of different regions and how these interact with the administrative/political apparatus; and

- Lack of analysis of the impacts of rural migration, including immigration, emigration and temporary migration.

Indeed, there is a remarkable vacuum of knowledge of what is a uniquely heterogeneous rural landscape. This SA cannot hope to provide all the needed information as this would require a great deal more time and resources than was available for the process. The depth that needs to be reached in gaining a solid understanding of the history, regional specificity, factors associated with change, and molding the attitudes of national and local leaders cannot be accomplished without a broader focus. Equally important is the multi-faceted and extremely rapid pace of change that continually defines the local stakeholders and their relationships. Also, there are many aspects of rural development; the key stakeholders of land privatization, for instance, are not the same as those that concern the process of creation of new rural industries and agribusinesses.

Until 1991, the farming structure in Azerbaijan consisted of some 1,800 collective and state farms, cultivating a total of 1.46 million hectares. These large units provided comprehensive services from birth to death for rural communities in almost every aspect of life, from the economic and social to the cultural and individual. Non-farm workers, especially children, pensioners, and those providing social services, were a major part of collective and state farms. The members of the farm were essentially paid, specialized laborers.

In addition to the state and collective farms, other prominent rural institutions under the Communist regime were the State Agriculture Bank (Agroprombank), rural enterprises and the family. Today, farming units have become smaller and more voluntary, enterprises have largely collapsed, and Agroprombank is in financial and organizational distress. How the family functioned among the 26 different ethnic groups in rural areas is unknown, but judging from current patterns it appears that extended family relationships were strong among most of the Central Asian groups. The diversity of social organization must have been rich and is reflected in the current patterns of social change. These patterns are so diverse that it will do injustice to even attempt to illustrate it, but it is apparent that they affect patterns of coping mechanisms in important ways. Further research is needed to study the interrelationship of patterns of pre-transition social structure and coping patterns. We can provide an illustration of some elements of regional differences in the development of stakeholder groups.

Let us take two administrative regions in the Caspian Sea region. One of these used to be a major fisheries center with just one formal enterprise employing over 1,200 workers. It had developed a sophisticated fisheries industry, widely exporting well-known brands of caviar and other produce. It also produced a great variety of fruits. These were sold to Russia and other Soviet regions and the local people enjoyed a comfortable life that is visible from the housing stock and infrastructure that remains. The region was first hit by the Caspian Sea ecological disaster, the fisheries industries was completely inundated and the potential for continued production severely reduced. As a result of the armed conflict with Armenia, some families have left and many others were housed in the region.

After independence, loss of wage employment in the fisheries industries and non-payment of wages in the state farms pushed some families into poverty. Some started going to Russia for informal sector work. The local administrators and enterprise managers held their positions, hoping that the State would sooner or later revive some of the fisheries. A foreign firm moved in to drill for oil, established a well managed enterprise, and provided employment to several hundred families. Merchants from Turkey moved both to the Baku market and to some of the local markets, providing better quality agricultural produce at lower prices that competed with the crops from this region. Although the distrust against the key institutions and their managers of the previous era grew and farm restructuring reforms were accepted, not all farms proceeded with the process with equal speed. Those managers who had better connections were able to delay the process or force farmers into agreeing to leave the management of their land to them.

In another rayon that was once the center of silk production, the changes have been even more dramatic. The production process has been completely damaged as a result of marketing constraints. A severe reduction in local demand resulting from falling incomes was complemented by difficulties in competing with global prices of similar products. A great many of the lost jobs are women's, and the skill losses are large both in silk production and in production of fabrics and carpets. Neither the local influentials nor the people appreciate the advantages of farm restructuring, and there is an overall reluctance among stakeholders to identify with the reform process. The incentives for out-migration are large and new set of stakeholders has yet to emerge. The bonds between relatives and neighbors have been firmed up to cushion the adverse impacts of transition. In addition, small scale efforts of local women's NGOs to revive silk and/or carpet production is attracting attention to the needs of these stakeholders and giving a new visibility to their role. They have been particularly vocal in stakeholder seminars and are since calling the sector planners' attention to the special needs of women.

With the start of transition to a market economy in the early 1990s, the government decided to abandon the system of large collective and state farms, beginning with the distribution of land. In order to receive land, people were organized into land cooperatives, groups of 100-250 people that were smaller than the state and collective farms. While some farming cooperatives of this magnitude still continue, many of these land cooperatives proved unworkable as people preferred to group themselves voluntarily into smaller associations based on the extended family and groups of neighbors. These new associations do not have the administrative or management infrastructure of the collective and state farms, and any new form of assistance needs to take this into account.

The following are the most visible and readily identifiable stakeholders who are responsible for the proposed Agricultural Development and Credit Project and will benefit from it (Figure 9).

**The Central Government** is represented by the Ministry of Agriculture, the State Land Committee, the State Irrigation Committee and the Agrarian Reform Commission. The Central Government has passed a number of laws regulating land and farm property distribution, as previously described, and additional laws and regulations continue to be developed. It is also one of the largest purchasers of agricultural products from privatized and non-privatized farms.

**Local Authorities** comprise executive leaders in rural areas. These leaders are usually responsible for distributing land as part of the privatization process. Therefore, their cooperation is crucial to the success of agricultural reform efforts.

**Directors of State and Collective Farms** are usually tenured from the Soviet era. They are experienced in farm management and may exert profound influence on the privatization process. Many of these managers want to retain land and other farm assets under their control. There is evidence that land distribution has diluted their influence in some regions. But in most regions these directors have pooled the farm assets of their relatives and, in some cases, their neighbors, often through voluntary means.

**Former Workers of State and Collective Farms** comprise agronomists and other technical experts involved in both cultivation and non-production activities. Privatization has benefited those involved in cultivation because they have received land or are operating machinery on someone else's newly-privatized land. However, those involved in non-production activities have suffered under privatization because they lost their salaried positions after farm liquidation and do not have the experience to succeed in agricultural cultivation.

**Residents of Rural Areas** are legally permitted to obtain land through the continuing land privatization and distribution process. [21] A special group of these residents is the Internally Displaced People (IDPs) who live mainly in rural areas but are not entitled to land in their places of displacement. They are the former residents of Nagorno-Karabakh and its surrounding regions which are under occupation. They have been displaced from their home areas during the hostilities between Armenia and Azerbaijan. Since then, they have been living in tent camps, railroad wagons, public buildings and other makeshift arrangements. Their incomes are significantly lower than other rural residents; they are heavily dependent on foreign aid and are not entitled to receive land in their displacement regions. However, some of them work and earn incomes on newly privatized farms.

- **Smallholders** emerged after completion of the land distribution process in rural areas. They comprise former workers of state and collective farms, pensioners as well as non-agricultural residents of rural areas such as teachers, doctors and administrative workers. Most of them lack the know-how and equipment to participate in agricultural production. Nonetheless, the SA found evidence that these smallholders, in most areas, are joining forces in the form of informal production groups. The SA focuses on their current problems, coping strategies and possible methods to help them during the transition.

- **Women** suffer disproportionately from rural unemployment and lack of income. They are often the first to lose their jobs in collective farms and production facilities. Most work in individual garden plots and maintain households, while a few find administrative jobs locally.

- **The Elderly** are important in rural Azerbaijan. They lead extended families and often are the holders of land titles and other farm assets. Their pension income, when not delayed, contributes significantly to overall household income.

- **The Youth** are an important part of the population living in rural areas. They are, by and large, frustrated by the lack of opportunities in rural areas, and they are the ones migrating to Baku and other countries in search of employment. Remittances they send back to their villages are an important source of income for many rural families. There are concerns that outmigration from rural areas is a big threat to the future viability of the agriculture sector in the country.

**Private Enterprises** comprising cotton ginneries, small food processors, traders and farmers have emerged during the transition and contribute significantly to household income. Small "informal" private activity such as roadside marketing also adds significantly to incomes of

---

[21] It is also important to note that these groups are inter-related and there is a large overlap between the members of each. For instance, pensioners may at the same time be small holders due to land distribution, and there are female pensioners who are currently engaged in agriculture as well.

many families. As more land is distributed and credit and extension services become available, the private sector is likely to expand its importance within the rural reform process. **The World Bank** is financing the Farm Privatization Project and is working with **GOA**, in partnership with **Food and Agriculture Organization** (FAO), on the Agricultural Development and Credit Project and the Irrigation Project. Through these projects the Bank provides policy, technical assistance and financing for rural sector investments.

**Figure 9: Stakeholders in Rural Reform**

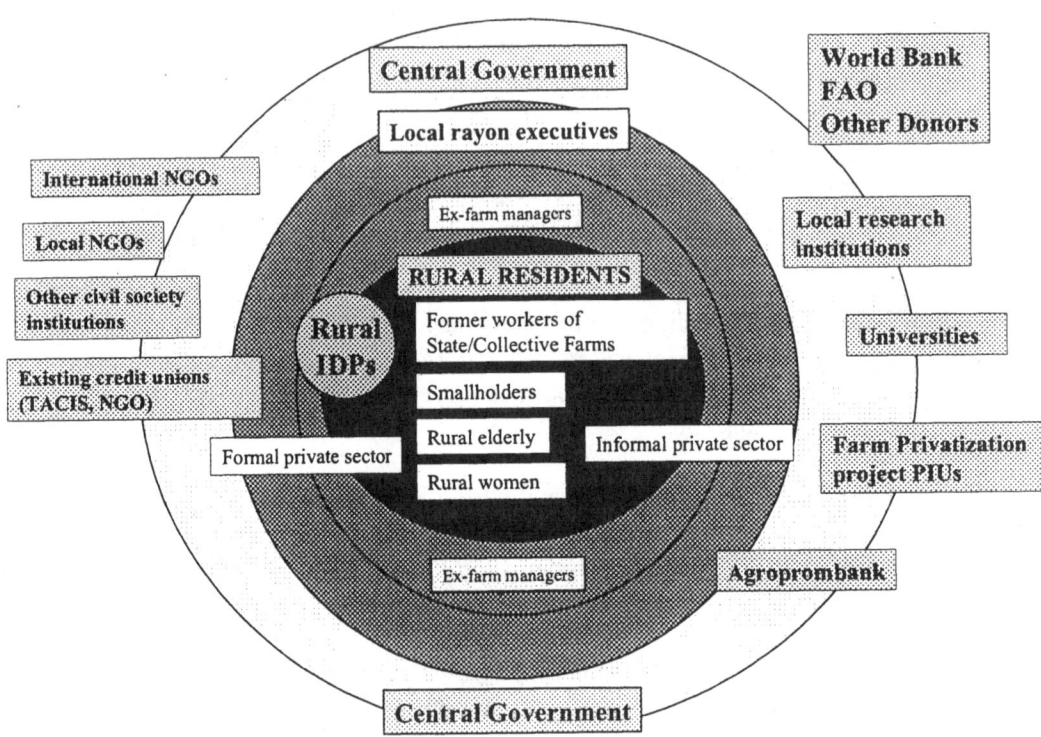

Among these stakeholders, several deserve specific attention. Various attempts have been made to reform, restructure and revitalize **Agroprombank**, which had a wide network of branches throughout the rural sector in Azerbaijan. The 1996 Farm Privatization Project financed by Government, the World Bank, and the International Fund for Agricultural Development included $13.4 million in credit for the rural sector using Agroprombank as the intermediary. Unfortunately, the restructuring of Agroprombank, partially financed under the Project, has not been successful yet, either in terms of creating a responsive organization or establishing financial soundness. In addition, Agroprombank is widely distrusted by farmers.

**Emerging Structures**

**Regional or local executive committees** are essentially new rural institutions. While these existed in the Communist era, they were overshadowed in practice by the collective and state farms. It is, unfortunately, too early to determine what role local governments will play in rural life in the long-term. As their governance structure moves from appointed to elected officials, and if officials, and if they are able to secure regular funding, they may become more instrumental in assisting people. Up to now, however, various social assessments have suggested that people have more trust in the national government to deliver services and provide assistance than in local government. In some cases, local government has even worked against the transition; one of the Bank missions monitoring the Farm Privatization Project in late 1998 noted that there was a general reluctance among farmers to seek credit because of the many taxes being levied and collected by local tax authorities. Obtaining a loan from an institution appeared to expose farmers to additional taxes which may not have been in the tax code.

New institutional structures in rural regions are developing at three levels. At the lowest level, families provide the social safety net, cushioning the pains of individuals as they transition from wage earners to farmers. These families have multiple breadwinners, many of whom lack job security; as such they are highly dependent on one another and have roles that are far more complementary than during the Soviet period. At the sub-community level, voluntary associations of relatives, friends and neighbors lead to the emergence of joint farms to facilitate reform by helping individuals leverage their resources. These are built on existing social capital and are being continuously rcdcfined. At the community or even sub-regional levels, credit unions, irrigation associations, extension services and land titling/registration systems are developing and are expected to anchor the rural reforms. Nonetheless, large-scale dissolution of the formerly existing organizational links in rural areas has hampered agricultural production and marketing. Success of agricultural reforms, to a large extent, depends on the ability of farmer/producers to sell their products in an efficient manner and at adequately wide margins to pay for operational costs and to provide profits conducive to further production.

---

**Examples of Production Associations**

*Imishli Region: An association received its land by the first lottery and individuals in the association received their land via the second lottery. Associations in Imishli tended to comprise larger groups bound together by formal, written agreements. After working together for the first few months, many individual families decided they did not want to work with other families. Consequently, the association is moribund as families work individually on an informal basis.*

*Sheki Region: In Sheki, a former farm administrator convinced 100 relatives and neighbors to work together under a formal written charter and registration. All have title and control of their individual land but work together according to written contracts. Led by an executive committee, decisions such as what and how much to plant are made through majority voting.*

*Ismayilli Region: In Ismayilli, individual land plots were too small to farm efficiently. In one association, a group numbering 500 individuals cooperates through informal, unwritten agreements. Half have not yet received titles to their land. A former state farm manager pays individuals for produce and markets it.*

**Source:** SA consultations, October-December 1998

---

Several models of farming have evolved during the privatization and land distribution process. Some reflect the values of the post-Soviet era while others maintain the values of the Soviet period. Some directly promote positive economic development while others may obstruct it.[22] Three primary models dominate the new economic order in rural Azerbaijan; they are

---

[22] The situation regarding the variability of the types of distributed land is a common characteristic throughout the former Soviet Union. In several of these countries, the early phase of privatization was rather spontaneous and often inequitable. For instance, in Georgia, even before the formal land privatization process was initiated, some land fell into the hands of private owners although no "formal" or "official" data exists on the extent of such activity.·

characterized below. Differences exist in the working order of these associations, as explained in the box below. These associations are still evolving and farmers may switch from one to another as they see fit, although the results of such switches are not always visible in official records.[23]

**Family Farms** are two to four hectares in size, owned by a single family that has significant knowledge about agriculture. Such families often worked in the kolkhoz during the Soviet period but do not want to work collectively in the new era, usually to savor independence and to avoid what they see as bureaucratic entanglements and complicated relationships with family, friends and neighbors. Some families who were not farmers during the Soviet period received agricultural land through the land distribution program and decided to farm it. The experience of the post-land distribution years so far indicates that most family farms cannot support all the family's needs, so members may earn income through other activities and/or depend significantly on their pensions. This is discussed in the section dealing with poverty.

**Small Farming Associations**[24] usually comprise relatives and neighbors that govern themselves democratically. Associations can number from 10 to 400

> *A husband and wife, ten grown sons and daughters, and 13 children make up a family farming unit in the Ferzili Village in Jelilabad Rayon which received a total of 9 ha, including their household plots, through land distribution.[1] The two main areas of cultivation are separated by a river, although the five houses and their gardens are on the same side of the river. The family plants vegetables for their own consumption. While previously their land was used for grapes, the production is now wheat (60 percent) and potatoes (40 percent). The wheat harvest is distributed among the family sub-units, each one choosing how much to consume or sell. Produce is sold at local markets, where prices are low, because of high transport costs to other markets. As in the former Soviet Union, household garden plots, which grow mainly vegetables, are an important source of income. The unit keeps two cows and about 200 small livestock. No one in the village has received land certifications and most people work on their land individually—only this family cultivates collectively. The family unit has the typical problems: seeds are expensive and of bad quality; fertilizer and pesticides are not available; and equipment can be hired, but it is expensive and not available when needed. If available, credit would be used initially for small livestock.*

persons and farm between five and 400 hectares of land. Practices differ from association to association, but each family receives by written or verbal contract equity shares in the association equal to the market value of the assets it contributes to it. Family assets usually comprise land, buildings, farm equipment and the number of family members committed to the association full-time. Existing members decide democratically whether to admit new applicants. Members can leave the association voluntarily. Members appoint a leader and an executive committee which run the day-to-day business affairs of the association. However, all important production and administrative matters are usually discussed by the general assembly, often including the type and amount of production and the division of proceeds from the harvested products. An exception is the strictly family-based groups, in which older persons or most knowledgeable members may

---

[23]     Similar types of farming groups are also seen in rural Moldova, which is going through a similar transformation. There, various forms of organizations emerged from the distribution of land and assets among workers and former workers (retirees) of these farms. Each of these forms has advantages and disadvantages. They can be classified as smaller individual farms, larger-scale commercial farms and group farming with relatives and neighbors. (The Gap Assessment Report - Moldova Agricultural Support Services Project, 1999)

[24]     In Russia and Ukraine, there are regions and areas (such as those in proximity to large towns) where family farming has gained importance. In Russia as a whole, however, family farms account for less than 10 percent of the agricultural area; in Ukraine the figure is 2 percent (Csaki and Nash, 1998). In Moldova, "although land shares are reported to have been distributed in virtually all the farms, the process is still very much in the stage of 'conditional paper' shares." (Lerman et al. 1998:3).

make these decisions themselves. Association members usually receive most of their income through agricultural activity. There is a tendency in rural areas to concentrate small land holdings in voluntary small production associations which provide the flexibility to engage in agriculture as individuals deem fit while pooling resources for common activities such as land preparation and harvesting.[25]

**Large Farming Associations** are usually modeled after the former kolkhoz and often number 500 or more people. The few large farming associations that exist were started

*LG is a 49 year old female living in the village of Yeni Garadolag, part of the Agjabedi Rayon. With degrees from a pedagogy institute and an agricultural academy, LT was a teacher and worker in the former state farm. Since 1996, she had headed a production association established voluntarily among kin and neighbors who joined together during the process of land distribution. The association controls 50 ha of land belonging to 65 individuals from 16 families. Thirty members of the association are actively involved in farming. The association cultivated 30 ha of wheat, 15 ha of cotton and 5 ha of fodder. Wheat was not profitable due to price, and the cotton yields were low due to poor quality seed. Lack of credit prevented them from purchasing high quality seeds and the crops were damaged when rented equipment was not available when needed. The harvest was much lower than expected because of heavy rains. The association sustained itself on a herd of 300 sheep. Many of the association's problems can be cured by having access to credit but not if the association would have to work with people outside the present members.*

by former directors of state farms and are powerful. Members may work on their own land as salaried employees of the association, but the association leader usually dominates decision-making, deciding what to plant, how much to plant, and how to distribute the earnings. Leaders do not meet frequently with association members on these issues. These associations often control more than 80 percent of a village's land and may exert inordinate power over those who do not wish to join. Some large farm associations have obstructed the establishment of smaller farming associations by their neighbors to maximize their influence in a village. Many villagers do not have the expertise to farm so they join the large farming associations by necessity.

Types of production associations that households belong to largely affect their income levels. Individual smallholders find it increasingly difficult to engage in agricultural production by themselves due to lack of economies of scale and efficiency. As indicated, many single family farms join forces with their neighbors to generate agricultural income. Farming associations formed this way have greater scale economies and benefit from the knowledge and skills of former state farm workers. Some try out associations and become discontent; others try different arrangements. Some seek the guidance of ex-managers of state farms. Others claim they were forced into such an arrangement. However, whether as a result of trust, an inability to farm, or a realization that they cannot be effective alone, many households pool resources together— although it appears that this occurs less than it did at the outset of the privatization process. Despite ongoing marketing and financing problems, they usually enjoy more income than single family farms. Increased access to credit and extension facilities, reinforced with security of land ownership, would contribute significantly to their income-generation capability. The groups that have social capital will also have greater ease in getting access to credit for purchase of farm technology and other inputs.

---

[25] The SA team did not find evidence of collective marketing in these associations. Experience from other Former Soviet Union countries, nonetheless, show that this is not always the case. In Moldova, for instance, associations act primarily as marketing and/or purchasing cooperatives; their production functions are secondary but nonetheless exist. Over there, an average association of peasant farms manages around 200 ha and has a membership of 100 farmers (Dudwick, 1997).

The people are coping by expanding social capital and the change is dynamic, and thus the tentative trends noted above need further confirmation. In Azerbaijan the precise relationship between land privatization and agricultural production, especially efficiency gains and losses of

*AH is a farmer in Qaralar village of the Imishli Rayon. He is 61 years old and considered the aksakal of his family; he leads his family's small association. Together with the other farming group leaders in their village, he decides what to plant this year. After that, he discusses this with the members of his group and then proceeds to work. He complains that before, everybody knew what his job was and did it. Now, people have to make decisions about everything and this is harder. During and after the distribution of former state farm lands in the village, AH was leading a significantly larger farm association. However, after the distribution, members of this group wanted to work on their own land. Now, he is still listed on official records as the association leader; however, the members work on their lands as they please. He thinks that the distribution of the land was fair, and created more opportunities for hard-working people. But at the same time, people who are not used to farming are in a worse position. He says that they need a lot of new information on farming techniques, and they get it partially from their neighbors who may have more experience in farming. Overall, though, he is pleased with the new situation, which gives him a chance to work independently as he wants*

different types of farm enterprises in the various parts of the agricultural cycle, is unclear beyond very small samples and anecdotal evidence. In fact, with respect to yields, it appears that village location, proximity to urban areas, and the climate play more important roles than the level of privatization. Since the data are inconclusive, systematic impact monitoring and evaluation of the proposed Project is of particular importance.[26] This will ensure that the Project does not end up focusing on items such as privatization, advisory services and credit to the detriment of other more important factors that are not understood fully at present.

Livestock production has become more important for the livelihood of small farmers especially. Livestock provides opportunities to add value to self-produced crops such as maize or surpluses of perishable crops which could not find a market. Marketing animals is also easier and these animals are considered as a sort of savings which also produce income in the form of milk and meat. There is not adequate information on livestock marketing activities in rural Azerbaijan, however, the importance that rural people attach to their livestock is unmistakable and for many families, it is a major source of livelihood.

## *Participation*

### Land Distribution and Registration

Confusion and ambiguities in the land reform process are reported throughout the former Soviet Union. It is only fair to say that comprehensive reforms cannot be achieved overnight, and it is unrealistic to expect reform implementation to be completed to the same degree in all parts of a country, no matter how small, even if the legal frameworks were fully in place. In many of these countries, the legal framework is still being defined and sub-regional differences in reform implementation are substantial. The process involved is complicated and difficult to

---

[26]      Farmers need assistance in this transformation process. Some ways to provide this assistance include: 1) supporting the establishment of these associations when wished by farmers as a transitory way of separating from the collective while keeping access to some farm equipment and other assets; and 2) helping these associations improve their management and move into service cooperatives and smaller groups based on active management of individual members. (The Gap Assessment Report - Moldova Agricultural Support Services Project, 1999)

communicate, and it is difficult for ordinary citizens and farmers to keep pace with a large number of legal changes; access to understandable information on legal matters is limited and farmers have little opportunity to provide feedback on these issues.

In Azerbaijan, physical land distribution is largely completed; in most regions, the process was regarded by the local population as transparent and fair although charges of wrongdoing exist in some locales, particularly where privatization is not yet complete. These charges appear to increase with the length of time taken to implement the reform. In addition, the poverty levels tend to be higher in regions where land distribution lags and lower where land distribution is completed. Poverty is also related to the distribution other state farm assets; it is higher where assets have either not been distributed or, more often, where the distribution did not take place in a transparent manner with adequate information/communications support. Land title registration and the ability to trade assets, therefore, will enhance support for the reform process.

The land distribution process has the following general steps (specific steps may vary by region):

- The area to be distributed is surveyed using maps and "walk-throughs;" it is classified as fertile, non-fertile, municipal or state lands; and the number of people entitled to receive land and the average size of land plot per person is determined.

- Participants in the land lottery notify the land commission of their participation either as individuals or as part of a group; members of each group select representatives;[27] and lotteries are conducted to determine which groups and/or individuals are assigned to various land areas.

- A second lottery is conducted among individuals in each group to determine the size and place of land plots each member will receive; and the local land commission prepares a land certificate for each individual household or member and sends it to the State Land Commission for approval and registration.

In general, the livelihood of rural residents improved where land privatization and distribution occurred. No implementation or slow implementation of privatization has caused most of the problems (Figure 10). Land distribution also has had the effect of curtailing the influence of ex-farm managers and has helped empower the rural population. On the negative side, it adversely impacted the non-technical workers and population that did not have the know-how to engage in farming. The liquidation of state farms caused these people to lose their wage-based employment without providing them with the opportunity or the know-how to engage in agriculture, at least in the short term. It is therefore important to support the effective completion

---

[27] This step has been introduced more recently in response to feedback from the farmers who say that the spread of lands belonging to a kin group make it impossible for kin members of a community who occupy different households to cultivate their lands jointly. In early 1997, rural people were content with the lottery system but complained that fragmentation of kin lands prevented their ability to rent or purchase farm machinery. The current procedures allow families the flexibility that they need.

of land distribution and to promote private employment in small-scale rural enterprises while providing information and technical support to non-agricultural part of the population.

**Figure 10: Perceptions of Effects of Land Privatization**

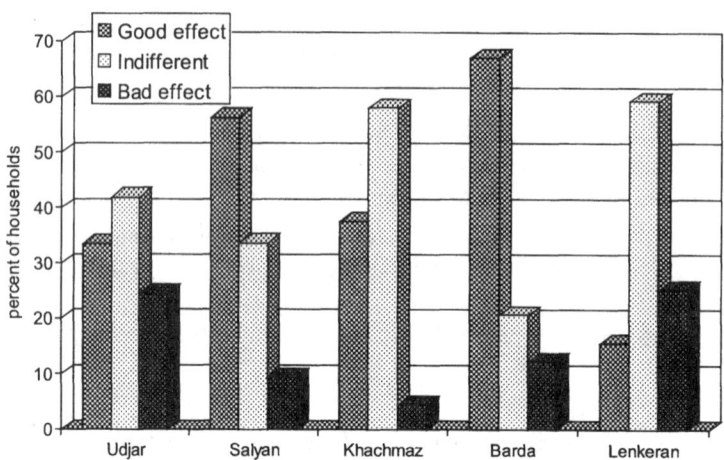

Source: Social Assessment Household Survey, 1998.

The Government intends to focus the subsequent stages of the land reform process on completing and legalizing the proceeds of land distribution and ensuring that the rights of individuals are not diluted through the legal registration process. To ensure the equitable integration of small-holders, especially rural women, individuals need better understanding of the legal and procedural changes of the process, and mechanisms for their participation should be developed. Official registration currently is time-consuming, impeding land trading and increasing ownership insecurity. Rural people receive their land shares, however, the registration of these shares at central offices in Baku slows down the process. Lack of regional land offices adds to the costs of land transactions, making these harder to complete for poorer people, and hampering the development of rural land markets.

Development of a land market needs to be encouraged since this will improve the efficiency of the agriculture sector. Informing farmers on different methods of land transactions and contracts is important to enhance the development of these markets. The Bank's support to strengthening and decentralizing this process will therefore further empower the rural people. The Information and Advisory Services component should also support this through dissemination of mass media of simple recommendations and advice on how to engage in land transactions. Decentralization of land registration would also make it possible for land offices to serve their region-specific needs better. Nonetheless, it is important to clarify the existing legal framework on land transactions and to ensure that the newer legislation put in place does not hamper the development of land markets and people's equitable access to these markets.

The Project will also include a unified cadaster and registration system for land and buildings to be set up in Nakhchivan Autonomous Republic. Real property information for all of the rural and urban land and all of the buildings will form the base of this system. The pilot approach is expected to make it possible to develop a cadastral model and a sound national policy that can be implemented in the whole of Azerbaijan in subsequent phases. Nakhchivan, an autonomous region with six rayons and a mix of urban and rural areas, was chosen because it is representative of Azerbaijan as a whole and because it has its own Parliament which will facilitate the adoption of the appropriate legal framework. The component will pilot an appropriate approach to a unified national cadaster on a regional basis before devoting the substantial resources required for implementation of national system. This typically involves urban land and buildings as well as rural properties and experiences a far greater number of transactions then rural land registries.

### Access to Financing

There was a great deal of uncertainty at the outset of the farm restructuring process, and community visits showed that very few people wanted to borrow money except for livestock. They were concerned that investing money in other agricultural activities would not provide a reliable return. Indeed, the social science team was only able to identify a demand for credit in communities where the Bank Project was being implemented and where the villagers have been repeatedly told about the availability of credit. In other communities where no expectations have been created, there was little knowledge of or demand for credit. As with other dimensions of rural life, the attitudes rapidly changed.

**Figure 11**
**Access to Credit and Independent Borrowing**

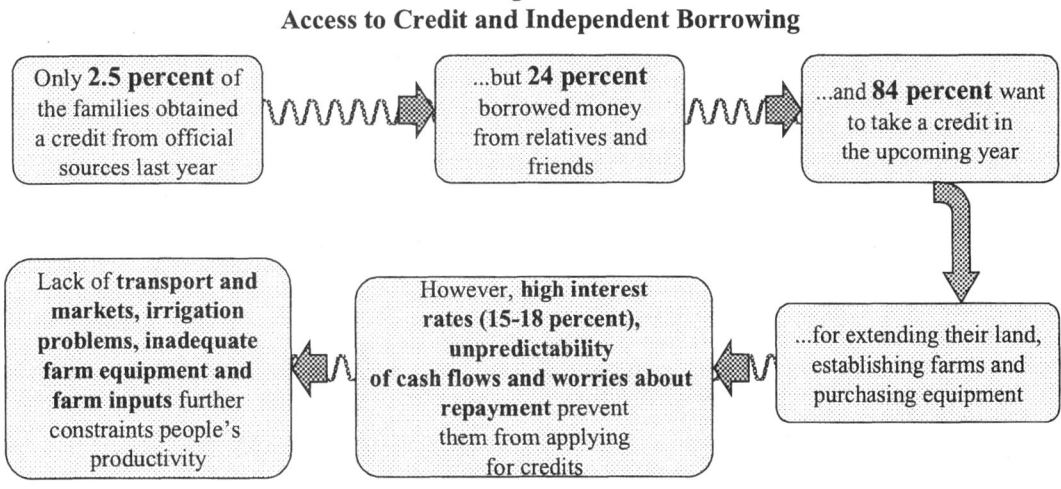

Source: Social Assessment Household Survey and Focus Group Discussions, 1998.

During the conduct of the 1998 SA household survey, a high demand for agricultural financial services was evident in rural Azerbaijan, provided that these services are affordable to emerging small holders and farmers. Depending on the region, 50 to 98 percent of farmers, wanted to borrow money to improve their farms, and almost all of them were willing to use land and equipment as collateral if favorable conditions exist. However, many of these people could not obtain adequate financing, preventing sufficient marketing and access to inputs (Figure 11). Only two percent of respondents reported obtaining formal credit in 1998. Small-scale informal borrowing, on the other hand, was more commonplace, although the availability of funds from friends and relatives is an important obstacle. By end of 1998 and early 1999, difficulties in raising demand for credit were reported from international NGOs working in limited geographical areas; this appeared to be associated with two factors. First, the credit was offered for small amounts and short maturity, which did not coincide with the agricultural needs of the rural people. Secondly, the focus was on areas, such as newly liberated communities, experiencing multiple problems.

There are several factors that contribute to the inadequate availability of financial services in rural areas:

- Inadequately-performing rural financial institutions and the unwillingness of existing urban financial institutions to expand services to the rural sector;

- The slow implementation of legal reform for land in many locales, hampering the use of land as collateral;

- Favoritism to the powerful and informal payments;

- High rates of interest and short maturity charged for existing loans which hamper investment in agricultural assets such as machinery; and

- Lack of information coupled with cumbersome bureaucratic procedures for obtaining loans, resulting in delays.

The most important financing problem in rural areas is systemic; few financial institutions have a network in rural areas and their lack of liquidity has effectively halted their operations. GOA has little money to infuse into the state banking system. There have been

> *"...Last year, we did not know how to get seeds, how to prepare our land, how to plant and irrigate... This year, we learned everything and we are ready to work. But now we don't have any money to buy seeds and pay for land preparation. There are those who offer money, but they ask back for it in three-six months, or they want high interest rates. It may be possible to get official credit, too, but we don't want to pay the "hat" of 50% on it..."*
>
> **Source:** A farmer from Imishli Rayon, December 1998.

regional efforts to enhance access to financial services. For instance, some credit unions started in 1996 with assistance from the EU/TACIS Program. But confusing regulations have delayed their operations until 1998; even though they have been able to operate on a limited scale since, the sustainability of these unions, in which the borrowing members also have equity stakes in the union, is highly questionable. Financial institutions in Baku do not have the capacity or the willingness to engage in the rural sector, especially when the use of rural land as collateral for credit is legally cumbersome. As reported by an EU/TACIS credit union manager, even those unions have asked for assets ownership in Baku as collateral while joining the union.

Since rural reform efforts have started, the nature of problems experienced with inputs seem to have changed. At the start of the reforms, the most important production constraint was access to good quality agricultural inputs, in particular seed, pesticide and fertilizer. However, as reforms have progressed and the private market developed, access to inputs increased while financing to facilitate this access became a crucial factor. According to people, agricultural credit is not completely unavailable, although available financing (both formal and informal) is often too expensive and has to be paid back in an unrealistically short period of time. There is also a strong element of distrust among the population toward existing institutions; stakeholder consultations point out the need to "bribe" bank officials to obtain credits. Consequently, farmers tend to pay for as many of their inputs as possible—machinery, equipment use and labor—with production.

Informal borrowing is often from relatives and no interest is charged; rather, reciprocity governs these relationships. Some borrow from informal lenders; people have reported borrowing 10,000 manats at the beginning of the month and paying back 12,500 manats at the end of the month, bringing monthly interest to 25 percent. It is important, however, to stress that these high rates are dependent on the small amount of the money lent. In addition, the incidence is neither systematically confirmed nor is it high, and the amounts are usually small, even by Azeri standards.

Nonetheless, the current extent and outreach of informal borrowing and efforts to combine financial resources is highly limited. Most available financing is small-scale and at short maturities. Collateral use is not common, and there is reluctance among rural people to use their house and land as collateral for expensive credit, granted on unfavorable terms. Large-scale credit, which would allow farming households to purchase equipment and inputs and decrease their reliance on old equipment from former state farms, is not available. In addition, the SA found little evidence of formal mobilization of savings in rural areas. The ownership of privatization vouchers, increased ownership of livestock and the extent of informal borrowing point that rural people have some savings, albeit not always in cash.

Other obstacles to credit also exist. Currently, the main information sources on credit programs are the World Bank's foreign and local representatives, local TACIS staff, some international and local NGOs and the State representatives. These sources of information do not suffice to inform the general public on lending programs and procedures. Prospective borrowers often must complete cumbersome and time-consuming state registration forms and loan applications without adequate instruction. Many farmers are also reluctant to borrow without improvements in inputs, transport facilities and marketing facilities. Any intervention in the rural financial system has to take into account these factors to achieve maximum impact and sustainability over the long-term.

In the Farm Privatization Project areas, farmers perceive charges on lending to be unfair, wondering why they must pay interest at much higher rates on loans than GOA pays the World Bank on IDA loans. The documentation required for obtaining loans under the Farm Privatization Project also proved excessive, particularly in view of the fact that most loans requests are for about $2,000. As a result, by November 30, 1998 a total of only 205 short term loans had been approved under the Farm Privatization Project, amounting to about $510,000, of which only half had actually been disbursed. In addition, Agroprombank, which is responsible for executing the credit scheme, could not develop adequate capacity for loan administration and is in the process of restructuring.

In general, rural households are hesitant to borrow when the maturity of the loan is less than 12-14 months, which is their current crop cycle. For livestock activities, families think at least two years is necessary to have a reasonable return under current conditions; therefore, households are not willing to borrow for livestock unless loan maturity matches their revenue stream and/or cash flows.

The Agricultural Development and Credit Project's estimation of the demand for credit is based on the SA survey finding that 84 percent of rural households would be willing to obtain credits for agricultural operations. The reasoning is that "...these data, combined with the approximately $290 million in annual revenues being generated by rural entrepreneurs, suggest significant unmet demand for loans. It is therefore assumed that 65 percent of entrepreneurs *might* have an effective demand for loans (i.e., they have the means and the desire to borrow), of which only 4 percent would actually obtain loans from a CC initially." The issue, therefore, is whether this utilization of a general willingness to borrow as the basis for establishing demand is supported by more detailed findings of other parts of the social assessment and stakeholder participation process.

There is some contradiction between the results of the social assessment with regard to the terms and conditions people are willing to accept for credit and current practices based on calculations of conditions necessary to establish a sustainable market-based rural credit system. The strongest point of such a system would be its flexibility to provide financing for a diverse range of activities at sustainable interest rates. Financial calculations show that the monthly rates currently charged by international NGOs—three to four percent—will be necessary to sustain these financial intermediaries in the short-term; as their operations become more diverse and efficient, they will align themselves with what their market demands and adjust their lending rates accordingly.

To establish adequate flexibility and outreach, the proposed Project will finance two different types of local financial intermediaries. The first, credit cooperatives, are expected to be formal, sound and competitive financial institutions with a wide ownership and clientele base. They will be incorporated under the "law on enterprises" as limited liability enterprises and will provide comparatively simple savings and credit services to their owner-clients. Their credit services will be funded with equity contributions and deposits from their members and IDA funds. They will provide financing to their members at maturities ranging from 6 to 24 months, based on business plans. The second group of LFIs will involve groups of jointly liable borrowers with 10 to 20 individual entrepreneurs in each group. These groups, also known as "village banks," will be used as additional channels for component funds in order to accommodate the great diversity in credit demand. They will be particularly important for micro-entrepreneurs in the smallest rural localities where there may not be a large enough population to support the fixed-cost structure required for the proper functioning of credit cooperatives.

Table 4. Proposed Uses of Credit (percent)

|                              | Barda | Lenkeran | Salyan | Udjar | Khachmaz |
|------------------------------|-------|----------|--------|-------|----------|
| Extension of land            | 28    | 39       | 78     | 11    | 6        |
| Establishing farming business| 28    | 12       | 14     | 40    | 26       |
| Purchasing new capital goods | 23    | 30       | 4      | 30    | 48       |
| Agrotechnical services       | 17    | 14       | 3      | 19    | 18       |
| Paying salaries              | 4     | 0        | 0      | 0     | 1        |
| Other                        | 0     | 4        | 1      | 0     |          |
| N                            | 120   | 69       | 153    | 150   | 77       |

As Table 4 illustrates, the purpose for which people say they will borrow money varies among localities. But the majority of purposes are for purchasing land and obtaining working capital. The designed system of credit cooperatives has inherent flexibility to address the various needs expressed by the people. However, in the event that there is a mis-match between what people seem to want and what they can "reasonably expect," the Project needs to implement a communications strategy that will reduce the gap between what people want to use credit for and what they can reasonably borrow for in a commercial environment. Similarly, the NGO implementing the initial credit operation needs to assess the credit needs of rural people on a continual basis and make necessary adjustments. Otherwise, the result could be disillusionment on the part of cooperatives. "You told us to be independent, responsive and self governing, but when we try to be, you tell us that this is not appropriate" could be the reaction of cooperatives if the gap becomes reality.

The social impact monitoring (SIM) process will thus need to focus very quickly on the way in which loan marketing justifies the higher interest rates and the purposes for which the loans are used. This focus has to be region-specific, since credit cooperatives will be formed in a number of regions initially and are likely to serve their immediate area, at least at the beginning. The danger is that people in rural areas will borrow short term for long term needs at interest rates which they perceive as unreasonable and for activities which may not generate sufficient early cash flow, possibly leading to high rates of default and the failure of the Credit component of the Project. There is an expressed need for long-term credit to the farm sector to facilitate investment in machinery, equipment, and other assets for small-scale processing facilities. Therefore, the progress of the Rural Finance component should be monitored closely and as need arises, support to such enterprises could be incorporated into the subsequent phases of the Program.

**Extension and Communications**

In many countries rural populations are inadequately informed of changes in rural policies because direct consultation with affected populations in designing rural reforms is a rare practice, and policy makers are not under pressure to provide rationale for reform.[28] This is a major problem for the population's equitable inclusion in rural reforms and their adaptation to a life style in which peasantry is the dominant form of existence. There are information gaps in two main areas in rural Azerbaijan. First, the general public needs reliable information about general rural policies, regulations and current events. Second, people working on farms and related operations need specific and focused technical training to manage their work properly.

Technical advice seems to be limited in its quality and its suitability to current conditions. Sources of general information are diverse (Figure 12). Rural residents gain information through district and national newspapers, radio and television programs, village and farm cooperative meetings and visits by government leaders. But the information flowing through these channels is often limited, outdated and sometimes inaccurate. Rural residents believe only official sources (Figure 13). However, meetings often fail to clarify issues, access to newspapers is limited, the language used in some publications is incomprehensible to many of the less educated poor, and

---

[28]      For example, in Uzbekistan, a significant portion of the population is not convinced that any significant reforms took place. Kolkhoz leadership had an influence on the flow of reform-related information to the population.

electricity shortages limit the usefulness of television and radio for information dissemination. As a result, agricultural reform efforts currently appear to be limited to land distribution; people have difficulties in being able to anticipate the broader benefits of privatization.

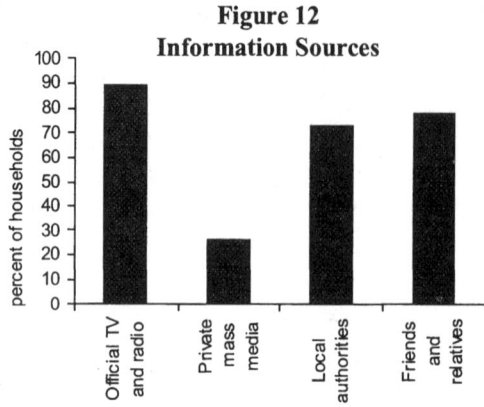

Source: Social Assessment Household Survey, 1998.

**Figure 13: Which information source do you trust?**

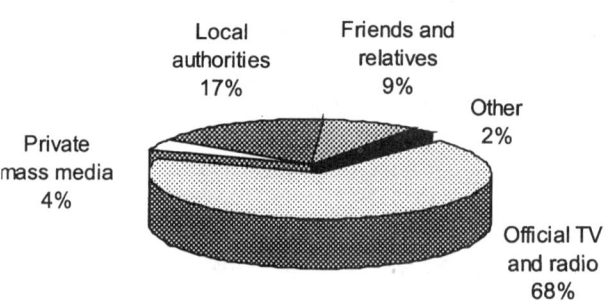

Source: Social Assessment Household Survey, 1998.

During the Soviet period, wage-earners working in state and collective farms required specific, job related knowledge. However, former wage-earners who are now farmers need information on a whole range of functions and skills to ensure proper management of an entire farm. Most of them do not have the knowledge base to engage independently in a wide range of agricultural activities. In addition, those who were formerly involved in non-agricultural activities, such as accounting and machinery operations, are not knowledgeable about agricultural production.

Current extension needs of the small holders, therefore, are quiet different than previous extension needs. Farmers engaging in production need information on crop cycles; the use and selection of appropriate fertilizers and chemicals; farm business planning and management; preparation of business plans and legal forms; farm records keeping and accounting; marketing pportunities; and animal health. To some extent, these needs are met informally by former specialists of the state and collective farms. In addition, especially in smaller villages, farmers continuously exchange information and experiences to have a better understanding of newly emerging agricultural issues. They consult each other and the specialists for the most immediate information needs. Nonetheless, the value of these experiences and the knowledge of the former specialists is limited; they reflect conditions that used to exist in the past or current problems. Farmers seek better yields by a process that is trial-and-error for the most part.

A targeted extension system is necessary to facilitate the transition from a wage-based to a more peasant-like subsistence. Smallholders also need information on their legal rights and responsibilities to make well-informed decisions regarding their land, equipment and livestock. The results of the social assessment have been incorporated into the Agricultural Development and Credit Project both with regard to the substance of information needed and the mechanisms for communicating messages to intended audiences.

The Project will rely on locally recruited specialists who are contracted on a part-time basis and paid on the basis of outputs. In addition, no attempt will be made at cost recovery for group activities such as training and demonstrations. The capacity for farmers to meet the costs is at present limited, and their willingness to do so will depend on the demonstrated value of the service. Instead, it is expected that the specialists will progressively develop their own line of business, and the amount of paid work will be reduced accordingly.

Connected with this, early efforts to meet extension needs have been well received. Extension organizations have been established in the Farm Privatization Project areas. Training covers farming equipment, land preparation, crop cycles, planting mixes, breeding, grazing and selling livestock, irrigation, and marketing. Feedback from farmers has been favorable where this training has been conducted. Additionally, more informal farmer-to-farmer information exchanges are carried out by some of those who worked in the agricultural research centers of the Soviet era.

A dedicated communications and public relations strategy is an important part of project implementation and of the process of permanent social assessment and stakeholder consultation. The availability of project benefits to the rural population, ensuring fair access to project services, and assessing the appropriate transfer of resources in a transparent and accountable fashion can be greatly enhanced by funding a communications and public relations strategy as part of the investment projects. In addition, the best evaluation of project outcomes and the most appropriate assessment of how to improve project performance often comes from beneficiaries themselves through a process of consultation and assessment that is objective, systematic, quantified, and carried out by locally trained experts in the profession. The local network of social scientists created with the assistance of the World Bank and with the financial support of the Swiss Government has proven a valuable instrument in the Agricultural Development and Credit Project.

In addition to more general information campaigns and publications, various stakeholder groups will be targeted through the Information and Advisory Services component of the proposed Project. Among the groups to be encouraged will be water user associations (WUAs), with the objective to improve and maintain on-farm irrigation networks, reduce conflicts among users and negotiate adequate water delivery with the water authorities. Other groups include women farmers and associations of rural entrepreneurs. Short, topical training courses will be provided to groups of farmers on the basis of a needs assessment and training material prepared by the Extension Support Center in the Ministry of Agriculture. Some of the modules will pursue themes already addressed by information campaigns. Other modules to be offered will include accounting, business management, gardening, home food processing, dairy management, fruit tree pruning and seed selection. The courses will also provide opportunities for advisors to meet with prospective clients and discuss their needs. Advice will be offered on farm management, post-harvest handling and marketing, and preparation of business plans for commercial farms and rural enterprises. Such group work is expected to lead to requests for specific advice to individual farms, which would be directly contracted between the advisor and his client. Technical advice and technology demonstration will be offered to groups of farmers sharing a common concern or interest, on the basis of an agreed program. Depending on the technology, this will consist of on-farm demonstrations and trials to test and compare different varieties or techniques, as well as of teaching of practical skills.

### Access to Agricultural Inputs and Equipment

Disrupted access to agricultural inputs has largely contributed to the drastic drop in agricultural output in rural Azerbaijan. Farmers have limited access to seeds, feed, fertilizer, equipment, fuel and agricultural services such as plowing, sowing, and harvesting; this is one of the main technical reasons of low crop yields. During the Soviet period, the state was the dominant supplier of most of these inputs. In recent years, inaccessibility to inputs has been one of the major causes of declining agricultural output and productivity. However, as new farmers gained experience during the past year, the ability to pay for these inputs has become the main problem.

> *"...I am the head of a farming group which has 14 families and 40 ha of agricultural land. Most of us are family and old friends. We have 12 ha of garden plots next to our houses. But life is not easy for us... We don't have our land certificates yet, we cannot get any credit for seed and pesticides and we cannot pay cash; fuel is also very expensive... Sometimes we borrow equipment to work on our land, but even then, it is a big problem to pay for fuel and lubricants. Last year we tried selling fruits in the rayon market, but people want to buy cheaper imports so we did not even harvest our fruits this year. My son in Russia sometimes sends money, and thanks to him, we can still survive here.."*
>
> **Source:** In-depth interview with farmer, Devechi Rayon, October 1998.

The problem of inputs has two aspects: the availability of machinery and equipment for land preparation, and the availability of seed. Most farm machinery and equipment has not yet been privatized and distributed. Several interests are claiming the equipment, which in most cases is still controlled by former state farms. Many former state farms are indebted to the Azerbaijan State Agricultural Bank (Agroprom) and use the equipment as collateral for the debt. In addition, former employees of some state farms have not been paid for significant periods of time and they have claims on the equipment to guarantee back-payments. The claims from these two constituencies complicate GOA efforts to implement the equipment distribution. In the meantime, a significant portion of the available equipment has deteriorated and become non-usable.

The utilization of existing machinery and equipment varies from one village to another. In some villages, the equipment has completely deteriorated and is almost impossible to use. In others, equipment has been distributed to villagers and they make use of it, taking turns. In still others, the former operators have "privatized" the machinery and they provide contractual services to local farmers for a fee. Even when equipment is available, there are problems with its use. Machinery operators, who also have farms, use the machinery on their own farms and their relatives' farms. Fuel is expensive for local people, and delays for use of the machinery are sometimes so long that yields are significantly low. It is a challenge to find spare parts for some of the existing machinery. There are also reports of former farm managers taking ownership of this machinery. In general, both individual farms and small farm associations have contractual access to the equipment; however, larger farm associations, which control considerable sources within a village, usually have direct access.[29]

Women appear to have far less access than men to farm equipment and other assets although gender specific information is scarce. The criteria of property share attribution implicitly discriminate against women. Property shares are calculated according to workers' length of service and wage levels. Since women retire five years earlier than men and their average wages are lower than men's, the value of their shares is lower.

The second aspect of the inputs problem relates to the availability of good quality seeds, feed, fertilizers and pesticides. During the Soviet period, farmers obtained their seeds from state farms and did not have to make any choices between price and quality. Availability was not a problem. At the initial stage of rural reform, gaining access to these inputs was indeed problematic (Figure 14). However, as time progressed and land was distributed to more people, households started to learn how to obtain necessary inputs, and the problem has become one of lack of financing to purchase them. Usually people buy wheat from their local markets if they have the money to do it. Otherwise they use the wheat that was produced last year, resulting in low yields and low marketable produce.

Fertilizers and other inputs are also available in local markets, although at high prices. Regional specificity is important in gaining access to these inputs. For instance, there are reports of nitrate fertilizer being purchased from a factory in Georgia. Farmers closer to the Iranian border report receiving sugar beet seeds from Iranians, growing them, and giving a certain amount of the produce back to the seed suppliers. Cotton ginneries in the southwestern part of the country provide seeds and fertilizers on credit to farmers, although the real interest rate on these inputs is high and the offered price of cotton is low, resulting in low interest in cotton production. Nonetheless farmers continue to obtain smaller amounts of inputs from the ginneries, and some subsidize their wheat production through fertilizers obtained from the ginnery, effectively reducing their potential yields but hedging their family's subsistence.

---

[29]     The Gap Assessment in Moldova (1999) found service cooperatives that are being formed by farmers to be able to use their right to value share of the collectives from which they separated. The SA, however, did not come across such cooperatives in rural Azerbaijan. The distribution of assets and equipment in Azerbaijan is uneven and it seems that this unevenness, for the time being, prevents potential service providers from taking such initiatives. Nonetheless, the evolution of such initiatives should be closely monitored through the Project's M&E system; as such formations develop, they could serve as an efficient channel for supplying their members with appropriate machinery services through long-term credit or leasing agreements. The Project could also extend the scope of their activities by developing their capacities for input supply.

## Figure 14
## Access to Inputs at Initial Stages of Land Privatization

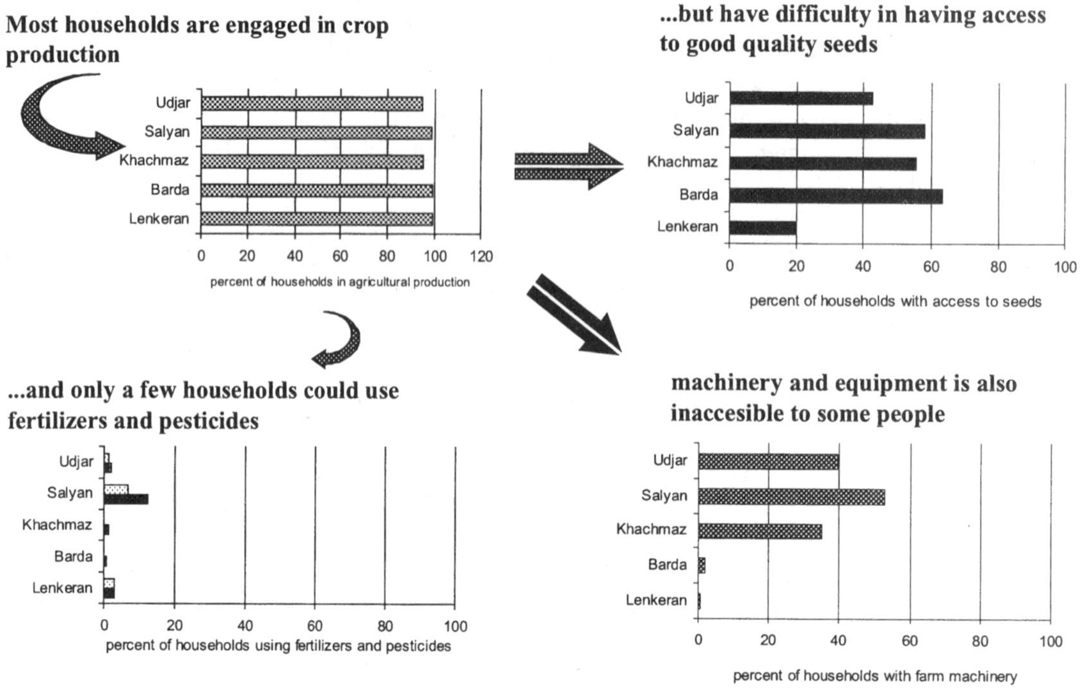

**Most households are engaged in crop production**

**...but have difficulty in having access to good quality seeds**

**...and only a few households could use fertilizers and pesticides**

**machinery and equipment is also inaccesible to some people**

Source: Social Assessment Household Survey, 1998.

### Rural Infrastructure and Social Services

The inadequate and unreliable electricity supply is a major problem in rural areas, disrupting livelihoods, access to information and the functioning of the already dilapidated irrigation systems. Most households have electricity for only a few hours a day (Figure 15). Other rural infrastructure, such as water and transport, is also in bad shape. Adequate transport facilities do not exist between rayons, contributing to marketing problems. In addition, irrigation systems are often too large for smallholders to use efficiently.

**Figure 15: Dissatisfaction with Electricity Supply**

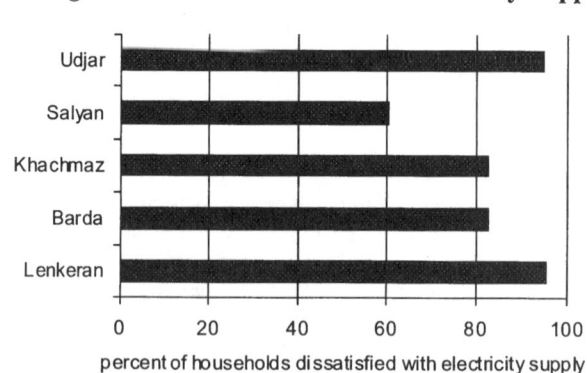

Source: Social Assessment Household Survey, 1998.

Currently, most social services such as health care, education, and kindergartens do not function properly in rural Azerbaijan; people usually make under-the-counter payments to receive these services. The state of social services imposes a burden on women and the elderly in particular. Lack of kindergartens increases the workload of women at home; their child care responsibilities distract them from engaging in more remunerative activities (Figure 16). The household survey found that about half of the interviewed households have kindergartens in their villages. Health centers existed in these villages, but the majority of residents want to see improvements in their services.

**Figure 16: Deterioration of Social Assets**

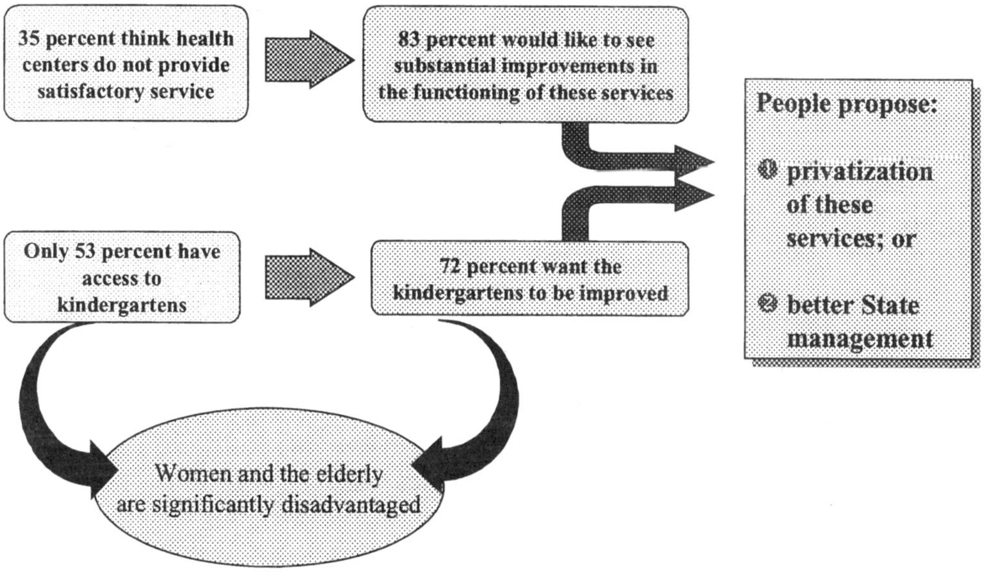

35 percent think health centers do not provide satisfactory service → 83 percent would like to see substantial improvements in the functioning of these services

Only 53 percent have access to kindergartens → 72 percent want the kindergartens to be improved

People propose:
① privatization of these services; or
② better State management

Women and the elderly are significantly disadvantaged

Source: Social Assessment Household Survey, 1998.

Most rural infrastructure problems are outside of the scope of the Agricultural Development and Credit Project. Nonetheless, they are important factors contributing to the potential impacts of the Project. However, irrigation issues influence the agricultural output and there is a high level of dissatisfaction with the functioning of the irrigation system (Figure 17). However, these concerns are being addressed through another World Bank project on irrigation.

**Figure 17**
**Households Dissatisfied with the Irrigation System**

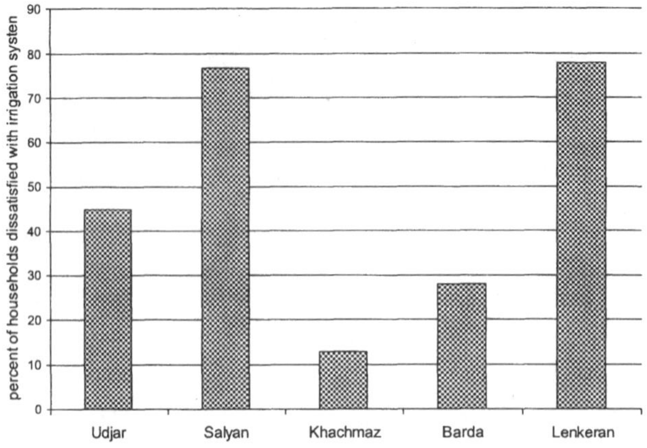

Source: Social Assessment Household Survey, 1998.

During the Soviet period, approximately 1.45 million hectares in rural Azerbaijan were developed for irrigation—about 80 percent of all cultivated areas—with little or no attention to economic costs and viability. Major irrigation sector issues are:

- Deterioration of the irrigation infrastructure, caused by insufficient maintenance;
- High reliance on pumped irrigation—over 500,000 ha—which in many instances may prove uneconomic as many pumps are not maintained properly;
- Negligible user contribution to the operation and management budget;
- Inefficient water distribution and application due to deteriorated inter-farm and intra-farm irrigation facilities—the latter is also less suitable to the new farm structures that are emerging from privatization. In addition, farmers lack experience and remain largely unorganized; and
- An increase in areas affected by soil salinity and water logging caused by inadequate or poorly performing drainage systems and inefficient use of water.

The proposed World Bank project in irrigation would contribute to the rehabilitation and completion of major irrigation systems to bring critical infrastructure back to fully operational standards. It would provide support to improve the collection of water charges and provide capacity building of the State Committee for Irrigation in project planning, design and implementation according to international standards.[30]

Life in the rural areas is a complicated set of interrelationships and factors which change over time, particularly with regard to rural infrastructure and social services. There may be a difference of perspective between what the people believe is necessary to improve their lives and what the Government and the donors are prepared to fund as priority. For example, the Agricultural Development and Credit Project will be complemented by an irrigation rehabilitation effort later. It represents the common view of the donors that assistance programs are most successful if they are selective—choosing a small number of interventions—if they are simple,

---

[30] Azerbaijan Irrigation II Project. Project Concept Document. August 1998.

and if they focus on the legal and credit framework rather than on broader areas of rural infrastructure and social services.

Social assessments in Azerbaijan and other areas have shown, however, that people in rural areas support a more comprehensive program of support in a larger number of sectors. The advantage of the Adaptable Program Loan (APL), which is the instrument the Bank will use for the Agricultural Development and Credit Project, is that project strategies, components and implementation mechanisms can change over time with the lessons of actual experience. It may be, for example, that access to markets is a more serious bottleneck than presently conceived or that people will not remain in rural areas in Azerbaijan if the present system of social sector support is not restored to what it was previously. The permanent or continuous SA process makes it possible to adjust the Project in midstream in light of findings. If the findings show that a more comprehensive program is more appropriate, then adjustments can be made. If this is a point of contention, it can be resolved through the SA process.

### Marketing Bottlenecks

Poor access to marketplaces, insufficient and expensive transport, roadside corruption, lack of storage facilities, low market prices, poor credit availability, and lack of proper information hinder the marketing of many rural products. Villages closer to rayon centers and rayons closer to Baku, therefore, have an advantage in marketing opportunities. Farm outputs are no longer marketed through the state farms but by individual farmers. The restructuring or closure of agribusinesses removed key outlets for the marketing of farm produce. Most households have access to local markets, but many of these markets do not function well. Most markets are along roads, where roadside corruption disrupts trade. Access to larger markets that function better is blocked by the high cost of transportation, roadside corruption and extortion in these markets. The institutions that are required to replace the planning ministries of the Soviet period have not yet fully materialized. These include physical marketplaces as well as the legal and regulatory system to operate and enforce transactions.

Small farmers have a number of marketing problems. Production associations, as they exist, do not necessarily cooperate in the marketing of their produce. Even though the members usually work together at the harvest, each member takes his share from the produce and either tries to sell it or keeps it for household use. Farmers who initially were willing to market produce cooperatively soon dropped the idea because it was too cumbersome and everybody wanted to use the produce in a different way.

So far, much of the effort in agricultural privatization has focused on farms and agribusinesses and little attention has been paid to the myriad of other players in a market-based agricultural system. Other important elements of a market—transport, storage and processing— are either still in the hands of monopolist enterprises or inadequate. Consequently, the fate of both large and small farms, whether restructured or retrograde, is to a great extent determined by institutions that also must adapt to a market economy.

Figure 18 illustrates the regional variation in marketing bottlenecks. In Lenkeran residents complain about inadequacy of transport in marketing their products while in Salyan the lack of proper marketplaces is the most important constraint. Small holders' marketing problems partially relate to the lack of information and contacts in larger markets. In addition, lack of financing for small holders prevents them from transporting their produce to larger markets at affordable prices, effectively reducing potential income levels. Producers overall also complain that their current production cannot compete with the better quality packaged imports. Qualitative assessments showed that large quantities of the same fruits and/or vegetables appear in the market at the same time, and there is a lack of proper storage space and processing facilities. This results in low prices in the domestic market.

**Figure 18: Marketing Bottlenecks**

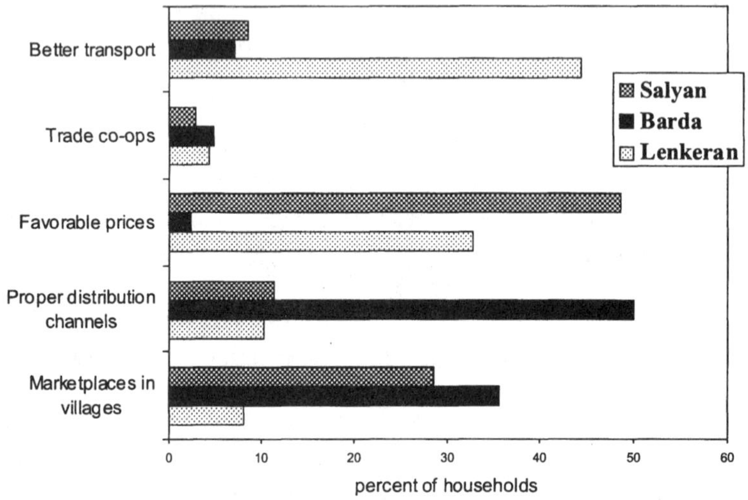

Source: Social Assessment Household Survey, 1998.

The economic crisis in Russia during the summer of 1998 has halted the border trade in the northern regions. Baku traders have access to large transport; they buy produce from villages at low prices and sell it in Baku markets. Farmers do not have access to large trucks and they cannot enter the Baku markets because of high penetration costs and extortion. In addition, roadside corruption increases the cost of transport significantly for small farmers who use their cars for transportation, and it is difficult for them to compete with traders. However, the traders' capacity is also limited. As a result, most produce spoils on the ground without harvesting. Table 5 illustrates how informal payments on the road, coupled with higher transport costs, make it difficult for smallholders in Qusar rayon in the northwest part of the country to market their produce independently.

Table 5: Basic Cost Calculation for Producers

| Items | Small farmer (personal car, can take 500 kg of produce) | Trader (truck, can take 2 tons of produce) |
|---|---|---|
| Sales price of 1 kg of apple in Qusar | 250 manats | 250 manats |
| Transport per kg | 1,000 manats | 350 manats |
| Roadside bribes per kg | 250 manats | 250 manats |
| Total cost | 1,500 manats | 850 manats |

Traders from Baku and other urban areas go into rayons periodically to buy produce from villagers. Sometimes people from neighboring villages also engage in local trade. People with their own cars sometimes work as traders, although it is not very profitable. Barter is commonplace. Traders usually do not work on commissions from other buyers; they facilitate sales themselves. According to farmers, these traders and people participating in Baku markets protect their places in these markets jealously and exclude others from participating in these markets. There is strong protection against outsiders and it is not easy for a farmer to penetrate into these areas. However, farmers' business with traders is not regular or dependable. Usually, due to their lack of access to vehicles and lack of cash, farmers cannot make much money from sales of their products. There are also rayon-level differences in access to traders' services.

Farmers and traders are willing to engage in exports but the current rules and regulations as well as corruption largely prohibit exports and exclude those who cannot deal with those costs. First, taxation and customs are seen as the most important obstacles to international trade. People claim that informal charge for crossing the border into Russia is about $50 while the driver pays an additional $100 to cross Dagistan region to go into Russia. Big trucks pay about $300 to cross the border into Russia when they are empty; depending on their cargo, they pay more when they are loaded. If these payments are not made, the vehicles are kept waiting for days and sometimes for weeks at the custom stops. The official border-crossing fee, traders and farmers claim, is $15. Second, police harassment and corruption is an important deterrent. Transport problems constitute the third priority obstacle to exports. Currently, most international traders are either Russians or Turks and villagers are more willing to do business with these traders than local traders.

The results of the surveys and the qualitative investigations in the social assessment and stakeholder participation process suggest that access to markets remains a major obstacle to agricultural improvement. The Agricultural Development and Credit Project has chosen to give priority to other aspects of agricultural reform, but on the basis of current evidence, careful, systematic surveys of market access need to be done at local levels to provide data rapidly. Measures to improve access may have to be added quickly to Government's reform program which the Bank and other donors are supporting.

### *Monitoring and Evaluation*

The Azerbaijan Agricultural Development and Credit Project is designed to be an Adaptable Lending Program which will span 10 years in its implementation and require the achievement of pre-determined benchmarks (triggers) for succession into future phases. It is, therefore, crucial to build into the project participatory monitoring and feedback mechanisms to shape the implementation of the subsequent components of the project as well as to ensure their sustainability. These indicators should focus mainly on the expected development impact of the Project, and less on the progress of implementation. The objective is to link project goals with expected outcomes using the minimum number of indicators, which provides a reliable estimate of efficacy and efficiency, Project impact and sustainability. The social assessment indicates that proximity of regions to Baku and other regional characteristics are crucial determinants of household income levels and poverty. Therefore, the monitoring and evaluation process will track the indicators with a regional and community focus as well as by gender, age and other social characteristics of the beneficiaries.

Performance indicators can be classified as Project implementation indicators and Project impact indicators. Indicators specifically link a project's inputs and activities with quantified measures of expected outputs and impacts. Project implementation indicators can be classified as input, process and output indicators. These measure the extent to which various physical and institutional dimensions of the project are implemented as envisaged. To that end, the indicators are multidimensional and cover such aspects as procurement, disbursement, appointment of key staff, management performance, adherence to covenanted actions, availability of local funds, and technical assistance progress. Such indicators may point out that remedial action may be required to get implementation back on track.

The healthy utilization of these indicators depends on having reliable baseline data at the beginning of Project implementation. Therefore, the first responsibility of an M&E system should be the establishment of the baseline for future monitoring of the Project. Once the baseline is established vis-à-vis the pre-defined monitoring indicators, it will be possible to track quantitative and behavioral changes as a result of the Project and make revisions in the implementation as well as indicators themselves as it is required.

**Project input indicators** show what the Project intends to accomplish in terms of physical improvements and measures the quantity (and sometimes the quality) of resources provided for project activities. They monitor the project-specific resources to be provided for each project item. These measures have the advantages of being readily measurable and directly under the discretion of implementing bodies, but they only indirectly reflect the Project's success in achieving its development objectives. The input indicators include, but are not limited to, funding—counterpart funds, Bank loan funds, cofinancing and grants. In addition to funding, they aim to keep track of provision of physical goods and services during implementation. These inputs will develop over time through various processes (which will be tracked through process indicators) and produce physical outputs. They will also have a development impact on the target population living in the project areas. The input indicators may include: the procurement of equipment for various project components, recurrent costs, training activities, and human resources—number of person-years for members of the implementation unit, consultants, and technical advisers.

**Project process indicators** measure changes that reflect either the achievement of particular dimensions of Project objectives, or of changes that are expected to lead to attainment of Project objectives. They are one step removed from the inputs that the Project provides, and one step closer to actual Project objectives. In project work, it may be convenient to divide the process indicators into financial, physical and institutional indicators, because they measure different aspects of project implementation. Monitoring project process indicators is a task for implementing agencies and supervision missions. In some cases, input and process indicators overlap, depending on what is being measured. For example, training extension staff is a process. The important output is the number of agents trained to assist farmers in agricultural production. The process indicators aim not only to measure how inputs are used but also to flag problem areas and issues which may come up during project implementation. For instance, cost effectiveness of project interventions is an important process indicator. So is the level of participation of the rural people in the implementation of the project components. It is also important that these indicators focus on trends emerging in the agriculture sector and provide timely feedback to the project implementation.

**Project output indicators** measure the output for each component in comparison to the project goals. These indicators aim at keeping track of the physical outputs produced by the Project by comparing them to the reliable baseline data obtained during the initial stages of project implementation. Output indicators will reflect the agricultural reform achievements in physical terms and also usually are readily measurable by the implementing bodies. They are normally complemented by measurements of longer-term impact, so that evaluation can take place. These will have to be monitored by region, settlement type and gender, where applicable. According to types of projects, they can include such factors as improved farming practices, reduced incidence of diseases and others. Collection of these indicators may depend on the existence and quality of national census or survey systems.

**Project impact indicators** illustrate the development impact of the Project interventions. Impact indicators should be linked to the overall objectives of the project. They measure the extent to which the project is currently expected to achieve the longer-term development impact that was expected at the appraisal and how it is affecting the various stakeholder groups involved in the process. Such measures may suggest remedial actions to safeguard or enhance the expected development benefits of the project. They are, therefore, of particular importance in guiding the implementation of the reform efforts in the medium term and in the identification of responses to the results of the monitoring efforts. A project's impacts may not manifest themselves as directly measurable numerical information but rather may show them in terms of the attitudes of beneficiaries. A purely narrative description of these effects, however, may be insufficient to measure results. It may, therefore, be necessary to devise ways of measuring impacts in quantitative terms, converting qualitative descriptions into quantitative information.

**Special studies**: Sometimes a project's routine monitoring and evaluation data do not provide sufficient information. If an unexpected problem arises, additional in-depth analysis through special studies can guide the way toward solving it—and avoiding it in the future. Special studies are formative evaluations of the fundamentals of problems and their origins, and in that way differ from monitoring indicators. For instance, project managers might need to learn more about the causal links among project outputs, outcomes, and impacts, especially when indicators reveal that the broader purposes of a project are not being achieved even though its planned outputs are being delivered. Special studies often provide important feedback for project redesign and higher-level policy debates.

An essential part of monitoring and evaluation is social impact monitoring (SIM) which aims at understanding the Project's impact on behavior to provide guidance for subsequent phases of the Project. SIM will build on the foundations laid by the social assessment process during project preparation by establishing a mechanism to periodically seek feedback from the affected populations which can be used to reorient the Project to ensure maximum, relevant benefits. The SIM process would have three broad categories:

- **Basic attitude and beneficiary surveys.** The main issue for these surveys and fieldwork is to understand the attitudes and expectations of key stakeholders with respect to the agricultural reform program and its implementation. The baseline survey and periodical updates will provide the necessary feedback to project implementation and will flag potential problems, especially at the village level, before they become serious obstacles. These surveys can be undertaken by consultants/NGOs.

- **Topical social assessments.** Throughout the life of the project, new issues emerge which were not considered at the beginning. In addition, there are a number of disadvantaged groups living in rural areas, in particular IDPs, which may not have an equal opportunity to achieve the full benefits from the Project. Two possible topics include the evolution of voluntary production groups and the changing role of women in the rural economy. These studies can be undertaken by consultants/NGOs.

- **System-wide and macro development impacts.** These Project impacts—caused by the energy sector, urbanization or changes in the importance of remittances for the rural sector—require a longer time to become visible; studies are usually carried out by national/international collaboration since they often require international comparisons and data gathering and analysis.

SIM will use a number of indicators to track the social impact of agricultural reform in the project areas. The methodology of SIM will include research into official data, household surveys and community discussions. The results of SIM will have a regional focus and be available by gender, age, and type of farm owned or worked by the household—individual, extended family, association, private company or joint stock. The following indicators need to be tracked by the baseline survey and subsequent activities:

- Household income sources, expenditures, and involvement in agricultural activity;
- Changes in the level of entrepreneurial activity, diversity and productivity of agricultural activity;
- Level of meeting demand for all agricultural land transactions in a transparent and timely manner;
- Formalization of land ownership arrangements and preservation of rights during registration and land transactions, especially for rural women;
- Trends in land consolidation, voluntary group formation, land trade, and obstacles to trade;
- Level of awareness of broad and basic subjects on agricultural reform among the rural population;

- Access to adequate and timely information on land rights and entitlements;
- Adoption rates, rationales for adoption and non-adoption in research and extension;
- Outreach of the mass media information campaign in rural areas;
- Satisfaction with advisory services and reasons;
- Suggestions for improvements in advisory services;
- Portion of farmers benefiting from veterinary services and their satisfaction levels;
- Access to credit through credit cooperatives/borrowers' groups for households and individuals;
- Obstacles for access to credit;
- Current uses of credit and emerging practices; and
- Barriers in input and output markets—number and quality of suppliers, access to supplies, pricing constraints, marketing constraints, and informal payments.

The results of M&E will be evaluated in a participatory manner at several levels. At the first level, the results emerging from the monitoring effort will be discussed between the Project Management Unit in Baku and The World Bank and feedback will be obtained. At the second level, these results will be discussed with the relevant stakeholders such as local administrations, municipalities, credit unions, farm association representatives, NGOs and farmers themselves in stakeholder workshops and seminars. Indeed, joint evaluation of the results is expected to be an important step in the direction of creating local level capacity for Project monitoring and feedback of these results into project implementation.

Equally importantly, guidance and advice from an external Panel of Experts made up of a local group of advisors will be sought. It is suggested that the external Panel will consist of five to seven local experts specialized in different aspects of agricultural activities and social sciences. They will be reviewing the results emerging from the monitoring efforts and will attempt to make policy recommendations based on these results. The Ministry of Agriculture (MOA) and the Bank will decide jointly on the composition of these panels during implementation.

## SA Implications for the Project

In preparing the Agricultural Development and Credit Project, the Government of Azerbaijan and the World Bank relied on the process of social assessment and stakeholder participation as one of the inputs defining the Project's components and its mechanisms of assistance. In some areas the reliance was more than in others. On the whole, the Project reflects the peoples' priorities and concerns. In some instances, both the Government and the World Bank modified their policies, as in the area of cost recovery for extension services. In other areas, the peoples' concerns could not be dealt with directly, as in the case of removing bottlenecks to market access. The Project includes a process of permanent social assessment and stakeholder consultation so that peoples' behavior and wishes could influence implementation and lead to the restructuring of the Project where appropriate. In the final event what is important is not that the Project is implemented as designed but that the Project's development outcomes are realized and it has a positive impact on people's lives. The social impact assessment will contribute to this outcomes through a permanent process of learning, adaptation and renewal.

The SA analyzed rural institutional to ensure that the needs of the rural poor and the small holders could be met. During the transition period, institutions such as state and collective farms have been losing their effectiveness in the rural areas while new ones are emerge, albeit at a more informal level. The Project, by its design, is responsive to the needs of these emerging institutions and their effects in the reform process in the rural areas. While the socialist economy has lost its hold in the countryside, a transition to a market economy still awaits the establishment of functioning markets. Until then, and for the institution of markets, support of the Government will still be required. High expectations for cost recovery, whether they may related to the irrigation systems, advisory services, etc., may not be feasible in the short term. This is especially true since the past five years have not given an indication of reduced poverty within the country and, especially, within the rural sector.

All the Project components are designed to invite strong participation from the beneficiaries during implementation. The Real Estate Registration component offers assistance to the Government to strengthen its capacity in real estate registration. However, its functioning and people's satisfaction with the system will be closely monitored and changes in implementation will be made, especially with regard to the implementation of a unified real estate registration system. The Information and Advisory Services component has a strong participatory approach in terms of its inclusion of local people as extension advisors and contractors, its built-in mechanisms to obtain feedback from the population on relevance and importance of courses offered, and the mechanisms to change the content and/or delivery of these services as necessary. The Rural Finance component is based mainly on the SA findings that there are entrepreneurs in rural areas who would be willing to pool their resources under the guidance and support of an experienced financial institution to overcome financial bottlenecks. The cooperatives formed under the component would be fully self-managing. The small borrowers' groups will be solely based on the existing social capital in rural areas to provide support to small entrepreneur groups. The implementation of the Rural Finance component will be done together with NGOs to provide a outreach to the poorer parts of the population and has flexibility to obtain real-time feedback from its clients and can make changes in its operations within the proposed framework. The Agricultural Strategy Unit is the Project component with least built-in participatory capacity, although the implementation proceeds of other components will undoubtedly affect the proposals and policy alternatives emerging from this Unit.

Project implementation involves a highly participatory monitoring and evaluation system designed to ensure participation of the rural people in Project implementation through a series of surveys and stakeholder consultations. The evaluation of the results emerging from the monitoring effort will be evaluated by a group of Azeri experts who will then feed their views into the subsequent stages of Project implementation and design of the subsequent phases of the Program.

# Chapter Five

# Republic of Kazakhstan Agricultural Post-Privatization Assistance Project: Farm Privatization and Restructuring in Akmola and Former Taldy-Korgan Oblasts

### Stan Peabody, Leonid Gurevich and Dinesh Aryal

## Background[1]

### *Privatization and Restructuring in the Agriculture Sector*

The Republic of Kazakhstan requested funding from the World Bank to undertake a Post-Privatization Assistance Project which seeks to support the development and commercialization of privatized farms and agro-enterprises in two agricultural areas of the country and to improve rural productivity and incomes. The changes required at the enterprise level to increase productivity and achieve and sustain commercial viability may involve: splitting-up existing enterprises; changing ownership, management, corporate status, and accounting practices; new investment; and changes in product mix, input use and production practices.

The Government attaches high priority to the reform of agriculture which accounts for 25 percent of the labor force, 12 percent of GDP and significant export revenues. As a result of reforms, the farming community in Kazakhstan has undergone dramatic changes since 1991 when the sector consisted of about 2500 state and collective farms, and 1997. The first peasant farms were formed from sovkhozes and kolkhozes. The official privatization process affecting the proposed project sites, Akmola and former Taldy-Korgan Oblasts, started in 1992. State and collective farms were turned into private enterprises, primarily "joint stock companies" and "collective enterprises," with ownership transferred to members of the former sovkhoz and kolkhoz and the district administration. The ratio of member/administration ownership varied according to the type of new organization. Administrative structures remained unchanged on the district and oblast levels, however, and the privatization was largely formalistic. In 1994, the new agricultural enterprises began to adopt different types of legally recognized structures. In 1996, Kazakhstan adopted a new Civil Code and all economic enterprises were required to conform to one of the legal forms recognized by the code, ranging from joint stock companies to producers cooperatives and various partnerships. "Collective enterprise" and "private enterprise" were not recognized, thus most of the new agricultural enterprises had to restructure and re-register. Practically all of the recognized legal entities are present in both oblasts.

[1] This work was undertaken by the Kazakhstan Social Science Network, led by Dr. Leonid Gurevich of Bilesim International Kazakhstan. Members of the Kazakhstan Academy of Sciences and institutions in Akmola and Taldy-Korgan also contributed to the work. Ilya Lipkovich, University of Delaware, prepared a draft survey instrument. Leonid Gurevich prepared the final survey instruments and the sampling frame, managed the field studies and analyzed the data. Dinesh Aryal conducted additional data analysis with Can Adamoglu. From initial design to final report, the social assessment was the responsibility of Stan Peabody, a member of the Social Development Team in the Environmentally and Socially Sustainable Development Sector Unit in the Europe and Central Asia Region (ECSSD). Roy Southworth is Task Team Leader in ECSSD and Laura Tuck is Sector Manager.

The privatization and restructuring of the agriculture sector had direct consequences on the people, particularly the rural population. A social assessment (SA) was undertaken to understand the impact of privatization on the rural economy and to formulate mitigation measures to eliminate or minimize hardship on the rural population. It concentrated primarily on issues related to farm privatization and restructuring. It was designed to build on the SA undertaken in 1995 for the Irrigation and Drainage Improvement Project (IDIP)[2], which also focused on privatization and restructuring. The earlier SA looked at the management status of about 30 state and collective farms in 10 oblasts, which were candidates for inclusion in the project. About half of the farms were privatized at the time of the SA; others were scheduled to be privatized by the end of 1995. The SA involved extensive semi-structured interviews with managers of the 30 farms and a wide range of officials in the oblasts, a household survey on a sub-set of ten farms, focus groups on five, semi-structured interviews with a wide range of officials and an institutional analysis of the water sector.

The IDIP SA revealed great variation in the management structure of farms that were privatized or about to be privatized. Except in South Kazakhstan Oblast, very little restructuring had taken place. It also found a very low level of knowledge about the privatization process and options; high levels of uncertainty and apprehension regarding family economic status; and evidence of the concentration of land and property shares in the hands of farm managers. The last essentially disenfranchised farm workers of their shares in the new farm enterprises. At the same time, however, it also appeared that the role of farm manager might be in the process of changing in response to new ownership patterns, with increased accountability to farm members. Based on the results of the SA, the project adopted a five-step process which required major decisions affecting the design, cost and repayment of project investments to be made by a general assembly of farm shareholders.[3]

## *Project Area*[4]

The social assessment of the Post-Privatization Assistance Project was carried out in Akmola and former Taldy-Korgan Oblasts in March and April 1997. After field work for the SA was completed, oblast boundaries were changed and the national capital shifted from Almaty city to Akmola city which was subsequently renamed Astana. The oblast names were not changed, however. The two oblasts, Akmola and Almaty, have very different climates, agriculture and

---

[2]     The World Bank. 1996. "Kazakhstan: Irrigation and Drainage Improvement Project." The World Bank, Environment Division, ECA/MNA Regional Technical Department (EMTEN). Washington, D.C.; and The World Bank. 1996. "Republic of Kazakhstan: Irrigation and Drainage Improvement Project." Staff Appraisal Report No. 15379-KZ, Europe and Central Asia Country Department III, Agriculture, Industry and Finance Division, Washington, D.C.

[3]     The World Bank. 1996. " Republic of Kazakhstan: Irrigation and Drainage Improvement Project." Staff Appraisal Report No. 15379-KZ, Europe and Central Asia Country Department III, Agriculture, Industry and Finance Division, Washington, D.C.

[4]     The World Bank. 1996. "Republic of Kazakhstan: Agricultural Post-Privatization Assistance Project." Staff Appraisal Report No. 17789-KZ, Europe and Central Asia Country Department III, Agriculture, Industry and Finance Division, Washington, D.C.

population. They share one characteristic, however, the capital city of the oblast also served as national capital.

Akmola Oblast is located just west of the center of Kazakhstan in the northern third of the country on the open steppes. Almaty Oblast is located in the southeast corner of the country, north of the Zailiyskii Alatau in the Tien-Shen Mountains, which form the border between Kazakhstan and the Kyrgyz Republic. The boundaries of both oblasts are new; in April, 1997, Almaty Oblast, which excludes the city of Almaty, absorbed most of Taldy-Korgan Oblast, located to the north and east; and Akmola Oblast was expanded to include most of Kokshetau Oblast. Akmola Oblast now covers 224,200 square kilometers and has an estimated population of 880,500. Almaty covers 121,700 square kilometers and has an estimated population of 1,632,900. Almaty Oblast is predominately rural (70%) and Akmola Oblast is predominantly urban (45% rural). Despite considerable emigration of Germans, Russians and Ukrainians since independence, Kazakhs are still a minority in Akmola (31%); in Almaty Oblast, however, Kazakhs account for about 57% of the population.

## Akmola Oblast

Akmola Oblast is one of five oblasts in Central Kazakhstan with a combined population of 4.4 million people. The climate is continental, with warm summers (19-20° C) and cold winters (-17-19° C). Annual precipitation is 300 mm overall, with areas varying from 200 to 400 mm per year. Rainfall is most prevalent during spring and fall. Winters bring little snow (20-30 cm), many sunny days and strong winds. Summers are dry with hot, arid winds.

The steppe area is a natural grassland with soils that vary from fertile black chernozem soils in the north to brown forest and lighter soils in the southern part of the zone. Historically, this area was the home of sheep herders who moved about seasonally, with very little sedentary agriculture. In the 1950's, however, the land was opened for grain cultivation through the famous "Virgin Lands" program attributed to the leadership of Nikita Kruschev. Huge sovkhozes (state farms) and kolkhozes (collective farms) were built from scratch by immigrants from Russia, Ukraine and other parts of the Soviet Union. The program also drew in other local residents, including former gulag prisoners, and the sizable population of Germans who had been exiled to the area during World War II.

The principal crop in Akmola is rain-fed wheat, primarily spring wheat, which was left fallow every five years. Other crops are fodder (lucerne and grasses), maize, sunflower, and millet. During the Soviet period, this area also had a sizable animal population, particularly around Akmola, with dairy and meat production, as well as wool. Akmola was also a major center for manufacturing agricultural machinery and there were some food processing plants and light industries. Gold, coal, tin, and quartz sand were also mined within the boundaries of the expanded Akmola oblast.

Akmola is famous for the huge state and collective farms that literally transformed the steppes from grassland pastures to highly mechanized farms. The farms were as large as 300,000 ha, including crop lands and extensive pastures. Farm and community were one and the same. One farm could have from a few hundred to a few thousand farm workers along with their families. In addition, there were other community members which included medical and education professionals, mechanics, other skilled workers and pensioners. The farm would consist of a major settlement and a number of smaller satellite settlements, entirely managed by the farm administration. These huge grain factories provided wheat and animal products to much of the Soviet Union. Rural families enjoyed a standard of living similar to that of urban people, all elements of which were directly financed and managed by the farm with support from the State budget.

The collapse of the Soviet Union changed rural life. Since independence, agricultural production decreased dramatically in Akmola, as in all of Kazakhstan. The animal population was decimated as animals were sold by the farms and distributed to farm workers. The fallow period was often overlooked and little or no fertilizer was used, causing soil fertility and soil moisture to decline. The lack of spare parts and the high cost of available parts and lubricants resulted in poor maintenance of farm equipment and the cannibalization of much of the capital stock to keep some machines moving. Grain price controls were lifted in 1996 and prices increased to near world market prices, but electricity and fuel prices were also raised, giving farms financial difficulties and debt.

Similarly, the standard of living in the countryside decreased significantly. Wages continued to be unpaid, paid in kind, or paid after months and years of nonpayment. Some farm workers were reduced to half-time employment or put on leave, receiving basic food rations and fodder for payment. Increasingly, farm families turned to their home garden plots for subsistence and income. While some farms continued to fund social infrastructure (school and health care) their numbers were declining and local governments were still unable to assume responsibility.

Although the government's privatization program changed the legal status of most farms, many large units remained intact in Akmola. Data regarding the type and size of farms was inconsistent from one source to another, thus it must be taken as indicative, rather than absolute. Table 5.1 illustrates the trends at the time of the SA in farm management structures, as well as the average size of different types of units, both total land area and arable land. The data indicated (Table 5.1) that the number of partnerships and independent farms grew markedly, especially in the last year before the SA, and state and collective farms had almost disappeared. Farm size was large, with the exception of independent farms.

Table 5.1 Farm types and average size, Akmola Oblast

|  | Number | | | Average size (ha.) | |
|---|---|---|---|---|---|
|  | Jan-93 | Jan-97 | Jul-97 | Total | Arable |
| State farms | 256 | 17 | 13 | 73,000 | 4,691 |
| Collective farms | 33 | 1 | 0 | 34,690 | 15,285 |
| Production cooperatives | 11 | 173 | 144 | 43,841 | 21,735 |
| Partnerships | 0 | 83 | 169 | 36,938* | 15,123* |
| Joint stock companies | 0 | 32 | 43 |  |  |
| Independent farms | 1,177 | 2,670 | 3,629 | 327 | 103 |

\* Includes both joint stock companies and limited partnerships

Source: Goskomstat, 1998.

### Almaty Oblast

Compared to Akmola Oblast, much of Almaty Oblast had more favorable conditions, better access to markets and more diversified markets. Nonetheless, it suffers from many of the problems that beset Akmola. Almaty, has a continental climate. Compared to Akmola Oblast, Almaty has milder winters (-7-11° C), hotter summers (22-24° C), and less wind. The growing season is also longer. Snowfall is limited in much of Almaty Oblast (20-25 cm), but it is higher in the foothills in the south and east. Precipitation in Almaty Oblast varies from an average of less than 200 mm in the west and north to 1,000 mm in some areas of the south and east. Rainfall is more reliable than in Akmola, as well. The soils in much of Almaty are good quality former woodlands.

Almaty Oblast has been the site of sedentary agriculture for a long time, but it is not one of the ancient settlements in the republic. Cropping is diversified and intensive, with irrigation in many parts, drawing from snowmelt, perennial rivers and groundwater. The terrain varies in altitude from the southern steppes in the northern reaches of the oblast to mountains in the south, providing varied microclimates for many types of field crops, fruits and vegetables. Sugar beet, wheat, maize and rice are prevalent crops in the area that was formerly Taldy-Korgan Oblast. Winter wheat, barley and fodder corps are grown on the non-irrigated foothills and southern parts of the former Almaty Oblast. Temperate fruits are also grown in the foothills and valleys; indeed, the hills around Almaty still contain wild apple stock from which apples were originally domesticated. Vegetable production is prominent near the urban areas, as well as milk and meat production. Sheep breeding is prevalent in the northern and western parts of the Oblast.

As cropping is diversified, landholding is also varied in Akmola Oblast. Almaty had fewer state farms than Akmola initially, but now has more (120 in 1993 and 46 in 1997) (Table 5.2).

Table 5.2 Farm types and average size, Almaty Oblast

|  | Number | | | Average size (ha.) | |
|---|---|---|---|---|---|
|  | Jan-93 | Jan-97 | Jul-97 | Total | Arable |
| State farms | 120 | 59 | 45 | 3,422 | 788 |
| Collective farms | 91 | 20 | 1 | 15,723 | 1,213 |
| Production cooperatives | 0 | 375 | 525 | 22,613 | 1,834 |
| Partnerships | 0 | 765 | 780 | 10,571 | 1362* |
| Joint stock companies | 0 | 55 | 55 |  |  |
| Independent farms | 1,330 | 8,498 | 11,178 | 122 | 18 |

* Includes both joint stock companies and limited partnerships

Source: Goskomstat, 1998.

Farmers and farm families in Almaty suffered most of the same consequences of the collapse of the Soviet Union as the rural population in Akmola. Their standard of living decreased and services disappeared or became expensive; wages were unpaid, delayed or paid in kind; and the principal source of subsistence and income was the family garden plot. In other respects, however, many residents of Almaty Oblast had some distinct advantages over their counterparts in Akmola. To begin with, communities were much more dense and inter-related. Residents had more diversified survival strategies. Many farm communities near the cities of Taldy-Korgan and Almaty were relatively large and comparatively few of the residents were entirely dependent on the farms for their livelihoods. Many people commuted to the city for work or were engaged in other sectors of a diversified economy. Thus it was not uncommon for family incomes to come from different sources, only one of which was farm employment. Similarly, survival of the community was less dependent on direct financial support or management by the farm. Another advantage was access to diversified markets and a milder climate which enabled people to grow and market high value crops, as well as manage more animals on their family plots. This meant that the prospects were better for individuals to break away from the large farms and obtain a reasonable income from their small plots. A further advantage was the mostly homogenous population of Almaty Oblast. Kazakhs are still the dominant ethnic group in the countryside, at all levels of the hierarchy, so there is a lower likelihood of developing explosive conflicts during difficult times. Finally, the varied, intensive farming on smaller units decreased the need for major capital investments in farm machinery, thus reducing start-up costs.

Field investigators attempted to compile a comprehensive demographic, social and economic data set for the communities included in the study, but the effort was not fully successful. Nonetheless, from the most complete data set, the following information can be drawn. As almost every data set contained some discrepancies, the numbers presented must be seen as indicative, not absolute (Table 5.3).

Table 5.3 Comparison of sample communities

| Average | Akmola | Taldy-Korgan |
|---|---|---|
| Farm size | 28,201 ha | 11,354 ha |
| Farm population | 1,238 | 2,877 |
| Farm membership | 397 | 554 |
| Population actually farming | 18% | 21% |
| Pensioners | 17% | 8% |
| Doctor ratio | 1 per 530 people | 1 per 392 people |
| Teacher ratio | 1 per 9 young people | 1 per 18 young people |
| Secondary education | 10% | 32% |
| Youth (under 16) | 32% | 38% |
| Autos | 1 per 33 people | 1 per 39 people |
| Autos not functioning | 4% | 30% |
| Tractors | 1 per 33 people | 1 per 50 people |
| Tractors not functioning | 3% | 33% |
| Commerce | 3 shops per farm | 5 shops per farm |

Source: Household Survey.

## *Project Implications*

These characterizations of the two oblasts have a number of implications for the project:

- Farm size in Akmola will inevitably remain large. In this context, a small family farm must consist of several hundred hectares in order to generate enough income to support a family, and relatively large units are required to generate the income needed to buy or rent machinery. It is too early to estimate minimum or maximum viable farm sizes in the Oblast, but it can be expected that many of the existing small farm units will ultimately aggregate to form larger ones.

- The impact of reducing farm employment is likely to be greater in Akmola Oblast than in Almaty, as people have fewer immediate opportunities for alternative employment and impossibly great distances to commute to cities or other farms. Consequently, although the project will try to promote new income generating activities during its second phase, emigration is likely to increase over time.

- Divestment of social infrastructure is likely to be a significant part of business plans in Akmola, with potentially devastating consequences. Schools are the most likely to suffer. Unless local administrations are prepared to act decisively and quickly to fund rural schools, Kazakhstan faces the prospect of developing parallel and unequal educational tracks for rural and urban children. The project should not contribute to this trend by ignoring it.

- In the context of the diversified, intensive and competitive agriculture in Almaty Oblast, the Advisory Centers will need access to information regarding markets inside and outside of Kazakhstan in order to develop successful business plans.

The great differences between the two oblasts create very distinct challenges for the project in terms of communication strategies, complexity of restructuring, social consequences of attaining commercial viability, range of options available, and the amount of capacity building required, to name a few. Success in both contexts will offer very important lessons that can be applied when extending the project to other oblasts.

## The Social Assessment

### *Social Assessment Objectives*

The objectives of the social assessment were to:

- Identify key social impacts of farm restructuring and privatization processes and to ascertain the extent to which the rural population of Akmola and former Taldy-Korgan Oblasts participate in these processes;
- Identify key institutional constraints to equitable participation of people in the reform process;
- Ensure that a proper framework is established for the project that enables different stakeholders to benefit from different initiatives; and
- Contribute to the design of a social monitoring and evaluation program to assess the development effectiveness of Bank assistance.

*Key Stakeholders*

Primary project beneficiaries include farm owners and managers, farm workers, independent small farmers, commercial banks and the network of independent service providers. These groups participated in the project design process in the following ways:

- Field visits to each category of primary beneficiary were undertaken to consult the beneficiaries on priorities for project action;
- Systematic surveys and qualitative consultations with households, farm managers, workers, heads of farm enterprises and independent farmers were undertaken; and
- Consultative meetings were held with commercial banks, involving an explanation of project objectives. These meetings determined priorities to be addressed through the financing component and contributed to project design and the arrangements for implementation of the component.

Other key stakeholders include national, oblast and raion level officials, input suppliers, banks, the World Bank and other agencies. They participated in the project design process through:

- The working group, comprising representatives from all relevant Ministries and Departments, which influenced project design and implementation arrangements;
- Field visits to oblast administrations, where officials participated in determining project design and implementation arrangements; and
- Visits to input suppliers to identify bottlenecks and priorities.

*Social Assessment Methodology[5]*

The following social research methods were carried out to analyze the social development, participation and institutional issues:

---

[5]     The survey was conducted in face-to-face indoor interviews. A draft questionnaire was prepared by the World Bank and significantly modified by BILESIM International Kazakhstan with the help of well-known agricultural economists. The survey instrument was pilot tested in Akmola and Taldy-Korgan Oblasts from March 4 to March 7, 1997. The survey was conducted from March 18 to March 31, 1997. Altogether, 538 farm workers and 62 independent farmers were surveyed.     For in-depth interviews, 30 heads of farms (Directors, Chairmen of Boards of joint stock companies, partnerships and production cooperatives, their Deputies, Chief Accountants and Senior Specialists) were selected.     The rest of the sample consisted of workers of other categories corresponding to the project purposes distributed, approximately evenly, in the districts and oblasts surveyed.

In each oblast, four focus group discussions were held with workers on the base farms, two with independent farmers, and one with heads of agricultural enterprises.   The program and the tools of conducting focus group discussions were developed on the basis of the tentative results of the household survey and in-depth interviews.   The data was processed and analyzed using two programs: ANKETA, for the questionnaire; and SPSS. The data was further analyzed in the World Bank. BILESIM International prepared four reports and a statistical annex: Household Survey, Focus Group, and Summary Report in English and in Russian. The Summary Report was also translated into Kazakh. This report draws on all of the reports mentioned above as well as the results of subsequent data analysis.

- **Household Socioeconomic Survey**: The survey was carried out with 600 households in 20 settlements located in seven districts, using a stratified random sample of independent farmers and farm workers.
- **In-Depth Interviews**: In each oblast, in-depth interviews were held with 30 farm managers and 20 others in the agricultural sector.
- **Focus Groups**: Seven focus-group discussions were held in each of the two oblasts in the areas covered by the household survey.

The objectives of the household socioeconomic survey were to collect and analyze information and to develop recommendations for specifying the most effective policies of World Bank assistance regarding the introduction of new economic relations into agriculture of the oblasts surveyed. Four districts in Akmola Oblast (Atbasarky, Astrakhansky, Vishnevsky and Tselinogradsky) and three districts in Taldy-Korgan Oblast (Karatalsky, Koksusky, and Sarkandsky) were selected for the household survey. The basic criteria for the selection of districts were diversity of farm management structures and specialization; and distance from the oblast center.

One area was selected in the Vishnevsky district and three populated areas were selected in each of the other sample districts of Akmola Oblast. Four populated areas were selected in the Karatalsky district of Taldy-Korgan Oblast and three populated areas were selected in the other two sample districts of the oblast. The selection criteria included legal issues, specialization and the results of economic activity on farms in the villages. Altogether, the selection included 20 populated areas. The number of interview locations in Taldy-Korgan Oblast was greater than in Akmola Oblast, however, for the following reasons: i) the amount of farm restructuring Sarkandsky district of Taldy-Korgan Oblast; and ii) the revision of oblasts borders and newly established farms.

Sample households were selected on a random basis. In each small farm household, people whose demographic and social-professional characteristics suited the survey purposes were selected to participate. Those surveyed were 18 years or older and directly engaged in production. Workers of the social sphere and managers were not included. To obtain more significant statistical data, a special quota of 10 percent of the general selection was specified for peasant farms, even though they represent only 4-6 percent of the rural population in the two oblasts.

Among those surveyed, 83 percent were men and 17 percent were women. The wide discrepancy between the two genders resulted from household composition, rather than the selection of individual households. Age and occupation were the demographic criteria used to select households. If more than one person in a household met the selection criteria, the one who answered was the head of the family, usually a man. In addition, the gender disparity in the sample reflected the dominance of male employment in the sector.

Fifty-nine percent of the surveyed families had a single breadwinner, a person with a constant, paid job. This reflected the high level of unemployment in villages in Kazakhstan. Thirty-one percent of the families had two working members, which was typical in Soviet times. The average family had 4.5 persons, which approximately corresponded to the latest state statistical survey data of the rural population in Akmola and former Taldy-Korgan Oblasts. Seventy-two percent of households were married couples with children while 6 percent of them were married couples without children. Four percent of the households were single parent families headed by women and one percent headed by men. Eighty-three percent of the heads of households were between 20 and 50 years of age, the most economically active age groups.

The ethnic composition of the households corresponded to the ethnic structure of the population of the surveyed oblasts (Figure 5. 1).

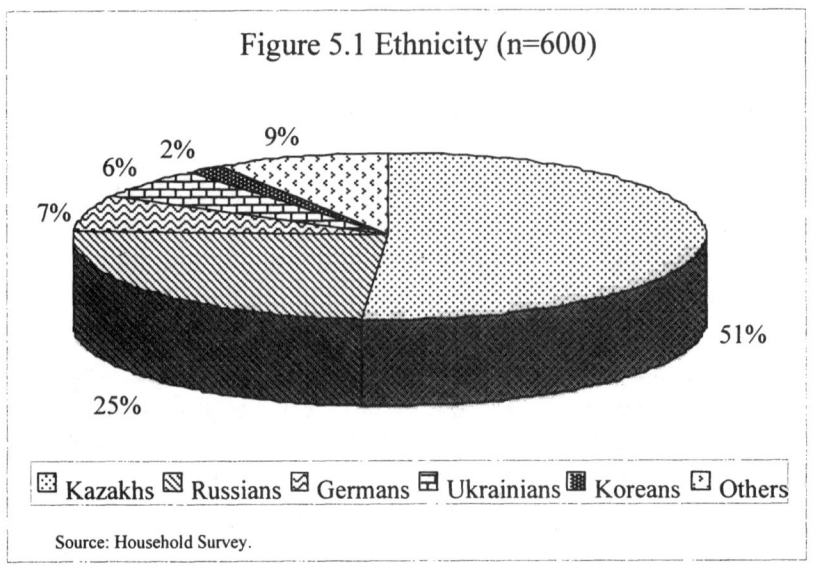

Figure 5.1 Ethnicity (n=600)

Kazakhs   Russians   Germans   Ukrainians   Koreans   Others

Source: Household Survey.

Samples from each oblast were representative in terms of ethnic parameters. In Taldy-Korgan Oblast, the households were 79 percent Kazakh, 14 percent Russian, 4 percent Korean and 3 percent German and Ukrainian. The people in Akmola Oblast were 35 percent Russian and 24 percent Kazakh. Despite intensive emigration, Germans still made up more than 12 percent of the rural population of Akmola Oblast and represented 13 percent of the sample. (In Soviet times, Akmola Oblast was one of Kazakhstan's main settlement areas of Germans and there was even a plan in the seventies to establish an autonomous German area there.) The remaining households were Ukrainian (11 percent) and Korean (0.3 percent) (Table 5.4).

Table 5.4 Ethnicity (%) (n=600)

|  | Akmola Oblast | Taldy-Korgan Oblast |
|---|---|---|
| Kazak | 24 | 79 |
| Russian | 35 | 14 |
| German | 13 | 1 |
| Ukranian | 11 | 1 |
| Korean | 0.3 | 4 |
| Other | 16 | 1 |

Source: Household Survey.

# The Four Pillars

The social assessment covered a wide range of issues and generated detailed data on socioeconomic status as well as experience and attitudes related to agricultural sector restructuring and farm privatization. It revealed a high level of dissatisfaction and discontent among people in the two oblasts as well as growing impoverishment. Except for farm managers and independent farmers, most farm workers were pessimistic about the future.

## *Key Social Development Issues*

### Privatization: Results and Prospects

Privatization and restructuring trends in Akmola and Taldy-Korgan Oblasts started in 1992. Since privatization and restructuring are a process and not a discrete action, it can be explained in a three-phase process.

> *"First there were state farms, then the names were changed and they became cooperatives. Then the cooperatives were dissolved. Now it is a private ownership."* Farm worker

Before perestroika all farming units were divided into state farms (sovkhozes) and collective farms (kolkhozes). While the difference between them was only nominal, both types were large, complex industrial farm communities made up of the households whose members worked on those units in various capacities. The director of the sovkhoz and the chairperson of the kolkhoz managed the farms. Many of the units were not profitable and regularly received government subsidies. During perestroika, the farming sector in Kazakhstan, which consisted of about 2500 state and collective farms, started to go through major changes. In 1991, the first peasant farms were formed by carving out parts of sovkhozes and kolkhozes, though otherwise farm management structure was unchanged. These independent farms, few in number, were generally created by technical staff of the sovkhoz or kolkhoz.

The official privatization process started in 1992 and ended in 1995. During this time, state and collective farms were transformed into "collective enterprises," and joint stock companies. This affected the proposed project sites, Akmola and former Taldy-Korgan Oblasts. The managerial structure remained intact, however. At this stage, privatization was largely a formality without much significance in day-to-day life. During 1994 and 1995, great numbers of these collective enterprises were transformed into private enterprises of various types. At this point, farm members were officially allocated land and property (capital stock) shares in the enterprise, related to their length of service.

The third phase started in 1996 after the Republic of Kazakhstan adopted a new Civil Code which required all economic enterprises to be reregistered in one of the forms recognized by the Code. Since the Civil Code did not recognize such formerly widespread entities as "collective enterprises" and "private enterprises," farms in those categories had to change their management structures to conform to the new code. Virtually all recognized forms of legal entities were found in each oblast. The most prevalent form was the production cooperative followed by joint-stock companies, which were somewhat less common, and partnerships, which were the least common. The results of the study did not allow assertions of whether one form was more or less effective than another.   The few advanced enterprises which successfully adapted to the new economic relations were found among partnerships, joint stock companies, and production and agricultural cooperatives.   Joint stock companies with predominately State participation demonstrated obvious inefficiencies.   For some time, State stockholdings were supposed to be sold, but buyers were not always found.

The stages and models of privatization and restructuring were clarified through in-depth interviews with the heads of the farms and representatives of the district and oblast administration. Analysis of the interviews indicated features specific to Akmola or Taldy-Korgan Oblasts as well as features common to both the oblasts, and unique characteristics of some farms. In Akmola Oblast, unlike in Taldy-Korgan, there was very little restructuring of the base farms and commodity partnerships were becoming more prevalent.   In Taldy-Korgan Oblast, especially in the Sarkandsky District, many small farms were formed with different legal status and the process of establishing peasant farms was also much more intensive.   Commonly, large, privately-owned farms with 100 or more workers were created on the basis of work teams (brigades) and branches of former state farms.

In Akmola Oblast there was essentially no partition of farms and farm areas were not reduced. Former Directors of state-farms, joint stock companies, and partnerships received a majority of shares and in a number of cases, former Directors leased shares for a period of five to 15 years.   There were also some unusual privatization arrangements.   For example, in 1994, the prosperous state-farm Rodina was transformed into a private enterprise with 51 percent of the property belonging to the Director and 49 percent to the workers.   Since the new Civil Code of 1996 did not recognize private enterprise, this farm was reregistered as an agricultural production cooperative with a majority of the shares belonging to the Director.   Karl Marx farm was turned into a joint stock company (avoiding the transformation into a collective agricultural enterprise) with 90 percent of the shares belonging to the State.   However, efforts to sell these shares were unsuccessful.

In Taldy-Korgan Oblast, interviews with heads of farms and oblast administration staff revealed serious conflicts between the farms, mainly over water.   Conflicts were especially strong in the rice growing area.   For example, in the Production Cooperative Pravda, base farms had to limit the areas for rice growing in order not to flood the fields of the peasant farms.   To resolve the resulting conflicts, the Agricultural Administration established special associations to regulate water allocation.   However, farmers had not yet recognized them. One of the unusual privatization arrangements was observed in the Aq-Yiq state farm which was turned into an open joint stock company.   Ten percent of the shares were given to farm workers, 20 percent were sold to members of the collective, and 70 percent remained State property.   In 1997, the State shares were sold by auction and now belong to private owners.

The SA revealed that the privatization process in Kazakhstan was developed and carried out without a broad social base and with little effort to inform rural people of their opportunities and responsibilities. Although 79 percent of the sample households responded that privatization was carried out on their farm, they were poorly informed about the economic

> *"At the time it was being decided which type of management structure we should have, there was an information vacuum; we were not informed or explained anything. For instance, I didn't know the difference between a joint stock company and other organizational forms. We even didn't see Charters of joint stock companies. Before March nobody knew how many shares are held by this or that person and even now we don't know this."*
> *Independent farmer*

and legal aspects of privatization and their rights and opportunities. For example, 39 percent of the households responded that they did not know what privatization meant. Twenty-nine percent of the households said that privatization options and implications were never discussed with them.

Changes in rural organization, as a result of privatization, are complex, prompted to some degree by the decrees and legislative acts of 1994. One decree gave a large share of the land of former state farms (up to 20 percent) to the heads of farms if they had 20 years of service. In some cases, this may have prompted the managers to acquire additional shares, but the extent to which the decree was implemented before it was revoked in 1995 is not well-documented.

> *"How the people let the director have their shares is another question. We signed empty papers. For several years the people live without electricity, wages. The only thing we get is bread from the bakery. They forced us to our knees. We were presented with an ultimatum: either you sign a paper or you'll not get bread, or you must pay in cash. We don't have cash. Of course, the people signed the papers." Independent farmer*

Of the 355 households who knew that they had land and property shares, more than 20 percent had transferred their land shares, and sometimes their property shares, to someone else (Figure 5.2a and 5.2b). In 75 percent of the cases, the transfer occurred free of charge. Formerly, such transfers were considered to be voluntary, but the survey revealed that compulsion, blackmail and fraud quite often were involved in the process. The most common ploy seems to have been for managers to promise to pay wages once farm members transferred their shares to them and threaten to fire farm members who did not comply.

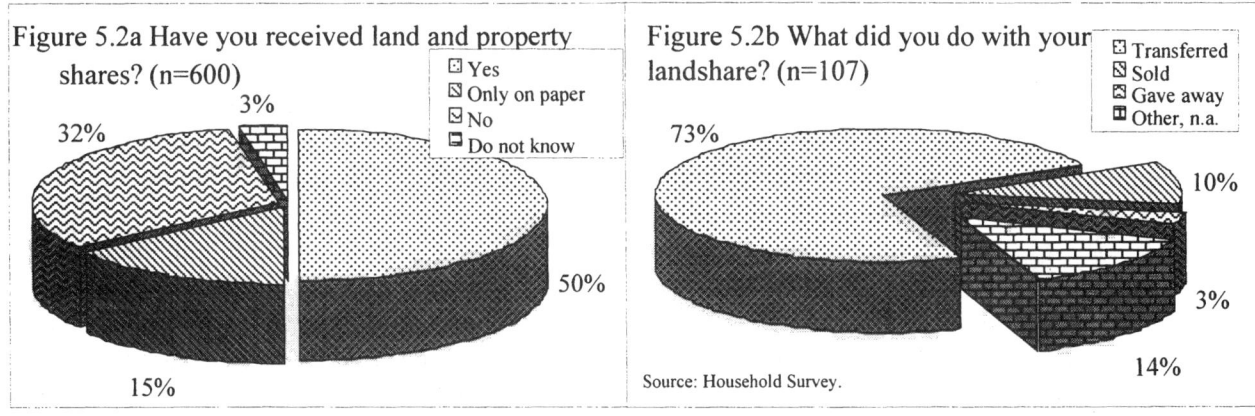

Figure 5.2a Have you received land and property shares? (n=600)
- Yes
- Only on paper
- No
- Do not know

3%  32%  50%  15%

Figure 5.2b What did you do with your landshare? (n=107)
- Transferred
- Sold
- Gave away
- Other, n.a.

73%  10%  3%  14%

Source: Household Survey.

Interviews with farm managers indicated that land and property shares in both oblasts were allocated to farm workers according to standard methods common to the whole republic. In this case, land and property shares were assigned to individuals according to a formula set for each farm, without reference to specific land or property. No significant local peculiarities were found in this process. In neither oblast, however, was the right to obtain shares translated into practice. The actual land share certificates were generally not distributed to the new owners. Rather, they were kept in the offices of the base farms. No cases were found in which property shares were distributed outright to members. Two-thirds of the households who separated from the base farms were unable to obtain their property shares because of one or another excuse. Farm members generally received promises of ownership, rather

> *"We are not real independent farmers, we are owners only on paper. As an independent worker I can't do anything by myself. Am I an independent worker if I am dictated what to do from below and from the top? Still we have plans what crop to sow. "Private" is only the name." Farm head*

than concrete proof, thus this privatization did not significantly change the status of the farms or their workers. Consequently, a major part of the rural population still considered privatization to be merely a formal act. They did not understand the potential value of their land and property shares and did not really consider themselves to be joint owners of former State property. A final consideration was the reallocation of the land and property shares of former farm members who repatriated to Germany, Russia, Ukraine, and other countries after independence. The importance of this reallocation process and its impact on equitable distribution of farm resources varied from one farm to another, depending on the previous ethnic composition of the farm. If the process was concluded without transparency, it could be manipulated by farm managers to undermine the principle of equitable distribution.

The social assessment revealed that few people actually received their land and property shares. One-third of the respondents in Akmola had received land share certificates and only about 11 percent had received property share certificates; in Taldy-Korgan the distribution was 41 percent and 8 percent, respectively. The data indicated that success in acquiring land and property shares is associated with level of education. For example, in Akmola, only 27 percent of the people without secondary schooling had land shares while 57 percent of those with higher education possessed land shares. In Taldy-Korgan, 40 percent of the people had land shares while only 13 percent did not. There was a wide difference between households with regard to gender and ethnicity. Forty-three percent of the male households in Taldy-Korgan had land shares as compared to 21 percent of the female households. Another noticeable difference was that only 38 percent of the Kazakhs, the dominant ethnic group in Taldy-Korgan with 79 percent of total population, had land shares. In contrast, Russians were the second largest ethnic group with 14 percent of the total population in Taldy-Korgan and 64 percent of Russians owned land shares. The number of households with property shares was very low compared to those with land shares. For example, only 8 percent of Kazakhs and 12 percent of Russians had property shares in Taldy-Korgan.

The data indicated that people were hesitant to become independent farmers for various reasons. Financial constraints, lack of machinery, and inexperience were some of the major factors that prevented farmers from separating from the main farm or extending their own farm. Other factors were the perception of risks associated with being independent and bureaucratic obstacles. For example, 27 percent of independent farmers did not want to extend their land because of financial reasons. Of those who wanted to increase the size of their farms, 68 percent responded that funding impeded them. Among farm workers, only 21 percent (12 percent in Akmola and 30 percent in Taldy-Korgan) expressed a desire to separate from the base farms. When asked if they would separate if credit available, however, 53 percent (36 percent in Akmola and 70 percent in Taldy-Korgan) said they would like to start their own farms. Similarly, 78 percent of those who wished to separate responded that financial constraints prevented them from doing so. In short there is still a great number of potential independent farmers remaining on the base farms.

Bureaucratic obstacles were another major factor which impeded farm workers from becoming independent farmers. One of every four farm workers who did not want to become independent said they were discouraged by bureaucratic obstructions. Data indicated that the income and the age of the households might have influenced their response. For example, nine percent of the high income (over 30,000 tenghe per year) households were concerned about the obstruction of farm administration as compared to 50 percent of the low income (less than 15,000 tenghe per year) households in Akmola. Only 29 percent of those households 31 to 40 years of age said they were discouraged by bureaucratic obstructions while 67 percent of those who were younger than 30 years of age felt discouraged.

People had gloomy expectations with regard to the impact of privatization. For example, 48 percent of the households expected privatization to decrease the level of social protection, 42 percent expected it to cause a decrease in salary, and 49 percent expected it to worsen education possibilities. In addition, the data indicated a growing pessimism regarding the level of economic risk (Figure 5.3).

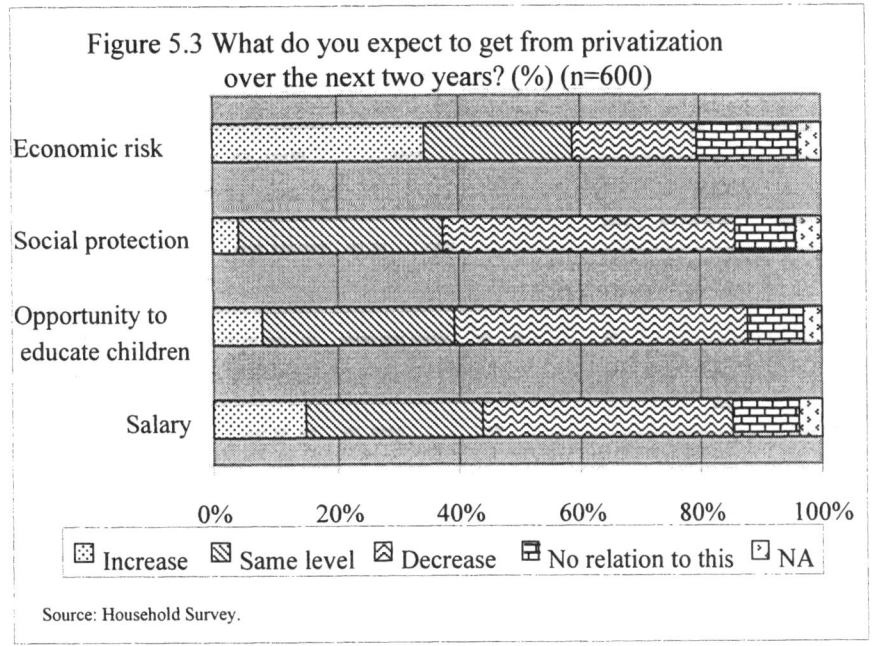

Figure 5.3 What do you expect to get from privatization over the next two years? (%) (n=600)

Source: Household Survey.

The expectations about the future of privatized farms was predominately low. For example, 44 percent of the households thought that irrigation would worsen on privatized farms. Forty-six percent thought that harvests would decrease and 51 percent thought that livestock production would decrease (Figure 5.4).

The level of dissatisfaction with the outcomes of privatization was very high among people. Thirty-eight percent of them were dissatisfied with the results of privatization and 21 percent were somewhat dissatisfied (Figure 5.5). The portion of the households completely satisfied was only 15 percent and the portion of those who were somewhat satisfied was 13 percent. Responses were similar in both oblasts. Essential differences could be seen in the attitudes of the workers on the base farms and the owners of the peasant farms to the results of privatization.

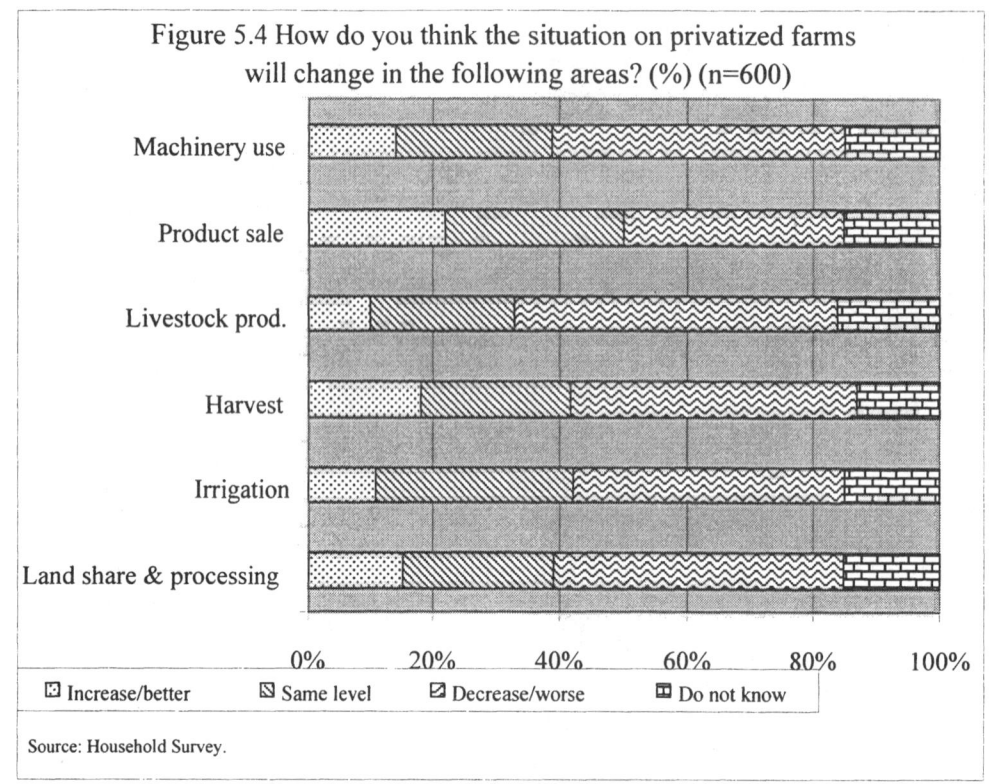

Figure 5.4 How do you think the situation on privatized farms will change in the following areas? (%) (n=600)

Source: Household Survey.

Figure 5.5  Are you satisfied with the results of the privatization in your farm?

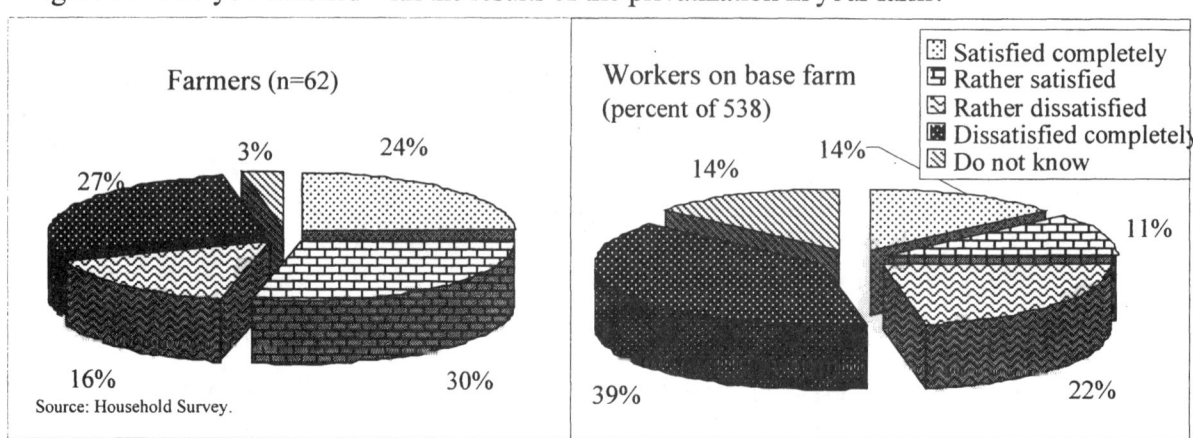

Source: Household Survey.

Educational and professional status strongly influenced a respondent's attitude toward the results of privatization. People with higher education were more likely to be satisfied with privatization than those with less education. Among occupational groups, groups, agricultural specialists were mostly satisfied with the results of privatization and unskilled farm workers were the least satisfied. This indicated that the educated and qualified individuals of the rural population were more likely to adjust themselves to the new economic relations (Figure 5.6a and 5.6b).

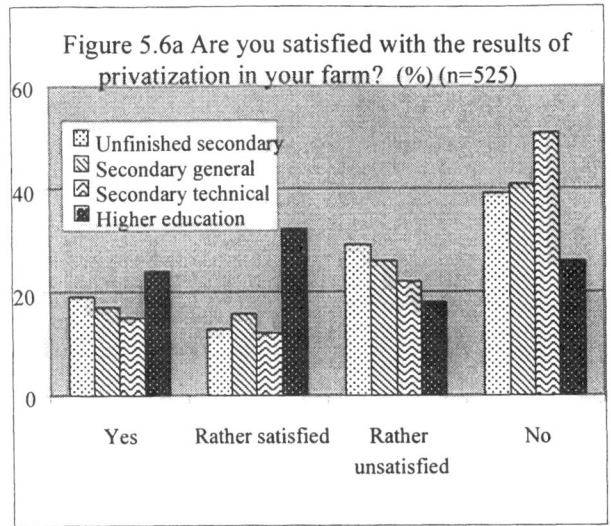

Figure 5.6a Are you satisfied with the results of privatization in your farm? (%) (n=525)

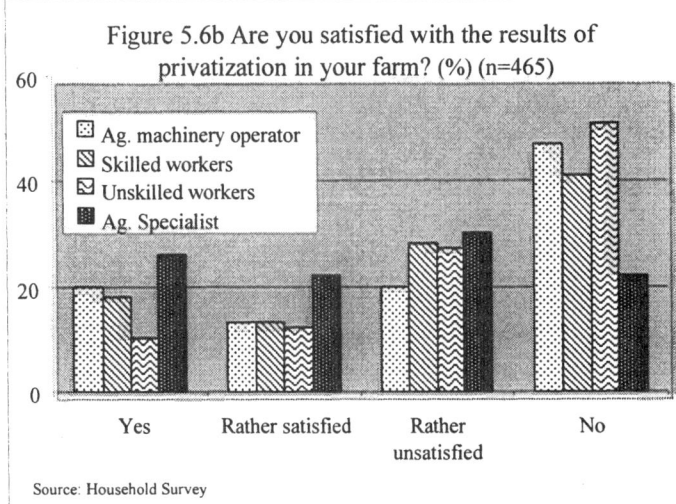

Figure 5.6b Are you satisfied with the results of privatization in your farm? (%) (n=465)

Source: Household Survey

One of the major reasons for dissatisfaction was an unfair distribution which did not meet expectations. Primary reasons for satisfaction were: a chance to become independent; privatization was fairly carried out; and receiving his/her own land. The major reasons for dissatisfaction with the privatization process were inequitable distribution of assets, followed by the failure to meet expectations. The primary reasons for satisfaction were a chance to become independent, equitable implementation and receipt of land (Table 5.5).

Table 5.5 Why are you dissatisfied (or satisfied) with the results of privatization? (Multiple answers, top six responses)

| Dissatisfied (%) (n=355) | | Satisfied (%) (n=170) | |
|---|---|---|---|
| The distribution was unfair | 39 | Everybody was given an opportunity to work for himself and depend on himself | 19 |
| We did not get what we had expected to get | 30 | I got my own land and property share | 24 |
| Privatization has led to a decrease in living standard | 12 | Privatization was carried out fairly | 15 |
| Only managers gained from privatization | 10 | Privatization has led to an increase in living standard | 8 |
| There are no positive economic results | 5 | Privatization was carried out in accordance with laws and rules | 7 |
| Privatization has led to the ruin of the base farm | 4 | There are more opportunities for economic activity | 7 |

Source: Household Survey.

The conviction that privatization was conducted in an inequitable way, regarding one or another aspect, was the greatest reason for dissatisfaction. This reflects the widespread perception that only specific groups of people benefited from privatization. For example, of the 100 people who thought land and property distribution was inequitable, 49 percent answered that the administration of the farm got the best land and machinery (Table 5.6). However, of 336 people only 100 said it was unjust. The remainder said either that the land and property distribution was correct (46 percent) or they did not know (24 percent).

Table 5.6 Why do you think the distribution of land and property
shares was unjust? (% multiple answers possible) (n=100)

| | |
|---|---|
| Authorities received better lands and equipment | 49 |
| Some groups of people didn't receive anything | 22 |
| Farms which separated received the worst lands | 12 |
| Farms which separated received the worst or worn-out equipment | 9 |
| Other | 12 |
| No answer | 4 |

Source: Household Survey.

In addition, of 389 households who said that they received land or property shares, 72 percent answered that the administration of former farms got the most benefits or profits from privatization (Figure 5.7).

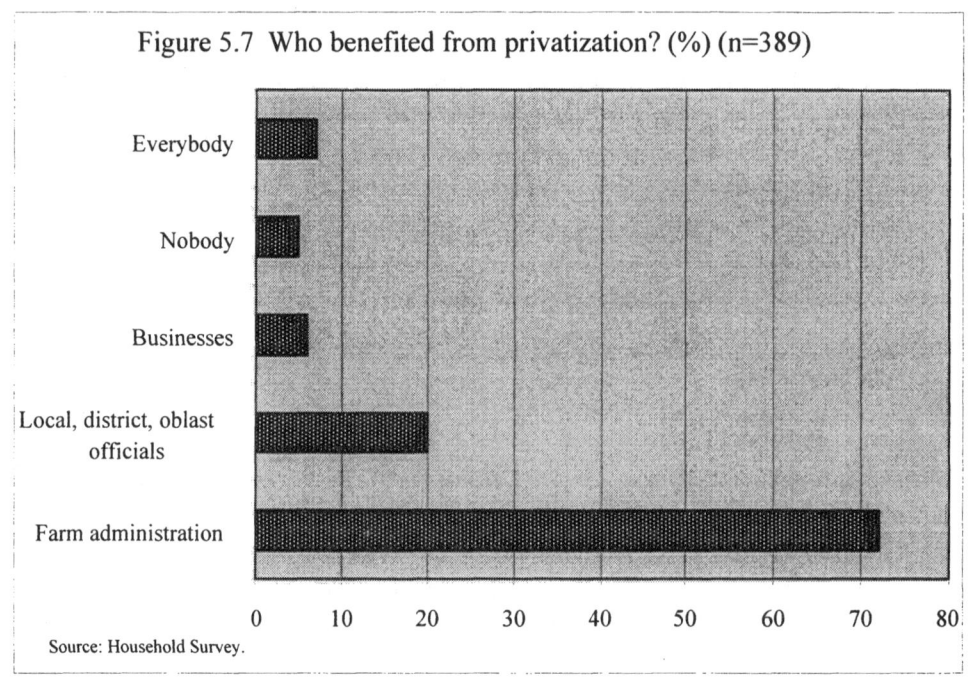

Figure 5.7  Who benefited from privatization? (%) (n=389)

Source: Household Survey.

In short, a process was taking place in which managers of some base farms were getting farm workers to transfer land and property rights to them, either temporarily or permanently, and the number of people who possessed land use rights and private property appeared to be decreasing. As the catalysts and bearers of new economic relations in the countryside, the new landowners, in many cases farm managers, were the principal supporters of further development of private ownership in the countryside. Independent farmers and owners of peasant farms, also supported the processes, but to a lesser extent. Among the employees of base farms, technical specialists viewed the new economic relationships more positively than did other categories of employees. Typically, workers on base farms had the most negative attitude towards privatization. Workers also had the most precarious and uncertain position, as many had either been deprived of their land shares or leased them under conditions that were ill-defined or clearly disadvantageous to them.

The overwhelming majority of households in the survey and participants in the focus groups assessed the results of privatization negatively. They associated the drastic decrease in production and standards of living with changes in the forms of ownership and economic management, which they believed only benefited officials and managers.

> *"Nobody is concerned about the development of agriculture. If we continue at this pace, there will soon be a collapse. We will have a starvation like in 1932 and then the main question will be how to survive." Farm worker*

They wanted to bring back the collective/state farm system (79 percent), without expecting it to occur (67 percent), and they expected further decreases in agricultural economic indicators and living standards in the countryside. Despite this, a considerable number of households and participants in the focus groups did not completely repudiate the objectives of the privatization policy; rather, they disagreed with the speed and methods used to implement the policy.

> *"Restructuring was conducted too fast. It should have been conducted at a slower pace or they shouldn't have changed anything at all." Farm worker*
>
> *"We are not against privatization. There is no way back, but we need real help. It is necessary to privatize the land. At least, we will be able to leave it to our children and grandchildren." Independent farmer*
>
> *"It seems to me that privatization should be continued taking into consideration of these concrete proposals: 1) financial support for the peasant farm; 2) help to organize mini-plants; 3) provide equipment - tractors K-700, mills, etc. 4) ensure that wages are paid on time." Independent farmer*

### Socioeconomic Conditions

The SA survey indicates that an overwhelming majority of the households, 93 percent, owned their house and about 91 percent owned the plot of land around their house. Eighty-one percent also had livestock. With regards to commercial goods, over 90 percent had a television set, 82 percent owned a refrigerator and 77 percent owned a washing machine.

Despite privatization of farms, the physical living conditions of rural inhabitants changed little from Soviet times. The total area of living space per person was 14.6 m$^2$ in Akmola Oblast and 12.9 m$^2$ in Taldy-Korgan[6]. These figures changed little in the last five years. The main source of subsistence for all families was the home garden. Ninety-one percent of the households had a garden. The value of these plots increased after products in the market became more expensive or disappeared. However, the home gardens were appropriate for vegetables, not fodder production, thus the base farm continued to supply an essential ingredient of home production.

The village was the weakest link in the process of market reforms. The transition to new economic relations proved to be more painful in agriculture than other sectors of the economy, and this was reflected in the status of the rural population. People's responses to the question "How has the level of your family income changed during the last two years?" are shown in Figure 5.8 and Table 5.7.

---

[6]     According to the State statistics of 1995, this indicator was 17.1 m$^2$ in the villages of Akmola Oblast and 12.3 m$^2$ in Taldy-Korgan Oblast (Ref. the Government of Kazakhstan. 1996. Kazakhstan statistical book "Agriculture in Kazakhstan". Almaty. p. 172.).

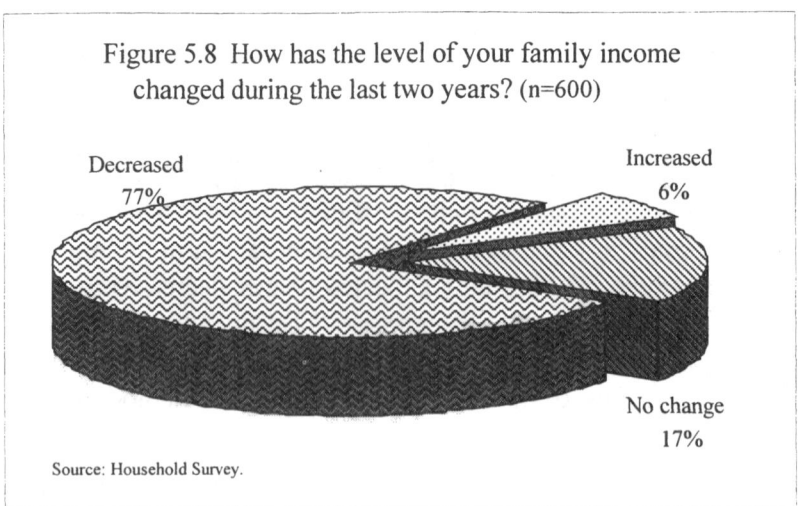

Figure 5.8  How has the level of your family income changed during the last two years? (n=600)

Decreased 77%

Increased 6%

No change 17%

Source: Household Survey.

Table 5.7  How has your family income changed during the last two years? (%) (n=600)

| Expected monthly income | Significantly increased | Slightly increased | Same level | Slightly decreased | Significantly decreased |
|---|---|---|---|---|---|
| < 2000 | 0.5 | 4 | 14 | 37 | 45 |
| 2000 - 6000 | 1 | 6 | 28 | 34 | 31 |
| > 6000 | 7 | 13 | 13 | 33 | 33 |
| Nothing | 0 | 4 | 10 | 26 | 60 |

Source: Household Survey.

The SA documents the hardships suffered by the rural population following privatization. For example, 77 percent of the households saw their family income decrease absolutely in the previous two years, while inflation was high. The average income was nearly the same in the two oblasts. Families with an income of less than 15,000 tenge (less than $200) per year made up 46 percent of the households in Akmola Oblast and 29 percent in Taldy-Korgan Oblast. However, residents of Taldy-Korgan Oblast estimated their incomes to be much higher than those of Akmola Oblast (Figure 5.9).

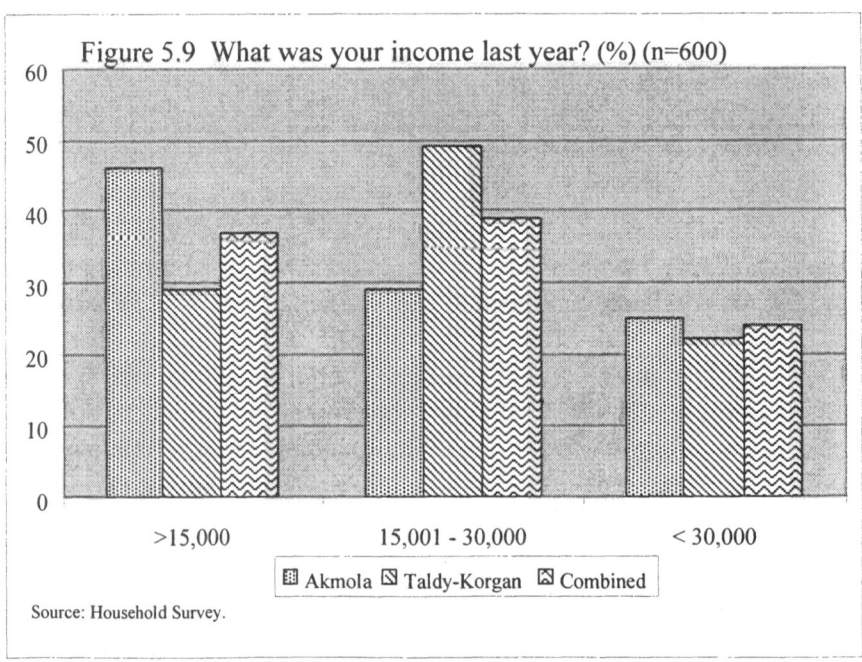

Figure 5.9  What was your income last year? (%) (n=600)

>15,000     15,001 - 30,000     < 30,000

Akmola  Taldy-Korgan  Combined

Source: Household Survey.

On the whole, the average income for the family of an independent farmer was almost twice that of a farm worker ($65 per month and $37 per month respectively). Since a significant share of farmers' income is derived from the profit at the end of the season, income from the sale of seasonal production etc., major differences were also observed in annual income. The average income of all households in the survey was $480 per year. The average income of an individual farmer was $1750 per year, whereas the income of a farm worker's household was $333 per year.

The analysis of income dynamics shows that the rich are getting richer, and the poor are getting poorer.

The SA indicated that rural inhabitants had lost traditional sources of income without developing new ones. The salary of family members was the primary source of income for an overwhelming majority (70 percent) of base farm workers followed by selling agricultural products they produced on their farm (28 percent). However, in the preceding year, 1997, only 23 percent of these households received wages, and more than one-fifth had received no wages for more than three years. Since a large majority of villagers' traditional source of income, wages of family members, was almost non in existent, they no longer considered wages as a source of family income. Officially unemployed households were not included in the study sample, as farmers are not classified as unemployed, regardless of whether or not they are actually employed.

Many independent farmers were able to generate a little income by selling products produced on their own farms (Table 5.8). Only 58 percent of the peasant farmers generated income from the sale of products from their farms. The same proportion of independent farmers responded that wages continue to be the source of income for their families. In other words, the hardships of independent farmers was still considerable as over 50 percent of them or members of their families continued to work for wages while maintaining their own farms. Nonetheless, the independent farmers appeared to have more viable income generating options than farm workers and had higher total family incomes.

Table 5.8 What does your family income include? (% multiple answers possible)

| | Independent farmers (n=62) | Workers of base farms(n=538) |
|---|---|---|
| Salaries of family members | 58 | 70 |
| Selling own farm's products | 58 | 28 |
| Share (stock) dividend | 2 | 7 |
| Income from non-agricultural business | 15 | 4 |
| Income from sale of land share | 13 | 5 |
| Income from property share | 5 | 2 |
| Other | 7 | 2 |
| Nothing | 2 | 1 |
| No answer | 3 | 7 |

Source: Household Survey.

In addition, only nine percent of the households received their wages in cash. Eleven percent got their wages more often in cash than in kind. Twenty percent got (Figure 5.10) their wages more often in kind than in cash and 46 percent got their wages only in kind. This data was supported by the survey of the heads of the farms who described various forms of payment in kind. These included special coupons for bread in Akmola Oblast, distributing fuel and seeds, and writing off public utility charges.

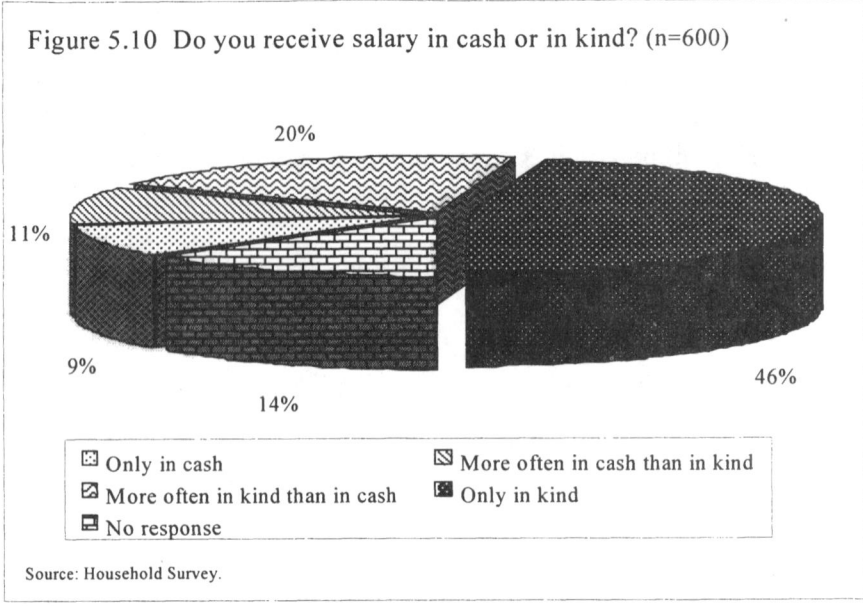

Figure 5.10 Do you receive salary in cash or in kind? (n=600)

20%

11%

9%

14%

46%

☒ Only in cash          ☒ More often in cash than in kind
☒ More often in kind than in cash   ☒ Only in kind
☒ No response

Source: Household Survey.

Seasonal employment did not provide significant material support for most of the households (Figure 5.11). Just 19 percent received wages for seasonal labor and only 13 percent had income from selling vegetables, fruits and grain. Sixteen percent of them received income from selling livestock and only one percent, 8 families out of 600, received a bonus at the end of the season. It should be noted that during the Soviet era, the end of season bonus was the most significant component of the material security of a rural family.

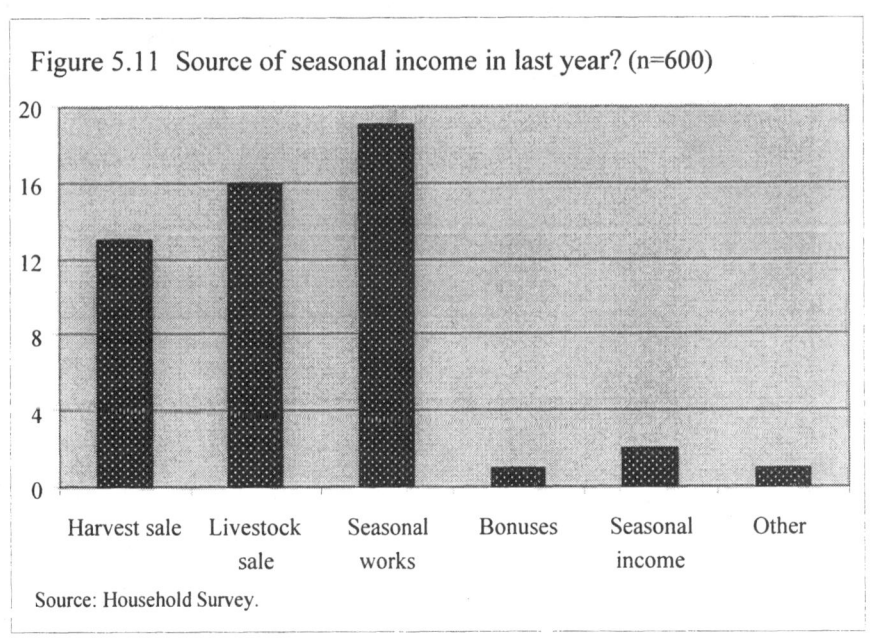

Figure 5.11 Source of seasonal income in last year? (n=600)

| | | |
|---|---|---|
| Harvest sale | Livestock sale | Seasonal works | Bonuses | Seasonal income | Other |

Source: Household Survey.

Traditionally the rural economy generated income from livestock. Eighty-one percent of the peasant families had livestock, the majority of which was cattle. Other types of livestock were much less common (Table 5.9) on independent farms. The data also reflected the traditional distinctions in the structure of stock breeding between the North and the South of Kazakhstan. Swine breeding was more developed in Akmola Oblast, whereas sheep breeding prevailed in

Table 5.9  What kind of livestock do you have? (%) (n=600)

|  | Akmola oblast | | Taldy-Korgan oblast | |
|---|---|---|---|---|
|  | have | don't have | have | don't have |
| Cows | 84 | 16 | 72 | 28 |
| Bulls | 28 | 72 | 17 | 83 |
| Sheep and goats | 16 | 84 | 34 | 66 |
| Pigs | 49 | 51 | 9 | 91 |
| Poultry | 62 | 38 | 53 | 47 |

Source: Household Survey.

Taldy-Korgan Oblast. Nonetheless, these sectors of animal production in both oblasts declined more than cattle husbandry on the household level. The data were also supported by national statistics.

In-depth interviews at Yntymak farm in Koksuisky district (Taldy-Korgan) revealed that the public herd decreased more than 99 percent. It was also found that none of the farms surveyed in Taldy-Korgan Oblast and only 50 percent of base farms in Akmola Oblast engaged in pig breeding. Only one of the base farms surveyed engaged in poultry farming. In addition, a significant part the livestock was moved from the public herd to individual subsidiary holdings. State statistics also support this finding (Table 5.10). Many individual home plots holders had no livestock and definitely no poultry although both were kept by every rural family until recently. The reasons for the decrease in herds was the same whether it was a state, public or independent farm. In particular, people pointed out that livestock became a means of barter and fodder was expensive.

Table 5.10  Total number of livestock in all categories of farms in Akmola and Taldy-Korgan Oblasts in 1995 as compared to 1991 (%)

|  | Akmola Oblast | Taldy-Korgan Oblast |
|---|---|---|
| Cattle | 64 | 61 |
| Pigs | 59 | 25 |
| Sheep and goats | 53 | 43 |
| Poultry | 40 | 27 |

Source: Household Survey.

The SA also indicated that the provision of public utilities was found to be extremely low according to western standards,. Only one third of the households were provided with a water-supply pipeline in their yard and only 20 percent of the households had a water supply in their houses. Less than 10 percent had bath-rooms or shower-rooms and a little more than 4 percent had access to hot water. Essentially no one had access to a centralized gas supply (Table 5.11).

Table 5.11 What kinds of facilities are there in your house?
(% multiple answers possible) (n=600)

| Gas stove with gas-cylinder | 94 |
|---|---|
| Electric power | 94 |
| Radio | 36 |
| Water supply in the yard | 34 |
| Water supply inside the house | 22 |
| Telephone | 21 |
| Sewerage | 12 |
| Bathroom, shower-bath | 9 |
| Water heater | 4 |
| Centralized gas supply system | 1 |

Source: Household Survey.

Although a majority of the households possessed durable goods, including television sets, refrigerators, washing machines, and sewing machines (Table 5.12), most of them were not used because of the unreliable electricity supply. Seventy-four percent of respondents said that lack of electric power supply was the second-most acute problem in the rural economy after wages. The data indicated a decline in living standard of the rural households in the post-Soviet years due to difficulties in meeting basic needs, including regular payment of wages and electric power supply.

Table 5.12 What kind of property does your family have?
(% multiple answers possible) (n=600)

| TV set | 90 |
|---|---|
| Refrigerator | 82 |
| Washing machine | 77 |
| Sewing-machine | 65 |
| Bicycle | 23 |
| Car | 21 |
| Motorcycle | 18 |
| Video | 11 |

Source: Household Survey.

Thus, it was not surprising that an overwhelming majority of the households noted a decline in their standard of living. Fifty-four percent of the households said that their family status was very low or low (Figure 5.12). Only about one percent of the respondents ranked their households as high or very high status. In comparing the two oblasts, the data showed a more pessimistic evaluation of living conditions with a decline in family income.

The attitudes towards privatization and family welfare were associated with ethnicity in each oblast. The rural population of Taldy-Korgan Oblast consisted mainly of Kazakhs who, as repeatedly seen in the SA analysis, were more tolerant of hardships than Russians and other European ethnic groups. This was also result of differences in the living standards of rural inhabitants during the Soviet period, when the living standards in Kazakhstan's northern oblasts were higher than the southern oblasts.

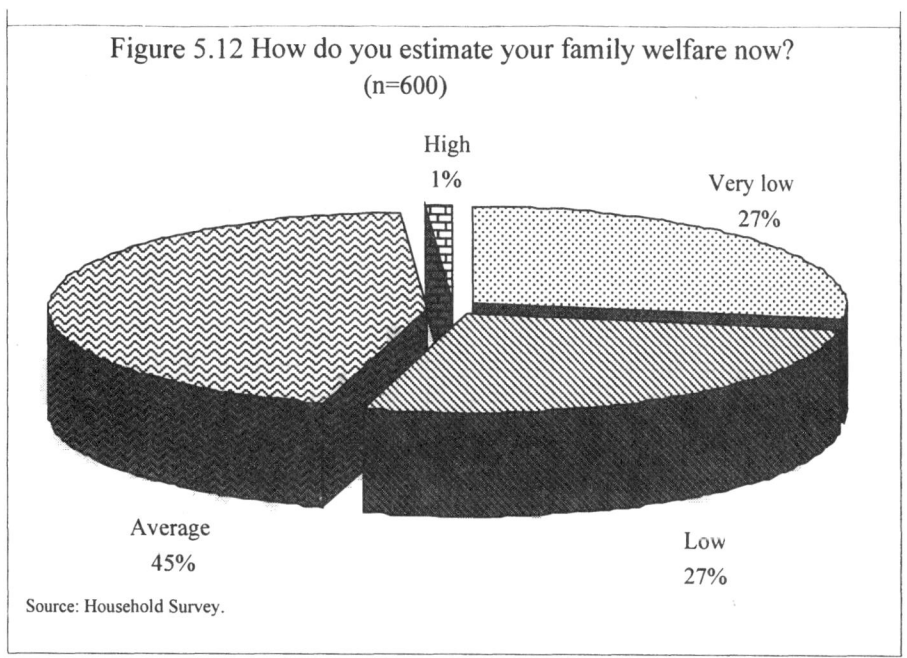

Figure 5.12 How do you estimate your family welfare now?
(n=600)

High
1%

Very low
27%

Average
45%

Low
27%

Source: Household Survey.

Attitudes towards privatization and family welfare were also associated with whether the household head was a farm worker or an independent farmer. A majority of independent farmers expected a decrease in their living standards. Thirty-one percent stated that their income had slightly decreased and the same percent stated that their income had significantly decreased. About ten percent stated that their income had significantly increased, but none of the farm workers claimed that their incomes had increased significantly.

The income level of the households had a direct relationship with their perceived family welfare. For example, only 21 percent of those with an annual income of less than $197

Table 5.13 Why did your family income decrease in the last two years? (% multiple answers possible) (n=462)

| Non-payment of wages, delay in payment | 50 |
|---|---|
| Unemployment | 12 |
| Increase in prices, inflation | 9 |
| Disorganization of the farm | 7 |
| Low level of wages | 5 |
| | 4 |
| Reorganization of the base farm, privatization | 4 |
| Guilt of local authorities | 4 |
| General complaints: 'Everything is bad' | 4 |
| Lack of machinery, gasoline and lubricants, fertilizers | 3 |
| Others | 18 |

Source: Household Survey.

characterized their standard of living as average, while 49 percent of those with an annual income of $197 to $394 and 57 percent of those with an annual income of over $394 characterized their status similarly.

Only a small number of households related the decrease in their living conditions to privatization and market reforms. Most of them thought that the decrease in their living standard was due to non-payment of wages. Unemployment, followed by inflation and disorganization of farms were cited as other conditions contributing to the decline. All these factors took place before mass privatization. Table 5.13 supports this finding. Despite the catastrophic state of the rural economy, the patience of the rural people had not yet been exhausted.

Privatization and restructuring were not the only reasons for the decline in the standard of living and the hardships for the rural population. The overall economy of the districts, oblasts and the country certainly contributed to the severity of the situation. Most of the households were aware of this reality. For example, 85 percent estimated that economic development in their district had worsened in the last two years. Fifty-four percent of the people forecasted further worsening of the situation in the agriculture sector.

The problems generated by the government's new economic policy were some of the main reasons for hardship on the farm (Table 5.14). Over 60 percent of the respondents said that non-payment of wages, deficiency in electricity, and low level of wages were the most acute problems on their farms.

Table 5.14  What are the most acute problems on your farms? (% multiple answers possible) (n=600)

| | |
|---|---|
| Non-payment of wages | 77 |
| Deficiency in electricity | 74 |
| Low level of wages | 62 |
| Out of date and depreciated machinery | 57 |
| Poor technical provision | 53 |
| High prices of goods | 50 |
| Unemployment | 36 |
| Inaccessibility of loan (credit) for farm members | 27 |
| Bad farm management | 22 |
| Inadequacy of school and kindergarten | 17 |
| Lack of drinking water | 15 |
| Lack of specialists | 11 |
| Insufficient watering | 10 |
| Low level of processing industry | 9 |

Source: Household Survey.

In addition, 90 percent stated that they lived better under the collective/state farm system. Seventy-nine percent of the respondents said wanted to bring back this system, but they had no illusions regarding the likelihood that it would occur. Only 21 percent thought that it was possible to bring back the Soviet system. Among pension-age villagers, 100 percent desired to bring back the past. Further, for all other age groups, the number of those who felt nostalgia for the Soviet system also exceeded 80 percent. Education level had significant influence on such attitudes (Figure 5.13).

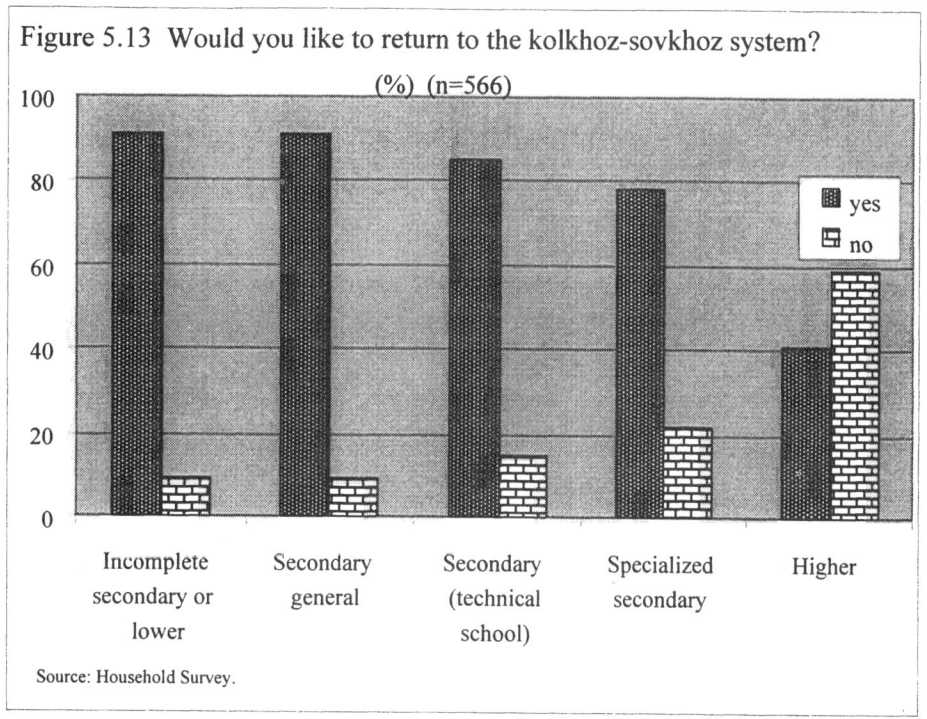

Figure 5.13  Would you like to return to the kolkhoz-sovkhoz system? (%) (n=566)

Source: Household Survey.

This growing pessimism could have a significant impact on social, demographic and economic spheres.  Potential migration from rural communities is significant.  For example, 43 percent wanted their children to leave the farm forever and 46 percent thought their children would actually leave the farm (Figures 5.14a and 5.14b).

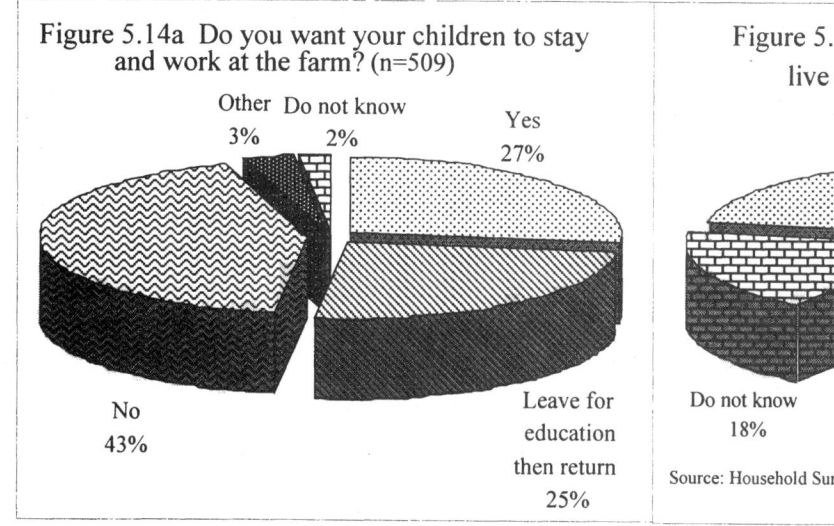

Figure 5.14a  Do you want your children to stay and work at the farm? (n=509)

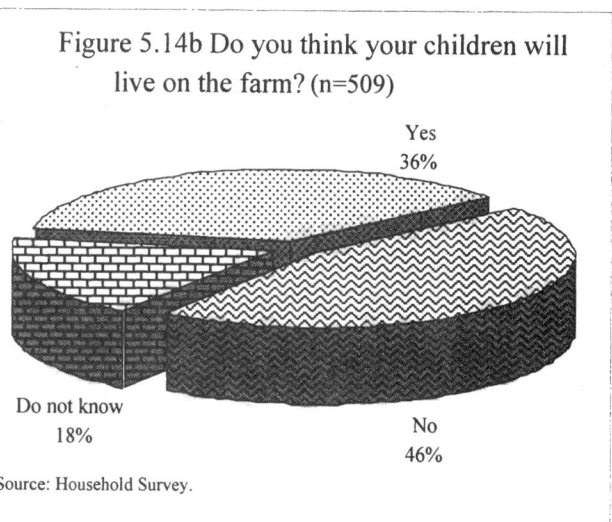

Figure 5.14b Do you think your children will live on the farm? (n=509)

Source: Household Survey.

The SA indicated that migration and ethnicity influenced the villagers' sense of social security.  Whereas 32 percent of Kazakhs would like their children to work on the farms, only 18 percent of Russians and 14 percent of Germans wished them to do so.  Thirty-four percent of Kazakhs, 57 percent of Russians and 56 percent of Germans expressed the desire for their children to leave their native village forever.  A similar result was seen in each oblast when dominant ethnic group was taken into account. Kazakhs are the dominant ethnic group in Taldy-Korgan Oblast, while Russians dominate in Akmola Oblast.

To the open-ended question "What is needed to improve your economic situation?" people responded in two main modes. Most respondents them asked for significant changes at the policy or leadership level. They also indicated a feeling of powerlessness, and thus a wide communication or consultation gap between themselves, stakeholders for whom the policies are intended, and policy-makers. As is seen in Table 5.15, the most common proposals were to improve employment conditions, especially wages. Administrative changes (change local and

| (% multiple answers possible) (n=600) | |
|---|---|
| Stimulate labor better (to increase wages, to pay wages on time, etc.) | 19 |
| Everybody must work better at work place | 15 |
| Change local management | 11 |
| Improve farm machinery, fuel and lubricants, seeds, fertilizers provision | 9 |
| Grant credits to peasants, to improve terms of credits | 9 |
| Change State leadership | 9 |
| Create conditions for effective work | 7 |
| Subsidize farms, to increase prices on agricultural production | 7 |
| Strengthen struggle against bureaucratic abuses, crimes and corruption | 5 |

Source: Household Survey.

national leaders, increase subsidies, regulate prices, strengthen law and order and fight against crime and corruption), together with the proposal to bring back the Soviet system, were supported by almost half of those surveyed. At the same time, less than 25 percent of the respondents supported w0rk place-related economic measures (to work better, create better working conditions, improve the credit practices, and change the taxation system).

Only one percent of the respondents thought that the key to improving the economic problems of the village was to speed up the privatization process and increase the number of independent peasant farms. Only two percent thought that it was possible to improve these problems by granting more independence to agricultural producers. These responses indicate that there was no social support for the existing model of privatization in the countryside. Having stated that, however, it is important to analyze the results with respect to two groups of households—farm workers and independent farmers. Only 11 percent of the workers on base farms said they did not want to bring back the Soviet system, compared to 58 percent of the independent farmers. In addition, 31 percent of the independent farmers responded that they lived better now than earlier, but only a very low number of farm workers said the same. Despite all the difficulties experienced by independent farmers, it appears that they are better able to make a social basis for privatization than the base farm workers. It also provided the basis for privatization which, if implemented together with social safety mechanisms, could alleviate or minimize the short term hardships to the rural communities and strengthen the local economy in the long run.

## Social Infrastructure

The quality of social infrastructure and access to it in the countryside have deteriorated greatly since independence and the subsequent economic crisis. The change has occurred in stages. Initially, practically all cultural centers stopped functioning and kindergartens were closed in the early 1990s. These social assets were state responsibilities, maintained by the sovkhoz and kholkhoz. When state budgets disappeared, and the farms lost their ability to maintain financial support, they were closed.

Most of the hospitals on the farms were also closed by the time of the SA. At best, ambulatory clinics and doctor's assistant stations remained open. The number of schools was reduced. The schools in less populated areas were closed or were in the process being closed. Children on settlements without school had to travel long distances to go to other settlements for school. In most cases, they lacked transportation.

Heating schools was another difficulty. In Akmola Oblast, most farms contributed to heating the schools and several schools were able to provide free food to school children. In Taldy-Korgan Oblast, schools were not heated and the schools were actually closed during the most severe winter months. A small number of farms were tried to find non-standard ways to maintain schools. For example, on Enbekshi, a six hectares plot of land was allotted to the school to farm. The school was responsible for buying textbooks from the income generated by selling the products gathered from this plot.

Education and health care were officially free, thus most of the rural people did not expect to pay for them (Figure 5.15). Nonetheless, patients had to provide themselves with basic needs such as medicines, food, and linen in formerly free medical institutions, as well as make unofficial payments to health care personnel. The charges for medical services were especially high for individual farmers. Fifty-six percent said that they had to pay for all their health services while only 45 percent of workers on base farms paid the entire amount.

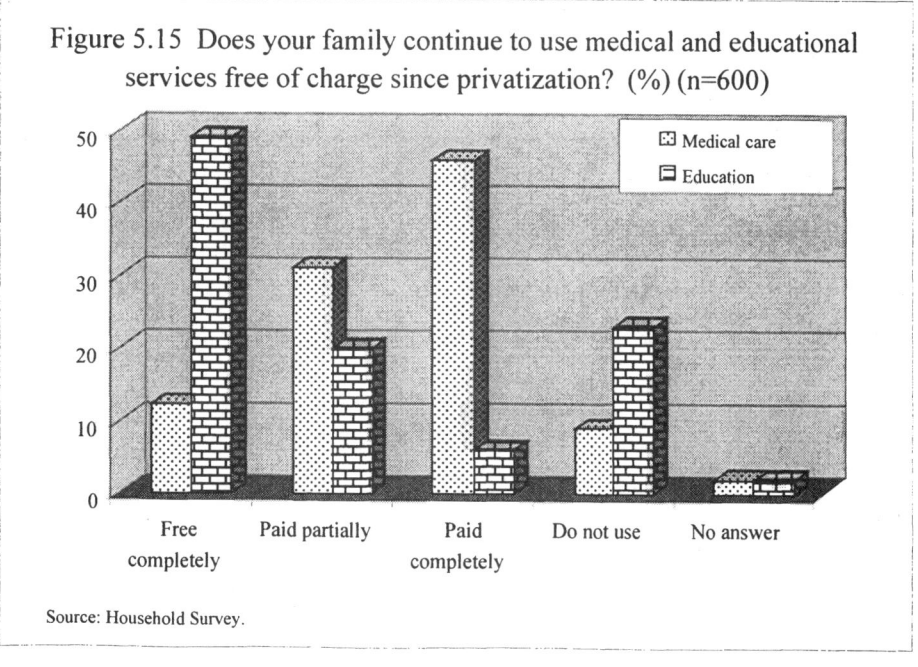

Figure 5.15 Does your family continue to use medical and educational services free of charge since privatization? (%) (n=600)

Source: Household Survey.

*Stakeholders and Institutions*

## Independent Farms and Farmers

Eighteen percent of the independent farms included in the sample were created before 1993—before the mass privatization of state and collective farms began. A number of respondents said that these farms got the most benefits when they separated. For the most part, they were close colleagues of kolkhoz and sovkhoz managers and their departure was not threatening. They received equipment and better lands on preferential conditions. Seventy-three percent of the peasant farms surveyed were created after the privatization process on the base farms began, however, and their fate was very different. A majority of them did not receive anything except lands, as the property shares were not distributed for various reasons. Only one of the farms surveyed was established later on the basis of acquiring additional shares, and another farm was leased by the owner.

> *"When the first farms were separated, the area of the land allotted was not limited. There were allotted the lands which they asked for."* Independent farmer

The heads of base farms almost unanimously stated that they allocated good or medium quality lands to the peasant farms. Among the owners of the peasant farms, only 7 percent thought that their land was good while 66 percent thought that it was medium and 23 percent thought that it was poor. Thirty-seven percent of the independent peasant farms consisted of only one family while 26 percent had two families, and 11 percent had three families. About half of the farms were made up exclusively of relatives. On farms consisting of more than one family, profits was generally divided equally. On five farms the profit was divided according to the shares contributed and on seven farms it was divided in proportion to the time spent working. The economic status of these peasant farms left much to be desired. For example, 23 percent of the farms made a profit in the previous agricultural season; 42 percent could only cover expenditures; and 23 percent had losses. The average turnover was $2,667.

Some of the most educated and qualified rural dwellers separated and started their own farms. For instance, about one third of the households with higher education were independent farmers while only three percent of those with incomplete secondary school were independent farmers. Forty-eight percent of the farms had no hired workers, 36 percent always hired seasonal workers and 13 percent had full-time hired workers. As a rule, managers of the base farms claimed that they did not hinder their workers from separating and creating peasant farms. Some managers, such as the Chairman of the Enbekshi Production Cooperative, said that they had actively promoted separation. However, during focus group discussions many managers, especially in Akmola Oblast, demonstrated hostility toward peasant farms. Quite often statements were made to the effect that the peasant farms were not necessary and had no prospects.

Other consultations also indicated that the separation of the peasant farms was carried out despite opposition from the managers of base farms. For example, 42 percent of the independent farmers stated that they faced obstacles, and opposition when they left the base farm. There was tension in the relationship between the base farms and the peasant farms. The managers of the base farms unanimously stated that the peasant farms did not depend on them. At the same time, the managers of Rassvet, Granit, and Stepnyak (Akmola Oblast) stated that the farmers were parasitic, and that they steal spare parts, seeds and fertilizers from base farms. Interview results, however, indicated that a majority of the independent farmers felt independent. They leased equipment and borrowed fuel and lubricants from base farms, then repaid in kind during harvest time. The household survey of the independent farmers supported these findings (Figure 5.16).

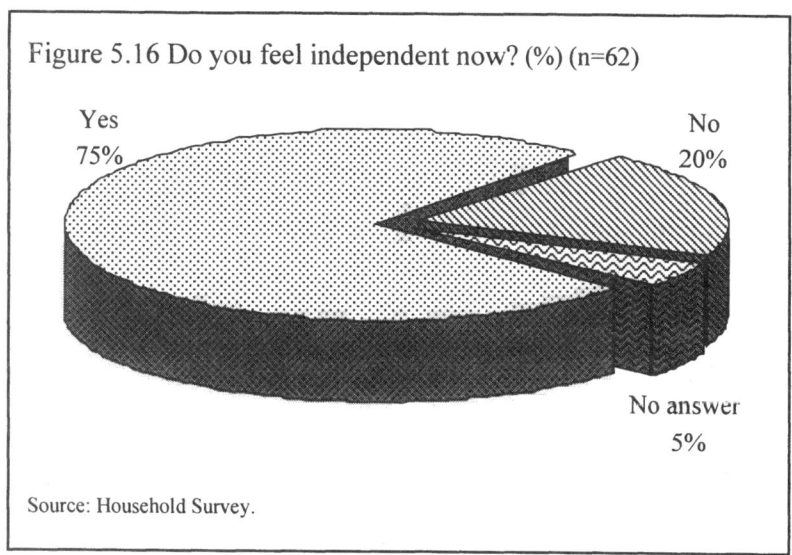

Figure 5.16 Do you feel independent now? (%) (n=62)

Yes 75%

No 20%

No answer 5%

Source: Household Survey.

Of course, it took a lot of courage to start an independent farm, as the farmers rarely had any start-up capital of capital stock, and they soon discovered that the enterprises which delivered inputs and purchased products were not able to deal with small farms and unwilling to change their practices. This institutional mismatch was reflected in high prices and low farmgate prices for independent farmers. The twenty percent of independent farmers who said they felt dependent attributed that to their lack of funds and equipment, not to interference by the base farms. Some degree of

| *Difficulties in managing independent farms (percent):* | |
| --- | --- |
| • *High price and inaccessibility of machinery and spare parts* | *61* |
| • *High price and inaccessibility of fuel and lubricants* | *37* |
| • *High prices of other inputs* | *18* |
| • *High prices and lack of seeds* | *10* |
| • *Lack of credit* | *10* |
| • *High taxes* | *8* |
| • *High prices and lack of fertilizers and chemicals* | *7* |
| • *Lack of assistance from local administration and the State* | *7* |

dependence was natural, given the local situation, principally leasing equipment from the base farm, sharing markets, or selling products to the base farms.

Despite all of these difficulties, only five percent of the farmers in Akmola Oblast tried to return to the base farms and another five percent would have liked to sell their farms. In contrast, fifty-five percent of the farmers would have liked to purchase more land and expand their production, though lack of funds and access to credit hindered them (Figure 5.17).

The attitudes of independent farms vary according to ethnicity. Among the Kazakh independent farmers, more than 55 percent said they want to expand their farms, whereas only 30 percent of the Russians wished to do so. Further, none of the Kazakh respondents wished to sell their farms but 20 percent of the Russians wished to do so. It should be noted that these figures may also reflected the desire of the remaining ethnic Russians to leave Kazakhstan. Although managers of base farms said that independent farmers were shrinking their farms and wanted to

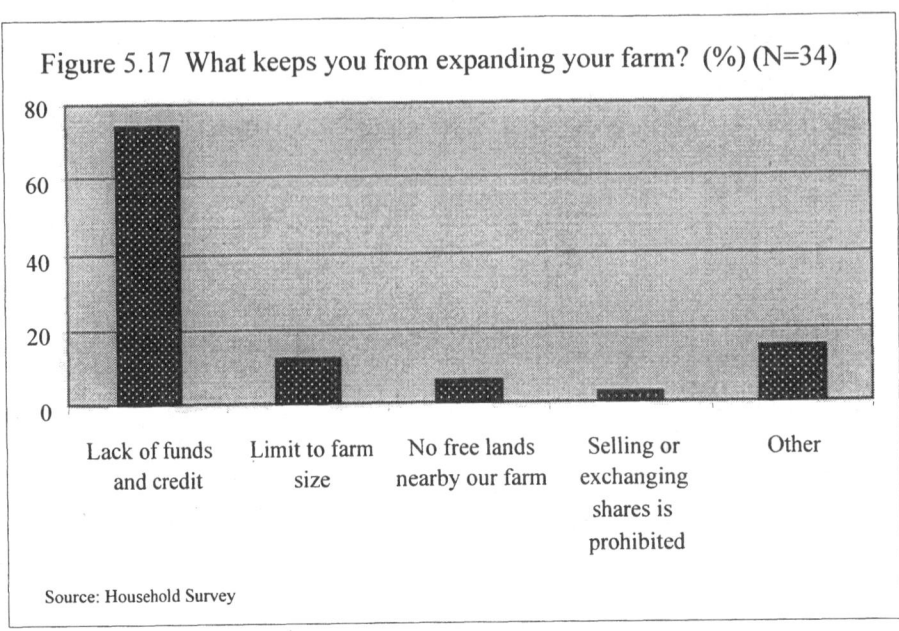

Figure 5.17 What keeps you from expanding your farm? (%) (N=34)

Source: Household Survey

return to the base farms, especially in Akmola Oblast, this impression is not supported by the response of independent farmers themselves. Indeed, State statistical data show a marked increase in the number of peasant farms over the few years before the SA (Figure 5.18).

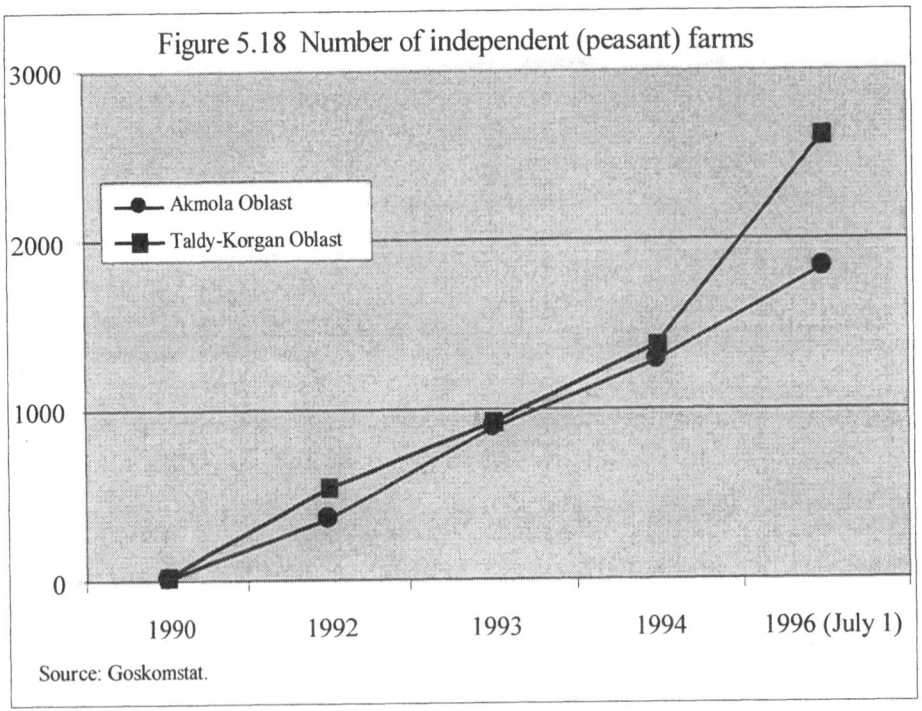

Figure 5.18 Number of independent (peasant) farms

Source: Goskomstat.

Among farm workers, 21 percent said they would to separate and establish their own farms. The SA revealed that two distinct categories of base farm members—agricultural specialists and unskilled workers—were the most interested in establishing their own farms.

Apparently, the former were motivated by their aspiration for self-realization and their confidence, based on better knowledge of their rights. The latter, unskilled workers, seemed to be motivated to leave the base farm out of despair, since their current economic situation was the worst of all social and professional groups and their prospects were dimming. The presence of the "despair factor" among workers on base farms who wished to separate was inversely associated with income level.

Ethnicity and a desire to be independent had some influence in the decision to separate from the base farm. The desire to separate was the greatest among ethnic Koreans (42 percent). A well-established agricultural tradition and the absence of an inclination to migrate may have contributed to their decision. In contrast, among ethnic Germans, considered to be exemplary farmers in Kazakhstan, only 14 percent wished to create their own farms and 18 percent of ethnic Russians wished to do so. In contrast, 25 percent of the ethnic Kazakhs wished to establish their own farms.

Those who did not want to be independent farmers (69 percent) were discouraged by lack of funds (71 percent), perception of associated risks (29 percent), lack of experience (28 percent) and bureaucratic obstacles (25 percent).

Seventy-one percent of the farmers surveyed were dissatisfied with their financial situation. Twenty-nine percent had debts. Most of those with debt (44 percent) owe it to private banks. Only a single household was indebted to the State, while the remaining were old debts of state farms, and debts to the base farms. The peasant farmers had almost paid off their debts to the State banks and the base farms and they paid 39 percent of their profit in taxes, on average.

Most independent farmers had used the services of private companies. The role of the base farms was to support farmers with equipment. In general, farmers who could not buy equipment leased it from the base farms. The role of State-supported organizations was still great and, as the in-depth interviews showed, these organizations were actually still considered to belong to the State, regardless of the fact that they have been privatized and today most had the status of joint stock company.

**Base Farms—Managers and Farm Workers**

After privatization, the brigade structure of most farms remained in place. However, on some farms, the number of workers was reduced by 40-70 percent. The staff in the district agricultural administrations was also reduced by 90 percent, from 40-50 to 4-5 people. They only controlled the most important technical norms. Their activities included verifying the correctness of maintained financial accounts and informing the farms of the changes in the laws and regulations affecting the agricultural-industrial complex. In interviews, none of the heads of the farms complained of any interference in their activities by district or oblast administrators. In most instances, farm level decisions were made individually by the managers. Most of them confirmed this in the in-depth interviews. As a rule, general assembly meetings of the production cooperatives were merely formal occasions, farm workers who hold shares in the base farms rarely participate in management.

The majority of the householders stated that they did not participate in any decisions on their farms (Table 5.16). The SA indicated that privatization had not changed the practice of making the key decisions regarding economic activity. Managers consult farm workers only regarding dates of the harvest, which requires maximum mobilization of human resources. Such situations were typical before and after privatization. Few workers participated in farm management because they thought that it was useless; they believed that the administration would solve all problems. Nonetheless, 54 percent of the farm worker sample said they would like to be able to participate actively in the decision-making process regarding the activity of the base farm, and nine percent would like to participate to some degree. The SA indicated that a systematic information campaign might encourage farm workers to seek roles in the decision-making process and motivate them to manage farms of their own.

| Table 5.16 Why do you not participate in decision-making on your farm? (% - multiple answers possible (n=538) | Percent |
|---|---|
| It is useless because managers solve all the problems | 40 |
| I am not allowed to do so; they prevent me from participating | 11 |
| I am unqualified, lack of experience, education | 7 |
| I do not want to | 7 |
| I am afraid to do so | 4 |
| I do not have time | 3 |
| I am not interested | 3 |

Source: Household Survey

Most of the base farms were unprofitable. While some isolated farms were prosperous, including Rodina in Akmola Oblast and Enbekshi in Taldy-Korgan Oblast, other farms were not. For example, Zarechnoye in Akmola Oblast and Pravda in Taldy-Korgan Oblast hardly made a profit. Farm workers were aware of the financial situation on their respective farms. Although 71 percent assessed their farms as unprofitable, 11 percent described their farms as profitable and 16 percent thought their farms would break even.

The profitability or unprofitability of the farms obviously influenced the financial position of the farm workers. Eighty-six percent of those surveyed stated that their wages would be less (or already were less) due to the increasing financial loss of the farm. Seventy-six percent were confident that if the farms became profitable their own financial position would improve.

One reason for the financial difficulties present on the farms was a huge debt inherited from the former state farms and the high cost of inputs, fuel and energy. The debt constantly increased due to the general financial disorder in the country and non-payments of wages. The amount of debt fluctuated on different farms from less than a dollar to $2.63 million. Most of this debt was the result of costs associated with obtaining spare parts, electric power, and fuel and lubricants. It was found that the indebtedness for wages and the pension fund were of secondary importance to the farms.

The debt completely paralyzed economic development on the farms. For instance, on the Rassvet Collective Agricultural Enterprise, all receipts from the sale of products were used to repay debts. According to the manager of this farm, which specialized in milk production, processed milk was delivered to a butter factory, but the farm did not receive money. Rather, the proceeds from the sale were transferred immediately to the electric utility.

Given the current financial situation, it was found that "direct compensation" rarely occurred. The barter system dominated trade in the marketplace. Most of the farms did not have cash or money in bank accounts to pay for fuel and spare parts. Grain and, more often, livestock were used to pay for agricultural inputs. This had a significant impact on livestock counts. The SA indicated that on most of the farms barter made up 80 to 100 percent of the total sale of production.

> *"We've already forgotten what money is. We don't use money at all. We work under the conditions of wild barter, i.e. everywhere everything is pledged for grain." Farm head*
>
> *"I acquired fuel oil and paid with livestock. Barter! It is not profitable for me. But what to do?" Independent farmer*

Some farms tried to increase their income by establishing processing facilities—mainly sausage or milk and butter production. In some cases it was successful. For example, the Rodina Production Agricultural Cooperative was successful in packaging milk products and supplying them to trading firms in Akmola under direct contracts. In most cases, however, the processing facilities were used below their production capacity or stood idle, primarily because of the irregularity in electric power supplies. In addition, the lack of fuel, including petrol, made it difficult to deliver products to customers.

In depth interviews with off-farm elements of the agricultural sector indicated that the overall demand for agricultural products was adequate for supplies. Managers of agricultural product consumer organizations indicated that they preferred to deal with large farms instead of dealing with many smaller farms to minimize administrative costs. At the same time, they often could not deal with larger farms because they lacked the means to pay for large deliveries. As a result, the needed products were not purchased systematically. The same situation was found to be typical for farm input supply organizations. Although the demand for agricultural input and output products remained high, markets in the countryside were not developed. The existence or nonexistence of market infrastructure was caused by the country's overall economic policies and the collapse of farm service and supply organizations, not by the privatization of farms, as such. Accordingly proximity to urban areas, and thus potential direct access to markets, was an important factor influencing the profitability of both base and independent farms.

### Participation

The Agricultural Post-Privatization Assistance Project will be demand driven. It is designed to help restructured farms become commercially viable by establishing advisory centers that give technical and financial advice and draft business plans, as well as a line of credit made available through commercial banks. The objective a participation strategy is to ensure that potential beneficiaries, especially independent farmers and entrepreneurial farm managers, are aware of project initiatives and use them effectively to revitalize the rural areas. To do this, the project must reach potential clients with information and services that they need, in ways they can understand and use effectively. The SA focused on several topics that relate both to demand and the design of communications activities and delivery mechanisms: credit experience and demand;

interest and use of advisory services; disposition toward organizing; and access to information and communications media.

### Credit Experience and Demand[7]

The SA revealed considerable latent demand for credit, but very little experience. Managers of the base farms said that they were able to obtain credit from 1992 to 1994. However, after 1994, access to credit was rare because the state and semi-state banks refused to give

> *"Those peasants who separated at that time were granted credit. They bought equipment and now live quietly." Independent farmer*

credit to unprofitable farms—in general most farms were in this category. Credit from private banks was also difficult because interest rates fluctuated from 50 percent to 200 percent. Respondents stated, however, that they would like to obtain credit at an interest rate of less than 30 percent for a period of three to five years, mainly to acquire equipment and/or expand production.

Only 23 of the 600 households (four percent) had received a loan. Fourteen of these households (61 percent) were independent peasant farm owners. Most of the loans were from State banks. Two farmers used the services of a commercial bank and only one obtained a loan from the base farm. Four farmers borrowed from private individuals. On average, people with higher education obtained loans. About two thirds of them obtained a loan between 1991 and 1993. The average size of the loan was 4.5 million rubles ($1800). The period for which the credit was obtained was less than one year and all who had received loans had already paid off their debts. Eighty-three percent thought that the loan was useful, and more than 40 percent would have liked to obtain a loan. More than 70 percent preferred state banks and organizations for obtaining credit. The average size of the loan they wanted to obtain was about $40,000. The majority of them wanted a three to five year credit period; only 30 percent wanted a 6 to 12 month period. People from some ethnic groups were more willing to take a loan than others. Members of different ethnic groups had a variable response to the question of whether or not they want to obtain a loan to buy machinery, raw materials, seeds, and fuel and spare parts (Table 5.17).

**Table 5.17 What is your aim in obtaining a loan?**
(% multiple answers possible) (n=240)

| | |
|---|---|
| To buy machinery | 58 |
| To buy raw materials, seeds, fuel, spares | 44 |
| To buy more land | 33 |
| Building | 13 |
| To repair machinery | 9 |
| Repayment of another debts | 1 |
| Repayment of another loan | 1 |

Source: Household Survey.

As expected, the desire to get a loan was almost inversely related to the relative size of the group (Figure 5.19). Most respondents said they would prefer to pay off their loans in installments, rather than at the time of maturity. They were not aware of the real conditions of loans. About 56 percent of the respondents who were ready to pay annual interest thought that it would be about 10 percent. This indicated that there is a need for information on how to obtain credit, current interest rates, and realistic payback times. Among farm workers, the lack of funds and credit were the major obstacles that prevented them from separating from the base farms and establishing their

---

[7]     To ensure that the credit component of the project is supporting a transparent and equitable restructuring process, a set of guidelines for the eligibility was established. Details of these guidelines are in Annex 2 and can be found in the SAR No. 17789-KZ.

own farms. More than 53 percent of the farm workers would have liked to become independent if they could have obtained a loan.

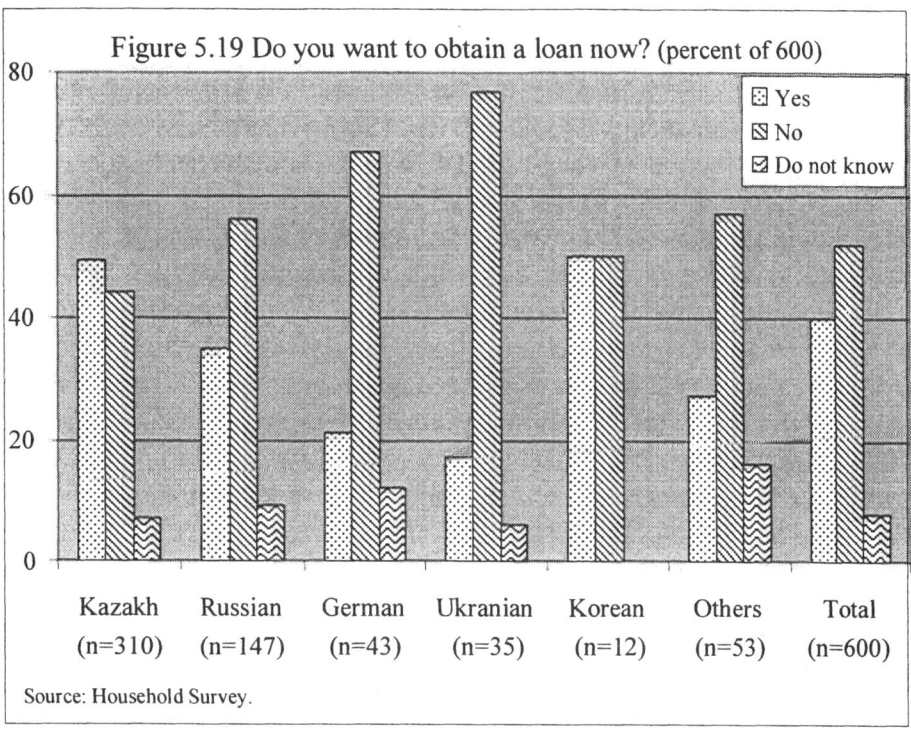

Figure 5.19 Do you want to obtain a loan now? (percent of 600)

Source: Household Survey.

## Use of Consultancy Services

In Akmola Oblast, there were no special consulting organizations in the countryside. In Taldy-Korgan Oblast, a marketing-consulting center was created attached to the Oblast Agricultural Administration with district subdivisions. Its functions included assistance to farms in finding clients and finalizing contracts. However, farm managers did not assess its activity positively. Managers of farms, supply organizations, and processing facilities would have liked to use the services of special consulting organizations. Their attitude toward establishing private

Table 5.18  What organization's (institution's) advice do you trust the most?
(% multiple answers possible) (n=600)

| Field | Local administration | Main farm administration | Foreign consulting firm | State consulting firm | Private consulting firm | I do not trust anybody | Do not know |
|---|---|---|---|---|---|---|---|
| Management and planning | 14 | 10 | 2 | 7 | 3 | 15 | 50 |
| Finance | 18 | 15 | 3 | 15 | 6 | 16 | 28 |
| Product sale | 12 | 12 | 2 | 8 | 7 | 15 | 45 |
| Technology | 12 | 10 | 6 | 5 | 4 | 14 | 50 |
| Access to information | 12 | 13 | 3 | 8 | 3 | 12 | 50 |
| Legal advice | 13 | 12 | 2 | 20 | 5 | 11 | 39 |

Source: Household Survey.

firms for such work was rather skeptical. Most of them preferred state structures. Although most of the households were willing to pay for consulting services, they wanted these consulting firms to initially demonstrate their value by rendering free or minimally priced services (Table 5.18).

People in both oblasts required advice on financial matters (42 percent) most, followed by legal matters (37 percent), sale of products (18 percent), and access to information (13 percent). Also, more households with higher education sought legal advice than those with unfinished secondary education. For example, in Akmola, 67 percent of those with higher education sought legal advice (61 percent in Taldy-Korgan) while only 36 percent of those with unfinished secondary education sought legal advice (32 percent in Taldy-Korgan). In Akmola, more households with high income (65 percent) than with low income (36 percent) sought legal advice. In Taldy-Korgan, more people with higher education (67 percent) also receive advice from specialist or information for their agricultural activities while only 17 percent of those with unfinished secondary school or secondary technical education received advice from specialists. A similar response can be seen with different income level households. More people (35 percent) who were younger than 30 years of age sought specialist advice than those over 50 years of age (18 percent) in Akmola. Nine percent of the total households in both oblasts took information on agricultural techniques. Farming and livestock husbandry technology was the highest among those who said they sought specialist advice or information for their agricultural activities. Agriculture and other techniques were the second highest. Specialists from the farm administration provided most of these services.

Workers on the base farms mainly showed an interest in consultations regarding technological matters. The owners of the peasant farms and the heads of the base farms sought consultations on marketing and financial issues. Eighty-five percent of those who consulted specialists received such consultations free of charge. However, 56 percent said that they were willing to pay for such consultation in the future. The question of readiness to pay for the services of special consulting firms was explored in more detail in focus group discussions.

### Farmers Associations

Of the farms included in the SA, cooperation or association of farmers and workers on base farms was quite limited. Only nine percent of rural people participated in voluntary associations established to manage production, financial and commercial problems. Seventy-two percent of the participants in voluntary associations were in Taldy-Korgan Oblast while only 28 percent were in Akmola Oblast.

Peasant farm associations or other voluntary associations could solve many problems related to farming. Since most farms face similar problems—acquiring and using equipment, irrigation and fertilizers, and applying new agricultural technologies—collectively, they may be able to address these problems more effectively if they had associations. Fifty-five percent of the farmers said

> *"We need the Association for the following purposes: first, to have a unified legal consultation center, secondly, to have an information bank on prices, suppliers, customers, thirdly, to defend our political and social interests. The Association should combine the economic, political and social interests of the independent farmers and to defend them."*
> *Independent farmer*

that they would like to unite with other farms to establish associations of peasant farms. These associations would be able to provide information and consultation on legal and financial matters,

> *"The Association should exist as a public entity. The contributions may be symbolic. It should be self-supporting, it should "feed" itself. Today, it is beneficial if we start to produce final products: flour, bread, sausage, force-meat." Independent farmer*
>
> *"We need a good leader to establish the Association. Also we need good economists, agronomists for the Association." Independent farmer*
>
> *"The institutions of social sphere are closed. All kindergartens are closed or sold. Bath-houses were sold. As for the schools, every day they say that they will be either paid ones or will be closed. The parents repair schools at the own expense. Free hospitals were closed." Farm worker*

which could have a significant effect on the farming sector. However, despite a need for associations and awareness of their purposes during the early years of privatization, lack of resources and experience have hindered their establishment.

### Access to Information

The flow of information in the rural areas is very restricted and very poor. Farm managers and some other educated people get information that allows them to protect their interests, but other rural residents are at the mercy of others. Few people know about their rights and responsibilities in such areas as the management of base farms, access to land and property shares and convertibility of land

> *"The basic mass of population (99 percent) didn't have any idea of the legislative acts which were issued. The State or the Parliament that issued these documents didn't inform the people. Therefore, only those people went to take lands who, due to their status, knew about this who worked in "the system." Independent farmer*

shares. Similarly, access to technical information, ranging from crop management to marketing and business management, is very limited. One major challenge for the project is to disseminate information about a broad range of topics in order to empower independent farmers, farm managers, farm workers and others in the countryside to take advantage of new opportunities and to protect their interests.

During in-depth interviews, practically all the heads of base farms said that they held meetings of farm members to explain the principles and processes related to privatization, as well as the rights and opportunities of farm members. However, 29 percent of the workers on the base farms, the only group to whom the question was asked, denied that such meetings were held. Thirty-nine percent of all households could not answer what privatization means (Table 5.19). More than 20 percent of the heads of households claimed that their farms had not been privatized or were unaware of the privatization. Those who were aware of privatization considered it to be a formal act which in reality had not changed working conditions.

Lack of information—legal, financial or technical—was one of the major issues that undermined people's trust in the privatization process. The SA revealed a wide information gap between farm workers and those with higher positions or specialized status. The information gap was manifested with regard to getting

> *"We need information on where we can find a good market, where we can buy good equipment, where we can obtain credit and what privileges will be granted. We need information on new taxes." Independent farmer*

land and property shares early in the privatization process, between 1991 and 1995. In 1997, however, the Ministry of Agriculture and Oblast Agriculture Department carried out an active information campaign to inform farm members about the privatization process and the requirements of the new Civil Code. As a result, a great number of farmers in each oblast took

Table 5.19  Which of the following describes the present situation on your farm?
privatized former sovkhoz, kolkhoz, etc. (%) (n=474)

| | |
|---|---|
| Many members of the farm got land and property shares which they applied into individual use, and later organized farmer association(s) | 2 |
| Many members of the farm got land and property shares, some of whom created private farms without joining associations | 26 |
| People work as in former times, but know that they have shares and get (or will get) dividends | 38 |
| There were no changes in production or in distribution. People work as they worked in former times, they do not see any impact of privatization | 30 |

Source: Household Survey.

their land share and broke away from the base farms to start their own independent farms. Thus, independent information and advice are important factors in the successful implementation of privatization.

Sixty-two percent of the people got information abut credit or other advice from friends and acquaintances. Other sources of information were radio broadcasts and television (40 percent). Radio and television were considered to be the most reliable information sources, but the shortage of electricity made them virtually inaccessible. Newspapers were a source of information for 29 percent of the respondents, but few obtain them directly. Rather, they are passed from one household to another, thus most of the news is out of date when they are read. Four percent read specialized magazines and only two percent sought advice from specialists. However, to solve economic problems, 24 percent of the households sought advice from specialists of one type or another who were not professional consultants. Among the owners of the peasant farms, 47 percent sought the services of specialists.

### Monitoring and Evaluation

On one hand, the project will have a relatively narrow focus--business planning and credit for agriculture and agribusiness. On the other hand, it has very high aspirations, as these initiatives are perceived to be key to both economic and institutional development in the transition to a market economy in rural Kazakhstan. If successful, the project will have four principal outcomes. First, the technical and business information and capital it provides will enable independent farmers and managers of farms and agribusinesses to make their enterprises commercially viable. Second, in doing so, it will significantly redress the existing gross inequity in access to information in the rural areas. Third, it will introduce new, client-centered institutions, and reorient the approach of existing institutions, notably banks. Fourth, by delivering information to rural residents about their rights and responsibilities, as well as alternatives and opportunities, it will encourage nascent entrepreneurs to break away from existing relationships; and by providing capital, it will empower them establish new enterprises that produce the goods and services needed for a vibrant rural economy.

The project will be implemented in Akmola and Almaty Oblasts as a prelude to a national program. The first phase will thus constitute a pilot program, for which systematic monitoring and evaluation are essential to maximize learning for application in subsequent phases. The monitoring program will focus on demand, delivery, social and economic impacts and emerging opportunities. The monitoring program is designed to assess:

- Changes in the composition, magnitude and content of demand for advisory services and credit;
- Equity and inclusiveness in the delivery of information and services;
- The adequacy, responsiveness, accessibility and effectiveness of the Advisory Centers and participating financial institutions, as well as their client orientation;
- Economic and social impacts on farm and community levels of public information campaigns, Advisory Centers and Networks, restructuring plans, investments and the divestiture of social infrastructure;
- The impact of small-scale investments in local commercial and service enterprises, and emergence of new opportunities.

## Project Implications

The findings of the SA have a number of implications for the design of a project intended to help farmers in the aftermath of privatization and restructuring. They can be summarized as simple directives:

**Spread Information Widely**. Too many rural people are still poorly informed about the privatization process, their rights, options and implications for their future. Without adequate information, they remain unable to protect their own interests. Project implementation will therefore be preceded by an intense public information campaign that covers the following topics in clear, understandable language:

- Privatization and restructuring options;
- Rights and responsibilities of farm managers and members in corporate farms of different types;
- Characteristics of profitable base farms;
- Opportunities and constraints in independent farming; farm management principles, including accounting; and production, marketing, processing and relative profitability of different major crops.

Drawing on materials prepared under the IDIP, and experience gained in the IDIP, the information will be presented in simple, but not simplistic formats, probably consisting of different brochures. Distribution will be widespread in the target oblasts, designed to reach all members of farms, as managers cannot be counted upon to pass out information that might limit their opportunities. Until farm members are well-informed, they remain subject to manipulation by managers.

**Regularize Land Share Transfer and Lease Arrangements**. Many farm members have transferred, sold or leased their land shares under arrangements that are open-ended, without specifying the reciprocal rights and responsibilities of the farms and farmer managers, and without bringing real benefits to the people who relinquished their shares, either temporarily or permanently. Before project funds are made available to any corporate farm, farm members will be informed of their rights and options, and an assessment will be made of the status of shareholdings. If it is determined that transfers have occurred through coercion, this must be rectified before a farm can be eligible for a loan. This means that transfers must be re-negotiated, lease arrangements must be regularized, and the re-allocation of shares must be carried out publicly and transparently to ensure equity.

**Promote Flexible, Dynamic Arrangements between Shareholders**. An uncertain number of farm members may eventually wish to exercise their legal option to become independent farmers, especially in Almaty Oblast. The project will thus promote restructuring and financing arrangements that preserve the option of farm members to separate from the base farms.

**Get Capital Flowing.** Capital demand is real. Farmers and farm managers understand very well that their opportunities are limited and declining due to the lack of short-term and long-term capital for operations and investment, respectively. Compared to the management of base farms, independent farmers appear to be responsive to market signals, however weak, and to adopt diversified strategies that enable them to experiment with new approaches in cropping and marketing while maintaining their solvency. They need capital to increase their responsiveness, stimulate their imaginations and intensify their operations. Base farms need capital to raise yields, rationalize sunk costs and transform their operations over time. In short, capital is needed to induce restructuring, to establish new farms, large and small, and to increase production, productivity and profits. Equity and effectiveness require the project to give priority to investments that will maximize overall impact. For the sake of equitable and broad based development, and to ensure that the credit demands of large units do not overshadow the needs of smaller units, a portion of the credit will be reserved for small independent farms.

**Make Business and Technical Advice Practical and Accessible**. Advisory Centers will need to prove their value to farmers who are justifiably skeptical of initiatives promoted by the public sector and unfamiliar with private sector operations. Responding to the findings of the SA, Advisory Centers will give priority to financial and legal advice. In addition, to the extent possible, they will provide access to market information or market information sources. The mobile teams will have a particularly important role to play bringing advice and resource people to the more distant rural areas. The Advisory Centers will monitor demands and utilization as well as respond to changing priorities and levels of sophistication over time.

**Prepare for Mitigation and Maximize Employment Opportunities**. To become profitable, many farms will have to divest social infrastructure and reduce their labor force. Although direct mitigation is beyond the scope of the Ministry of Agriculture, efforts will be made to help local administrations assume responsibility for social infrastructure and to provide start-up capital for local commercial and service enterprises. This will be done both to reduce unemployment and to serve local needs through the private sector. Loans to base farms will be preceded by an inventory of social infrastructure and the commercial and service structure of associated communities. The social infrastructure inventory will inform investment decisions, as well as discussions between the farms, communities and local administration. The commercial and service inventory will help assess unmet local needs which can be addressed by offering targeted training programs and credit opportunities, particularly to those who become unemployed as the result of Agricultural Post-Privatization Assistance Project.

**Monitor Demand, Supply, Impacts and Opportunities**. Systematic social monitoring will be an integral part of project implementation to maximize learning and make adjustments, accordingly. The project will be implemented through a three-phase program loan that includes two major review stages when project progress and impact are evaluated before deciding to move on to the next phase. A follow-up SA should be undertaken at each stage to take stock of progress and apply lessons learned from the ongoing monitoring program. The routine monitoring results and special studies, as needed, will identify strengths and weaknesses in project design and implementation arrangements and develop ways to incorporate lessons in subsequent phases. The principal indicators for ongoing monitoring are changes in demand for credit and advice; the responsiveness and effectiveness of Advisory Centers; the influence of public information and communications initiatives; the growth of community-level economic opportunities; and the fate of social assets and basic service delivery.

# Annex 1[8]
# Eligibility Criteria for Beneficiaries and Sub-Projects

*Eligibility Criteria for Beneficiaries*

The Project is intended to support a transparent process of farm restructuring where recently privatized state and collective farms with the support of advisory services undertake fundamental changes in the management, ownership structure, size, production mix, and technologies necessary to improve productivity and achieve sustainable commercial viability. The farming community is currently facing extreme financial difficulties and is in a considerable state of flux since the farms were initially privatized and property shares and entitlements to land were distributed among workers and farm managers. Much of the change that has taken place since the privatization is part of the painful but necessary adaptation that the farms must make to the new realities of the market place. But, as the social assessment has shown, there are also instances where the changes being made at the farm level are concentrating control, if not outright ownership of the farms into the hands of select individuals through transfers of shares and property entitlements in an atmosphere of mis-information and outright coercion. The project is intended to promote informed decision making on the use of property and land shares in the restructuring process and the advisory services will focus on issues of share ownership and transfers at the initial stages of restructuring.

To ensure that the credit component is supporting a transparent and equitable restructuring process it is necessary to establish clear guidelines for the eligibility of potential sub-borrowers. The PFI's will have the primary responsibility for checking for eligibility though a system of prior review will be implemented so that loan applications from sub-borrowers with large land holdings will be reviewed by either by the advisory centers or the Bank before being approved. To be eligible for loans under the project, farms must:

- be 100 percent privately owned and duly registered as a family farm or legal entity recognized by the legal code (joint stock company (JSC), production cooperative, or partnership);

- have a share holding structure that is transparent and fully documented. Where shares in a legal entity (JSC, production cooperative, or partnership) have been transferred to the farm managers by the farm workers or pensioners, the terms and conditions of those transfers must be fully documented in a formal lease or sales agreement that specifies the terms and condition of the transfer;

- have undergone or be in the process of restructuring and able to demonstrate in a business plan their financial and commercial viability and ability to repay the loan.

---

[8] The World Bank Staff Appraisal Report (Report no. 17789-KZ): Republic of Kazakhstan—Agricultural Post Privatization Assistance Project

The credit line would also be available to non-farm commercial activities in farming communities affected by restructuring including agro-processors, input suppliers, commodities traders, equipment hire companies, and farm community service providers such as retail outlets, canteens, recreation facilities or pharmacies. Such entities would need to be 100 percent privately owned and legally registered with a business plan that demonstrates their commercially viability and the ability to repay the loan.

### *Eligibility Criteria for Sub-Projects*

In order to achieve the Project objectives through supporting the targeted beneficiaries as outlined above, eligible sub-projects would include:

- Agriculture related activities—agro-processing, agro-services—that are directly linked to farms receiving a credit under the Project;

- Sub-projects with a self-contribution of at least 10 percent for rehabilitation or modernization, or 25 percent for new expansion;

- Sub-projects that can demonstrate a debt service coverage ratio of at least 1.3 over the life of the sub-loan, calculated on the basis of the entity's total debts.

- Investment sub-projects in agro-processing, agro-service or supply enterprises that can be completed within 18 months from the date of the signature of the sub-loan agreement.

The line of credit would not finance the purchase or lease of land, dwelling construction or improvement, or the refinancing of existing debts.

# Chapter Six

# Uzbekistan Agriculture Enterprise Restructuring and Development Program

**Nilufar Egamberdi, Peter Gordon, Alisher Ilkhamov,
Deniz Kandiyoti, and John Shoerberlein-Engel**

## Background

### The Uzbekistan Agriculture Enterprise Restructuring and Development Project

The Government of Uzbekistan (GOU) has expressed its commitment to promote a voluntary transition of farm management from the public sector to the private sector. This program will significantly affect families living in Uzbekistan's rural areas. Although the entire nation has been affected by extensive changes in the economy, farmers of the former state and collective farms (sovkhozes and kolkhozes) will be even more affected by the radical reorganization of their economic activities and changes in their access to social services. Cognizant of the far-reaching consequences of this program and of the need for this transition to be well organized if negative consequences are to be avoided for the rural population, GOU has formally requested the Bank to help design and support programs to assist these target groups during the course of the transition. The Farm Restructuring Program is the first medium through which the World Bank will provide such support.

Farm restructuring and privatization are intended to profoundly affect farmers' capacity to integrate themselves into the market economy; they present both challenges and opportunities for greater prosperity. This process also has fundamental implications for the provision of crucial social services. The success of the program will depend on the effective participation of the target population. Therefore, the participatory assessment of these target groups has been critical to the identification of strategies to ensure that farm restructuring involves the community at all stages of program design and implementation. This will guarantee sustainability by assisting target communities to make choices that meet their needs and, therefore, will improve their socioeconomic conditions in the long term.

### The Country Context

Uzbekistan lies in the heart of Central Asia between the Amu Darya and Syr Darya rivers. It has been a center of commerce and a sought-after prize in the struggle for political and economic influence in the region. Uzbekistan is the largest Central Asian state with a population of 23.7 million. It sees itself as a key country in the region, premised on its size, location, leadership, and abundant natural resources. It is relatively more homogenous culturally than its Central Asian neighbors. Surrounded by Kazakhstan, Kyrghyzstan, Tajikistan, Turkmenistan and Afghanistan, this semi-arid country covers 447,400 square kilometers. Although there are fertile oases and high mountain ranges in the south and east, respectively, almost four-fifths of Uzbekistan consists of flat lowlands, and the central and western regions are predominantly desert.

Uzbekistan comprises the Republic of Karakalpakstan and 12 provinces (viloyat), 124 cities and 157 districts.  The capital is Tashkent.  In August, 1991 the country declared its independence from the USSR as a multi-party republic with a Presidency and a unicameral Supreme Assembly or Oliy Majlis.  Representatives of the national executive are the Chairman of the Cabinet of Ministers in the Republic of Karakalpakstan, and the hakims (local administrators) at the regional, city and district levels.  Political parties include the ruling People's Democratic Party, Fatherland Progress Party (Vatan Tarakiyoti), Adolat (Justice) Social Democratic Party, and the Democratic National Rebirth Party (Milly Tiklanish).  It also has several political pressure groups such as the Bilrik (Unity) Movement, the Islamic Rebirth Party, and the Erk (Freedom) Democratic Party.

Uzbekistan is the leading producer of cotton among FSU nations and one of the largest in the world.  Agricultural production also includes sheep, cattle, silkworms, vegetables and a wide variety of fruits such as melons and grapes.  The country is the second largest gold producer in the FSU and has sizeable reserves of petroleum, natural gas, coal, copper, zinc, tungsten and various ores.  Industry is not well developed; during the Soviet period, however, Uzbekistan was the main producer of machinery and heavy equipment in Central Asia, but production has declined in recent years.  It also manufactures automobiles and aircraft.

## Population

Uzbekistan is a cultural mosaic reflecting its roots in the ancient Sogd, Bactria, Margiana, Shash, Khorezm and Turan civilizations and the influences of Persia, Arabia, China, Greece and the Great Steppe. It has more than 130 ethnic, tribal and linguistic groups, but Uzbeks comprise over three-quarters of the population and Uzbek is the dominant language.  Uzbeks belong to the Chagatai branch of the Turkik language subfamily. Russians, Kazakhs, Tajiks, Tatars, Karakalpaks, Koreans and Jews are other significant ethnic groups that have lived in Uzbekistan for centuries.  Rural areas tend to have higher concentrations of ethnic Uzbeks, Kazakhs, Tajiks and Kyrghyz (Table 6.1).

| Table 6.1  Ethnic composition of Uzbekistan (percent) ||
|---|---|
| Uzbek | 76 |
| Russian | 6 |
| Tajik | 5 |
| Kazakh | 4 |
| Karakalpak | 3 |
| Tatar | 2 |
| Korean | 1 |
| Kyrgyz | 1 |
| Turkmen | 1 |
| Ukrainian | 1 |
| Source: Uzbekistan State Committee for Forecasting and Statistics, 1995 ||

The current process of nation-building and de-colonization from the FSU has sparked ethnic tensions. [1] The borders imposed by Russia and later the Soviet Union aimed to impede the emergence of a Pan-Turkic nation; they respected neither the history nor ethnic composition of the local population.  For decades they raised great suspicion, territorial disputes and claims between the countries. Ethnic tensions have also appeared between the indigenous population and ethnic groups (Meskheti Turks and Crimean Tatars) who settled in the area by free migration or by deportation during and after World War II.  In 1989 and 1990 the region faced violent ethnic

---

[1]     Stefan Klotzli. 1994. The Water and Social Crisis in Central Asia – A Source for Future Conflicts?

conflicts in Ferghana Valley between Uzbeks and Meskheti Turks and Uzbek-Kyrghyz riots in Kyrghyzstan. The population explosion in Central Asia since the late 1970s, particularly in Ferghana Valley, has exacerbated economic, demographic and ecological stress. There is pandemic unemployment, shortages of agricultural land and housing, and increases in narco-trafficking, crime, and political crisis in the region.[2]

The Government of Uzbekistan (GOU) has a policy to promote the widespread use of the Uzbek language. Accordingly, Uzbek was declared the national language soon after independence. About 75 percent of the population speak Uzbek and 33 percent speak Russian. However, only 4.6 percent of ethnic Russians speak Uzbek fluently. Tajik is used as a second language by Tajiks, Persians and Uzbeks in Samarkand, Bukhara and some parts of Ferghana Valley and Zerafshan Valley. Kazakhs and Uighurs speak their own languages. Eighty-eight percent of the population is Sunni Muslim, 11 percent are non-religious, while only one percent are Eastern Orthodox (mostly Russians).

During the Soviet period the population of Uzbekistan and Central Asia as a whole grew rapidly. High birth rates and decreasing death rates fueled the growth. The quantity and quality of healthcare contributed significantly. Most rural families were large. Families with seven or eight children, and even 10, were not uncommon. However, infant mortality remains high, especially in rural areas. Areas around the Aral Sea have infant mortality rates as high as 100 per 1,000, perhaps caused by airborne salt, toxic dust and impure drinking downwind from the Sea and downstream from the major irrigated areas. The growth of Uzbekistan's population is summarized in Table 6.2.

| Table 6.2 Population (thousands) | | |
|---|---|---|
| | 1990 | 1995 |
| Total | 20,227 | 22,467 |
| Males | 9,997 | 11,138 |
| Females | 10,231 | 11,329 |
| Urban | 8,212 | 8,663 |
| Rural | 12,015 | 13,804 |
| Under working age | 8,687 | 9,656 |
| Working age | 9,948 | 11,083 |
| Over working age | 1,592 | 1,728 |
| Source: Uzbekistan State Committee for Forecasting and Statistics, 1995 | | |

In the last 25 years the population has doubled from 11.8 million to 24.2 million in 1998, increasing two percent annually. At this rate of growth, Uzbekistan's population will be about 50 million in 2038 and account for half of the population in Central Asia.[3] A significant portion of this growth will occur due to demographic momentum, as the large number of people born in the 1960s and 1970s now have children, even though they have fewer children than their predecessors.

Despite a recent decline in fertility in urban areas, growth remains high in the rural areas, which account for 90 percent of Uzbekistan's population growth.[4] Lower levels of income, education, and access to contraceptives, as well as pronatalistic cultural attitudes

---

[2]　　Anara Tabyshalieva. 1998. Central Asia's Quest for Regional Cooperation: Preventing Ethnic Conflict in the Ferghana Valley.

[3]　　Gulnara Kuzibaeva. 1999. Fertility Transition in Uzbekistan.

[4]　　Ibid.

toward women, all contribute to higher fertility rates in rural areas. Early marriages are becoming more widespread, which also contributes to higher birth rates and higher maternal mortality. In 1997 maternal mortality reached nine percent, compared to 3.2 percent in 1990. Infant mortality rates also are high, with estimates varying from 24.3 per thousand (Uzbek Ministry of Health) to 71 per thousand (U.S. Census Bureau). Infant mortality is caused primarily by poor health conditions for women, lack of health services in rural areas, extended participation of women in agricultural labor and widespread cross-cousin marriages.[5]

The most densely populated areas of Uzbekistan are the Ferghana Valley in the east and the Zerafshan River Valley in the south-central region. The dominance of irrigated agriculture and the recent out-migration of skilled Europeans have contributed to increasing the agrarian character of Uzbekistan since independence.

### Poverty and Living Standards

Uzbekistan is one of the poorest countries in the FSU despite its wealth of natural resources. A Family Budget Survey taken in the late 1980s estimated that 44 percent of the population lived in poverty. Rural areas are disproportionately impoverished with poverty rates of about 57 percent.[6] The 1997 UNDP Human Development Report concluded that in the early 1990s as much as 75 percent of the population was impoverished, while the middle class essentially disappeared and the number of wealthy declined.

Unemployment in rural areas, particularly in agriculture, reached about 27 percent in 1995 according to the 1997 United Nations Human Development Report.[7] Other reports put unemployment between 12 percent and 14 percent. Young people and women are disproportionately unemployed, as are residents of Karakalpakstan and the Ferghana Valley. The increasing population, regional economic disparities, and the out-migration of job-creating skilled labor and management contribute to unemployment. The rapid decline of the economy and growing political uncertainty in the early 1990s drove many workers with marketable skills to leave the country. The modest growth of the private sector has somewhat increased demand for industrial and service-oriented labor.

Income disparities increased for several reasons. Greatly reduced state budget revenues have hurt low-income people who depend on government social welfare programs. The collapse of labor demand by state industries and agricultural collectives has cut deeply into job opportunities for lower-skilled laborers. Many people who cannot survive on their wages have begun to engage in entrepreneurial activities and the sale of agricultural products grown on their own small household plots. To increase their earnings, highly educated professionals have flowed out of science, government and engineering into small-scale agricultural production, wholesale and retail trade, personal services and private transport.[8]

---

[5]        Ibid.

[6]        World Bank Uzbekistan Country Assistance Strategy, 1998.

[7]        In contrast, official data from GOU indicates unemployment is no more than 0.5 percent. GOU counts only those who register as unemployed.

### *Impact of Transition on the Rural Population*

Since the collapse of the Soviet Union, Uzbekistan has lost its subsidies from Russia and experienced social unrest, political turmoil, declines in personal income and employment and increases in poverty.

GOU has advocated gradual implementation of reforms, fearing that rapid privatization and dependence on the free market would engender excessive social, economic and political instability. It attempted to implement a national reform program in which state subsidies, price controls and gradual wage increases were used to shield hyperinflation. As a result, the state has retained control over foreign exchange, investment, and many state-owned enterprises and properties; however, the social safety net deteriorates under heavy government debt and incomes plummet under hyperinflation.

One million of the 3.5 million employed in the rural areas are under-employed according to the International Monetary Fund (IMF). There has been limited transformation of state collectives into joint-stock companies or cooperatives, but most rural people remain wedded to collective farms to access land, fodder, fuel, inputs and welfare. That is, one person in the household may remain a member of the collective while others pursue other income sources. Inflation, price deregulation, wage arrears, in-kind payments, state quotas for production and sales and the tattered social safety net further challenge rural households. Small-scale household plots have necessarily provided subsistence agricultural production and some also generate cash incomes for essential living.[9]

Uzbekistan has attempted to achieve rapid economic growth without major political changes. A central component of the approach is to attract foreign investment. Therefore, Uzbekistan has sought support from the IMF, the World Bank, the Asian Development Bank, the European Bank for Reconstruction and Development (EBRD) and other financial institutions. GOU has established three comprehensive objectives in conjunction with the World Bank:

- Strengthen fiscal conditions and balance of payments performance consistent with low inflation and prudential levels of external debt;
- Increase sustainable economic growth; and
- Develop human resources and a social safety net.

### Agricultural Sector

Uzbekistan depends on irrigated agriculture. The GOU policy is to maintain large-scale employment in agriculture and to attract investment in agricultural processing. It directly controls production and prices of inputs, production and processing, and it also

---

[8]     Ibid.

[9]     Ibid.

controls exports of cotton and wheat. This maximizes government revenue. Agriculture generated 30 percent of GDP in 1998, employed 44 percent of the workforce and generated around 50 percent of export receipts. Most agricultural and light-industrial output is related to cotton, which is produced throughout the country as a result of Soviet-era policies. Cotton production has declined steadily in recent years, and has remained below four million tons since 1994, compared to about five million tons in the 1980s.

| Table 6.3  Use of agricultural lands | |
|---|---|
| Type of production | Percent of cultivated land |
| Crops | 66 |
| Cotton | 30 |
| Grains | 15 |
| Animal feed crops | 12 |
| Vegetables | 3 |
| Orchards, vineyards | 6 |
| Livestock | 34 |

Cotton and grain are the most important crops in Uzbekistan; significant crops include fruits (apples, apricots, peaches and berries), vegetables cucumbers, tomatoes and potatoes), milk and silk. Cattle-raising, which takes up one third of the agricultural land, is especially important in Qashqa-Darya Province (Table 6.3).

About 53 percent of the arable land is pasture, fit only for sheep grazing; 36 percent is non-agricultural; and about 11 percent, 4,500 square kilometers, is cultivated. There are 2.7 people per hectare of arable land in Uzbekistan, compared to 0.5 people per hectare in the Ukraine or Belarus. Rural labor productivity is lower in Uzbekistan than in other Central Asian republics, but rural residents utilize the land intensely—98 percent of it is irrigated. Uzbekistan's two main rivers, the Amu Darya and Syr Darya, have been diverted for irrigation, causing an environmental catastrophe in the Aral Sea, now half of its former size.

Since independence, GOU has sought to become self-sufficient in wheat, using quotas, subsidies and directed credits. Wheat production increased ten-fold in the 1990s, at the expense of cotton, corn, fodder crops and vegetables. As production shifted from state/collective farms to private farmers, the decline of fodder crops aggravated livestock productivity. Other factors also contributed to declining livestock productivity, including the deterioration of herd quality, misuse of pastures, waste of animal feed and lack of proper equipment. Cattle and cow production has increased while poultry and pig production has decreased.

Yields declined during the 1990s, particularly for rice, but cotton yields have increased. Technical factors limiting yields and profitability include poor crop rotation practices, poor irrigation service, lack of farm equipment and lack of inputs and quality seed. Exports of cotton fiber account for about 10 percent of GDP, far more than any other commodity, including gold. Meat and potatoes also may be competitive products in a de-regulated economy. Uzbekistan's largest import is foodstuffs, nearly six percent of GDP in 1997.

Cotton is still the most important subsector, comprising 40 percent of GDP, 80 percent of government tax revenue and a significant amount of income, employment and foreign exchange. Uzbekistan has focused on producing cotton and exporting it to Russia for over 50 years. Cotton created a network of other industrial branches including irrigation networks, machine building plants, chemical facilities, hydro-electricity, cotton-processing and some textiles. Soviet investment focused almost exclusively on cotton production, with the bulk of value added generated outside of the cotton producing area, especially in Russia. Investment in industry and social programs remained very low in Uzbekistan during that period and since independence.

## Agricultural Reforms

After independence, Uzbekistan initially demonstrated a commitment to transfer management of farms and the rural economy from the public sector to private hands. GOU approved and began to implement a number of significant reforms:

- In 1991, collective farms, state farms and other agricultural enterprises were exempted from tax on profits and from the standard 11 percent deduction of wages.

- From 1992 to 1994, GOU gradually de-regulated prices for agricultural products, focusing on those less critical to development. The system of state-controlled purchasing was gradually modified to reduce the range of products covered and to transfer control of decisions about crop mix and quantities to state and collective farms.

- In 1993, state farms were transformed into collectives and joint-stock agricultural enterprises. The new organizations were forgiven debts owned to banks and suppliers. The state reduced its share of raw cotton to 80 percent of production.

- In 1994, a new "Model Charter for Collective Farms" was approved, creating ambiguity of control and ownership. It required democratic management of collective farms with farm chairmen elected by the farms' general assemblies once the candidates for chairman were approved by the district government (hakim). General assemblies could dismiss a chairman based "on the opinion of the district and province hakims, as well as the higher order agricultural institutions."

- In 1994, the President declared that those who are not productively engaged in agriculture are to be transferred to nonagricultural work such as industry and services.

Since the initial period of reforms, however, progress in implementation has been slow or non-existent. In late 1996, GOU reneged on key policy reforms and implemented highly expansionary macroeconomic policies. GOU states that it remains committed to economic reforms and market liberalization but wants to slow down the speed. However, over the last several years there has been a significant divergence between plan and implementation in Uzbekistan.

Current conditions result from the incomplete implementation of perestroika-era reforms, the collapse of the Soviet Union and, as described below, the partial introduction of market-oriented reforms. Living conditions have deteriorated severely throughout Uzbekistan, but rural areas have been hit hardest. State workers work without pay, social security benefits and public services have disappeared and infrastructure is decaying. Land, water and gas supplies are in short supply, population is increasing and educational levels are low. Processing capacity is low and located away from production, straining transportation and raising costs. Government officials support collective farms over private enterprise farming and household farm operations.

In many respects, the transition has yet to come. Rural people face an emerging configuration of new pressures and uncertainties in their environment as many elements of the soviet system are merely refashioned and reinterpreted in new transition context. Agricultural reforms have done little more than to convert some former state farms into joint-stock companies or cooperatives. This has introduced nominal rather than substantive changes in the ownership and control of resources. The dilemma for rural producers arises from the fact that they find themselves still locked into the collective farm system, for which formal membership is crucial to access essential resources such as personal land plots, fodder, fuel and welfare. However the drying of the public sector resources, inflation, price deregulation, arrears of wages, in-kind payments, state order on production crops and sales, and above all, lack of land ownership presents new challenges to rural households. Their small household plots play new roles as a source of subsistence to the household and as the source of a marketable surplus that partially compensates for declining cash incomes. For the rural population, sustaining livelihoods requires juggling a mix of wages paid in kind and cash, self-sufficiency in basic foods and petty trade, while social benefits decline.

The GOU and the World Bank share the belief that the successful restructuring of the agriculture sector will contribute greatly to the overall health of Uzbekistan and its ability to implement reforms in all sectors. As part of the reform of the agricultural sector, GOU intends to phase out monopsonistic government procurement of cotton and wheat and liberalize those markets with privatization and competition. It also plans to phase out subsidies of agricultural inputs and the state's monopoly on sales and distribution of inputs, agricultural production and agricultural processing. Accordingly, land will be distributed from the state to private individuals and groups. GOU also intends to improve agricultural support services, credit services, research extension and the quality of its cotton production.

## The Social Assessment

### *Social Assessment Objectives*

The Social Assessment (SA) for the proposed Agriculture Restructuring Program in 1996 had four objectives:

- assess strengths and weakness of the ongoing farm restructuring, as perceived by the stakeholders;
- assess how households are vulnerable during the restructuring (employment and social services);
- assess attitudes of primary beneficiaries toward the restructuring; and

- identify mechanisms by which stakeholders can participate in and resolve issues satisfactorily.

## *Social Assessment Methodology*

The SA included six of the thirteen provinces, selected on the basis of diversity and representativeness of varying agro-climatic conditions found in Uzbekistan. These comprise urban centers (Tashkent and Samarqand); areas of limited water resources but significant livestock breeding (Qashqa-Darya); areas distant from national centers with significant cross-border trade (Kharazm); areas with dense populations (Namangan and Farghana); and areas with significant cotton production (Kharazm). These regions cover the northwest, south central and eastern parts of the country and best represent various conditions throughout Uzbekistan. A team of local social scientists carried out the following components of the SA:

- A survey of 981 households across 65 communities randomly selected from six provinces in Uzbekistan;[10]
- Structured interviews with 51 local, regional and national government leaders;[11]
- Informal interviews and consultations with households, community leaders and government officials; and
- A review of literature, official documents and data.

The household survey addressed several subjects: 1) characteristics of households such as their composition, sources and levels of income and mechanisms of community support for the households; 2) the type of work done, 3) mechanisms for extending social services; and 4) channels of communicating information to the population and how policy

---

[10]     The provinces comprise about 80 rural districts. These districts were stratified into those with less than 80,000 people, those with between 80,000 and 105,000 people, and those with more than 105,000 people. Using probability proportionate to size (PPS) criteria, the quota (the number of surveys) for each district size was determined—250 for the small districts, 360 for the medium-sized districts and 440 for the large districts. Using PPS criteria again, the number of districts of each size to be surveyed were calculated—five for each size of district. Each of these districts received an equal-size quota of surveys—50 surveys per district for small districts (250/5), 72 surveys per district for mid-size districts (360/5), and 88 surveys per large district (440/5). Then PPS criteria was used to choose settlements in each district to survey. Ten households in five settlements in small districts were selected (50/5), 18 households in 4 settlements in mid-size districts were selected (72/4) and 22 households in 4 settlements in large districts were chosen (88/4). The persons surveyed in each settlement were selected randomly from the population of that district.

Of 1,050 surveys attempted for the household survey, answers were obtained from 981 households: 188 in Farghana, 132 in Kharazm, 106 in Namangan, 240 in Qashqa-Darya, 201 in Samarqand and 114 in Tashkent. In each case, the household member determined to be most economically active was chosen for the because they are likely to be the most knowledgeable person in the household. In most cases, these persons were men of the middle generation of the household. If that person was not available, the next most economically active person or the spouse was interviewed.

[11]     For the survey of officials, 51 interviews were conducted with local government officials who resided in the districts where the household surveys were conducted. Additional data were gathered from the Ministry of Agriculture, the State Committee for Planning and Statistics, the Tashkent State Economics University, agricultural economics experts and publications.

designers and implementers obtained feedback. For each of these subjects, respondents were asked: 1) the status of current conditions; 2) the degree of satisfaction with current conditions; 3) vulnerabilities under current conditions; 4) how the situation changed following independence and the reforms; 5) perceived options for improving current conditions and the feasibility of these options; and 6) the perceived roles of individuals and organizations in improving agricultural policies, programs and conditions.

Interviews of household members, community leaders and government officials augment the survey to confirm and elaborate on quantitative findings. The mixture of qualitative and quantitative methods helped overcome some government interference in the survey process. Additionally, given traditions of the Soviet system, some survey respondents may have given optimistic answers to some questions. In other cases, respondents fearful of taxation may have understated their income.

## Four Pillars of Social Assessment

### *Key Social Development Issues*

Rural Uzbekistan suffers from a number of problems that cause income and employment to decrease. Some existed during Soviet rule—rapid population growth and shortages in water, gas and credit. Others are problems resulting from the restructuring process—dilapidated social welfare and public goods such as education, extension/technical assistance, water and gas.

Another set of problems results from incomplete implementation of sound reforms—small land plots and land shortages, shortages in agricultural processing, poor communication of government reform efforts and little or no popular participation in developing and implementing reforms. These problems of incomplete reform are most critical. Implementing effective reforms will promote economic growth and increase both the tax base and government's ability to address other problems more effectively. The main findings are described below.

### Poverty

Average annual income is very low in Uzbekistan and has been falling over the past 10 years. Table 6.4 portrays household income by province and across the study group. Per capita rural income of between $76 and $125 dollars is very low. If the black market exchange

| Table 6.4 Average annual income, 1995 (U.S.$)* | | | |
|---|---|---|---|
| Province | Per household | Per capita | Per worker |
| Tashkent | 644 | 125 | 211 |
| Kharazm | 625 | 91 | 265 |
| Farghana | 584 | 87 | 193 |
| Namangan | 456 | 75 | 186 |
| Samarqand | 438 | 68 | 207 |
| Qashqa-Darya | 272 | 43 | 168 |
| Total | 476 | 76 | 211 |
| * Average annual official exchange rate of 30 soums to U.S.$1 used to calculate this income. Black market exchange rates averaged 40 soums to U.S.$1, making incomes 25% lower than indicated above. | | | |
| Source: Social Assessment, Household Survey | | | |

rate is used, per capita income would be about 25 percent lower than presented in the table.

Furthermore, rapid population growth in rural areas dilutes household income. The impact of rural population growth on income is illustrated in the differences between Tashkent and Qashqa-Darya. The ratio of per capita income of Tashkent and Qashqa-Darya is 2.9 ($125 : $43). But the ratio of income per working individual of the same rayons is 1.26 ($265 : $168). Qashqa Darya has much lower income relative to Tashkent when measuring by per capita income. This is due to the larger number of non-working children (and senior citizens) in the household, greatly diluting the power of household income in Qashqa-Darya.

Households spend a very large percent of their incomes on basic necessities such as food, clothing, housing and utilities (gas). The SA survey indicates that basic foodstuffs (84 percent of respondents), clothing and shoes (74 percent) and communal services such as housing, electricity and natural gas (63 percent) burden households the most (Table 6.5). Between one-third and two-thirds of the households are burdened significantly by expenditure on necessities or important social obligations. The far right column also indicates that many households cannot afford even these expenditures. Most households in Uzbekistan include three generations— children, parents and grandparents. Consequently, most households need medical care, child and elderly care, and education. One would thus expect a large proportion of households to spend money on these items. However, a significant proportion of sample households had not spent any money on these items in the preceding 12 months, indicating significant shortages of household income.

Table 6.5  Major household expenditures in past 12 months (percent of household respondents)

| Types of expense | Burdensome | No such expenditure |
|---|---|---|
| Medicines and treatment | 65 | 23 |
| Child and elderly care | 49 | 19 |
| Education | 38 | 34 |
| Weddings, funerals and community celebrations | 36 | 59 |
| Home repair and construction | 34 | 62 |
| Debt repayment | 18 | 80 |

A significant number of households spent no money on non-essential items or major purchases (Table 6.6). Most households do not have the power to purchase more than basic necessities. The fact that households are burdened significantly by the expenses of healthcare, child and elderly care, education, community obligations and home repairs and do not often purchase household appliances indicates that households have very little purchasing power.

Most rural citizens believe their income and standards of living have declined since 1986. Sixty-nine percent said they are in worse financial condition, 14 percent said their status is

Table 6.6  Households without major expenditure in the past 12 months (percent)

| Province | No purchase of appliances | No purchase of car or house |
|---|---|---|
| Namangan | 92 | 95 |
| Farghana | 85 | 86 |
| Tashkent | 84 | 91 |
| Qashqa-Darya | 79 | 84 |
| Samarqand | 72 | 77 |
| Kharazm | 54 | 71 |

Source:  Social Assessment, Household Survey

the same, and 15 percent stated they are in better shape. These opinions are held across all six provinces surveyed, with Qashqa-Darya being the most pessimistic (78 percent believe they are worse off) and Kharazm being the most optimistic (52 percent believe they are worse off).

**Social Capital**

Most citizens lack confidence in their ability to address their declining incomes by working in the private sector. Many doubt opportunities exist outside of the state/collective system and they count on government, not themselves, to fix their problems. There is pessimism that reforms and private sector initiatives succeed. At the personal level, the lack of experience, training, information, opportunities and start-up capital cause this pessimism.

The tradition of waiting for the government to solve problems makes this pessimism tolerable for many citizens. When asked how their income and employment problems can best be solved, citizens provided six common answers, four of which demonstrate their reliance on others:

- Create new workplaces—51 percent,
- Provide additional land—46 percent
- Get assistance from local authorities—24 percent, and
- Be provided with good jobs by kep people—20 percent.

Only two solutions require initiative by ordinary citizens:

- Professional training and education—33 percent and
- Capital to finance a farm or business—15 percent.

Less than five percent of respondents said greater "personal effort" is required to solve their problems. Such reliance on others presents a challenge to reforms and extension / training programs. Such programs will need to demonstrate clear benefits to motivate ordinary citizens.

> *"My husband got sick. He has some serious mental illness, so I had to leave him. He couldn't have a family. It's sad that it happened that way, but I had no choice. I am currently unemployed. Our kindergarten, the only one we have in the village, just closed down for the winter. There is no heating there. It is not the job I dreamed of, we work only several months a year and they don't pay us. For over a year I haven't seen any money although I used to get only 1,900 soum. They pay us in-kind, give us some foodstuffs. I also get my children allowances. It is only 1,050 soum per month for two of them, those are under age, and it's only once in three months. I wish I had a better job. I make cheese, yogurt, and sour cream and take to Tashkent. I take a bus twice a week to get to one of the city markets. It costs to travel back and forth, but I make some money to buy meat, potatoes, maybe some clothes. It's getting harder for me. My daughters are growing up, they need better clothes. Sometimes I cut on myself. If I need to choose between myself and my girls, I buy their clothes first. I just bought myself a scarf a few months ago for New Year's. My brother and his family help me a lot. They give us some food, even money when I need it. But I try to pay them back by helping them at the milk farm or do some garden work. I feel blessed that I have them. I know that they will help me marry my daughters. They know that unfortunately I cannot do it by myself. It costs so much these days to have a wedding or to prepare a dowry for a girl, and I have three of them." A divorced, under-employed mother of three daughters*

## Social Services

The difficulties rural citizens have in making a living are compounded by a parallel decline in the social safety net and provision of public goods such as education, healthcare and utilities. About 24 percent of the population age 16 and over has no source of income whatsoever. Of these, 41 percent are housewives, 36 percent are unemployed, 17 percent are students, five percent cannot work for health reasons, and one percent are pensioners who do not receive their pension. Of these groups, the unemployed, the sick, some women and a small number of pensioners are most vulnerable.

The unemployed do not benefit much from social safety net programs. Unemployment data are unreliable—official data show an unemployment level of 0.5 percent while findings of the SA suggest that unemployment is 10.5 percent nationally, with some provinces suffering unemployment as high as 20 percent. Of the 10.5 percent unemployed found by the SA, 84 percent did not receive welfare benefits (8.8 percent of surveyed household pool). About half of the unemployed are 23 years of age or older. About 83 percent of the unemployed have no more than secondary education.

Those who cannot work due to medical problems are also vulnerable. The government has traditionally provided medical care free of charge. However, 69 percent feel healthcare conditions have worsened in the last 10 years as doctors now charge patients for medical care and medicines due to their own income erosion. Other services such as drinking water and pre-school care are also no longer provided by the government.

Young, single women and those women working on state/collective farms are also vulnerable. The unemployment rate among women is 20 percent. Forty-one percent of the general population without personal income are housewives. Of these, 64 percent are aged 16 through 30 who stay home primarily to raise children. Many of the younger women have not had the benefits of working on state/collective farms such as pensions, maternity leave, etc. However the survey found that most women of working age are married (93 percent), and therefore can depend on spouses for social protection. It is the remaining seven percent

*A 45-year-old Russian widow, depends on two pensions to care for her eight children and extended family of five. In addition to her eight children, she cares for the three children of her oldest daughter, whose husband is in prison, and two other people. The family receives Lena's husband's pension, which is 8,000 soums per month, and the Afghan War pension of her older daughter's husband, which is 700 soums per month. In addition, Lena and her oldest daughter work at a state farm as milkmaids and receive 5,000 soums per month each, although their manager often pays them in-kind with meat, milk and flour. Altogether, the pensions comprise nearly half the family's income. Even with this income, the family subsists. They begin milking cows at 4:30 am and finish by 8:00 o'clock. They go to their jobs and work until 4:00 pm, and begin the afternoon milking at 4:30. This lasts until 8:00 o'clock, after which they prepare dinner and eat at about 9:00 pm. Then they go to bed. The family also grows potatoes, onions and tomatoes. Their earnings and pensions pay for care of the imprisoned son-in-law, gas and electricity bills, meat for two meals a month, flour, oil, sugar, tea and salt, and enough apples and candy for a family treat once a month. For breakfast they eat tea and bread. For lunch they usually eat tea and bread, but sometimes have soup. Dinner usually comprises pasta, but once a week they have a traditional Uzbek rice dish, palov. On some occasions they do not have enough to eat dinner. Her children cannot go to school in the late fall or winter because she cannot afford warm clothes for them. Sometimes neighbors will drop off rice, butter or flour to help out.*

of women who are most vulnerable. Additionally, most of the workers on collective farms are women. Although they have nominal benefits from working in state enterprises such as part-time employment, maternity leave, and pensions, they are vulnerable during the restructuring of state/collective farms.

Pensioners who do not receive their pension checks from the government are also vulnerable. However, the SA indicates this is a small proportion of the population, which is surprising given the troubles of other former Soviet republics in paying pensions to their retirees. These pensions often make a critical difference for the survival of households.

Local, regional and national governments have stopped providing public goods during this period of reform. Education is still free of charge. There are new fees for tuition and exams, however, and the government no longer provides for ancillary costs such as books, school supplies, transportation to and from schools, and housing for those who attend distant schools. Few households are able to afford these costs.

The SA indicates a significant decline in the quality of education in rural Uzbekistan. Even though ideological indoctrination is no longer carried out, 54 percent believe the quality of primary and secondary education has declined and only 24 percent feel quality has improved. Regarding universities, 42 percent feel their quality has declined while only 15 percent believe they have improved. The negative perception is stronger among those who have attended university.

Citizens traditionally depend on support from local institutions, neighbors and families in cases of domestic need, such as helping with an illness in the family or repairing a home. They view relatives (58 percent of respondents), neighborhood committees (39 percent), village councils (39 percent), collective farm/enterprise managers (33 percent), friends (30 percent) and neighbors (30 percent) as the most important resources of support. Only 5.3 percent believe they can rely solely on themselves in such situations. Table 6.7 shows that these support groups are decreasingly reliable for most citizens.

| Table 6.7  Trends in dependency on various support groups (percent of respondents) | | |
|---|---|---|
| Type of support group | Decreased role | Increased role |
| Friends, acquaintances and neighbors | 45 | 14 |
| Relatives outside the household | 41 | 18 |
| Managers of collective farm/enterprise | 37 | 18 |
| Mahalla committee/qishlaq council | 31 | 32 |
| District hakimiyat | 30 | 7 |
| Provincial hakimiyat | 28 | 4 |
| Local department of public organization | 23 | 2 |
| Central government | 22 | 6 |
| Central body of public organization | 21 | 2 |
| Mosques or people associated with it | 19 | 6 |

Source:  Social Assessment, Household Survey

Compared to the Soviet period, however, citizens now rely less on institutions or people that were important to them, including friends and relatives. The one exception is the local committees and village councils—institutions that gained as much importance as they have lost.[12] The difficulty in earning incomes and raising public budgets appears to be making it more difficult to help other family members, friends and constituents. In summary, Uzbek society is becoming increasingly atomized. This is worrisome, given declining incomes and reduced social services.

## *Stakeholders and Institutions*

### **Farm Stakeholders**

State/collective farms and cooperatives account for most of cultivated land, despite their relative small number. Cotton and grain predominate. A large number of smallholder and dehqan farms account for a much smaller proportion of cultivated land. These different types of farms share a large proportion of their labor with each other. About 37 percent of state/collective farm members also cultivate private plots as smallholder farmers. A significant portion of these workers also engage in non-agricultural service work. Many dehqan farmers also work on state/collective farms. Table 6.8 depicts the basic characteristics of the different types of farm stakeholders.

| Table 6.8 Characteristics of Uzbekistan farms in 1995 | | | | | | |
|---|---|---|---|---|---|---|
| Farm type | Total number | Total hectares | Percent of cultivated land | Average size (hectares) | Average size (people) | Products |
| State/collective farms | 1,405 | 2,281,000 | 57.0 | 1,500 | 2,500 | cotton; grain |
| Cooperatives (shirkats) | 866 | 705,000 | 18.0 | 1,200 | 2,500 | cotton; grain; feed |
| Independent dehqans (peasant farms) | 9,346 | 186,000 | 4.7 | 14.5 | 5.5 | livestock; cotton; grain |
| Leasehold dehqans (unregistered farms) | 12,168 | 207,000 | 5.2 | 14.5 | 5.5 | cotton; grain;feed |
| Smallholders (subsidiary plots) | 2,900,000 | 500,000 | 12.6 | 0.17 | NA | diversified |
| Private livestock farms | 1,600 | 104,000 | 2.6 | 75 | NA | dairy, beef, sheep, goats, poultry |

---

[12]    This finding reflects positively on the central government's efforts to make these institutions more important to the functioning of locales.

**Management and Workers on State and Collective Farms.** Under GOU's policy of gradual demonopolization, state farms have been re-organized into collective farms and receive no budget from the state. Because they have not responded to market forces well, GOU plans to transform collective farms into "closed-type" joint-stock cooperatives. However, they are cooperatives in name only since no shares have actually been distributed and many have yet to be transformed. Nonetheless, current leadership has found it useful politically to distinguish joint-stock cooperatives from the state farms and collectives of the Soviet period.

State and collective farms both organize their members through an internal lease. Uzbekistan law states that the property belongs to members but farm management controls the land, in practice. The terms of these leases do not favor members. Lease-holders give their production to the farm at prices set by the government (district administrations) in consultation with farm management. Lease-holders aim to fulfill state production orders, rather than respond to market signals.

State/collective farms focus on cotton and grain, which is totally controlled by the state. In Soviet times, lease-holders were able to market production that exceeded the state quota. However, now the state has monopolized the marketing of products such as cotton and there is only one processor per province. Therefore, all of a lease-holder's production is sold to the state.

State and collective farm employees comprise the largest labor pool in Uzbekistan, about 25 percent of rural employment. Of these, only 28 percent said they received income exclusively from the collective farm. Thirty-seven percent of collective farmers receive additional income from private plots as well. The average state/collective farm numbers about 2,500 people, half of which are under the age of 16.

The average annual income from collective farm work in 1995 was about 4,286 soums ($413), but many earn substantially less—40 percent earned less than $75 while only 20 percent earned more than $210. Consequently, most farmers on collective farms work part-time, earning most of their income from private plots and other activities. State and collective farm management controls access to land, water, agricultural machinery and social security benefits. Many who work on the private plots thus maintain their ties to the collectives to gain access to these resources.

State/collective farm employees have lower educational levels than other groups. Sixty-eight percent have secondary education but only 16 percent have education specialized in agriculture. Fifty-two percent are men and 48 percent are women. Most employees lack direct experience in organizing and managing a farm, lack entrepreneurial initiative and lack access to start-up capital.

**Cooperatives (Shirkats).** In practice, shirkats are very similar to state farms and collectives. Some shirkats resulted from the forced collectivization in the 1930s and others originated later through "bottom-up" entrepreneurial initiative. They generally have fewer regulations to follow than do state farms and collectives and, therefore, they have somewhat more freedom. The political leadership has promoted them as a way of distinguishing its rule from that of the Soviet Union. There are fewer cooperatives and they are somewhat smaller than state and collective farms.

**Independent Peasant (Dehqan) Farms.** Independent dehqans concentrate on producing livestock, cotton and grain. These are private family-based farms. They have registered with district governments, which recognize them as legal entities, approve the size of their land plots, and permit them to establish their own bank accounts. They can hire non-family members as workers. The lease of their land is permanent and can be inherited by heirs. Despite their apparent legal independence, these farms depend heavily on state/collective farms for irrigation, inputs and sales. The state/collective farms use mandatory cotton and wheat quotas to control them, as well as their own control of irrigation, inputs and customers.

Unlike workers on the state farms and collectives, most of these dehqan farmers are experienced in organizing small-scale production and have practical knowledge of a market economy. Although efficiency improved greatly between 1994 and 1995, the number of these farms is declining due to adverse conditions. Many received land in marginal areas located on the periphery of the state or collective farm. Generally, such land may be 1/3 pasture and half irrigated. These farms lack access to affordable credit, as high interest rates and bribery payments put loans out of reach of most farmers. In addition, land shares were given to all members of the former state and collective farms for free, thus even people without the requisite farming skills also received shares. However, the farms that survived have grown in size from an average of 6.7 hectares in 1991 to nearly 15 hectares.

**Leasehold Dehqans.** Leasehold dehqans are patriarchal based farms averaging about 15 ha. Until a new law was approved in 1998, these farms were not required to register with district governments; now they are required to do so as a means of control. Their lease terms are only for 10 years, a critical impediment to their optimal development. They cannot hire their own labor and do not have their own bank accounts. They depend on state/collective farms for irrigation, inputs and sales channels.

**Smallholders.** Household farms have permanent leases of land that are inheritable. They comprise about 24 percent of the rural population but only 13 percent of arable land. Their number and total land occupied have been growing rapidly from 2.5 million in 1993 to 2.9 million in 1998. They accounted for 160,000 hectares of cultivated land in 1960, 250,000 hectares in 1988, 491,000 hectares in 1993, 500,000 hectares in 1996 and 650,000 hectares in 1998.

Private household plots are limited by law to 0.25 hectares of land. About half of this land is permanently situated and usually supports a house, while the other half often is temporary, moving from location to location within a state/collective farm. The state/collective farms control about half the land of the many household farms, making the small farmer beholden to the state collective. Most smallholders are part-time private farmers, and they grow a wide variety of crops. Some cultivate for subsistence while others produce cash crops for income.

Smallholders generate income from a wide array of other sources—business, work on collective farms, and transportation or repair services. Thirty-six percent spend less than half a year farming their own plot. The average household farmer spends 7.5 months cultivating. Forty-two percent of them spend fewer than three hours a day on average working their plots. Forty-one percent of them work on collective farms as well. A small number also works in administrative jobs or marketing/sales. About 23 percent receive pension income from the government in the form of pensions, student aid, payments for military service and disability aid.

Overall, these smallholders are more likely to be educated than workers on state and collective farms. However, only seven percent have received specialized training in agriculture. The survey indicated average income was $120 per year, but this figure is likely under-stated because farmers try to avoid reporting taxable income. Eighty-five percent think income from household farming is quite reliable, higher than any other group.

**Private Livestock Farms.** These are the most independent farms in Uzbekistan. They do not depend on the state/collective farms for irrigation and other essential inputs. They occupy an average of 65 hectares and possess an average of 400 head of livestock.

### Other Stakeholders

**Non-Agricultural Labor.** This group comprises people working in services, industry, transportation, construction, education, medicine, as well as those who do not classify themselves as collective farmers, leased-land farmers or private farmers. However, many of them also engage in farming. This group is younger than the others, with 65 percent between the ages of 23 and 40. Sixty-nine percent of them are men, 29 percent have completed specialized secondary education, 19 percent have completed higher education, while only seven percent have completed fewer than nine grades of school.

Their income levels are relatively high, about $234 in annual salary; 77 percent feel their income is reliable, although 62 percent believe their income now is worse than it was 10 years ago. Many have multiple sources of income. Traders of agricultural products have the highest income. The SA found that only two percent of people are engaged in agricultural trade, but that statistic is likely to be understated because such activity was illegal and is now taxed heavily. Given its profitability, the true number is probably far higher.

**District Administrators.** District Administrators control the allocation of land and irrigation, as well as the purchase and distribution of agricultural products. One of their key objectives is to fulfill government quotas for agricultural production and food self-sufficiency.

**Mahalla/Qishlaq Councils.** These councils are an extension of regional and central governments into the local community. They have little capacity for initiative. They depend on the collective farms economically and on the district administrations administratively. They perform administrative functions such as registering the population and distributing welfare support.

### Access to Farm Inputs and Services

A small number of state-controlled enterprises dictate the terms and conditions of supplies of inputs and services such as fertilizers, pesticides, seeds and machinery. Due to traditional ties between the state enterprises and the state farms, the state/collective farms get the supplies first. Private dehqans and household farms then obtain their supplies through the state/collective farms.

**Fertilizers and Pesticides.** Fertilizers and pesticides are expensive and often in short supply. As a result, both are significantly underutilized. Current fertilizer use is perhaps 50 percent of recommended rates. The distribution of these fertilizers is also weak. These supplies are provided only on state/collective enterprise plans and credits, which is budgeted in advance of need rather than via the market. In some cases, large collectives may include the needs of dehqan and household farmers in their order, but these needs are often ignored.

Prices of fertilizers and pesticides are determined by the government. In addition a factory-to-farm service fee, other fees are added to the price along the way to the final purchaser; a 17 percent value-added-tax (VAT) was dropped in April, 1996, however.

> *Uzchimprom, a state-owned enterprise under the Ministry of Chemical Industries, is responsible for production of fertilizers and pesticides. It was privatized in 1995 with shares distributed to agricultural producers (35 percent), government (30 percent), private interests (20 percent) and its own workers (15 percent). All facilities and equipment were transferred to the shareholders. Uzchimprom also sold some of its branches.[1]*
>
> *Uzchimprom has six plants that produce pesticides and nitrogen and phosphate fertilizers. It has difficulty producing large enough quantities and delivering them on time, largely due to bureaucratic delays and limitations in mining phosphates. In addition, its spraying equipment is relatively old, reducing the effectiveness of the pesticides.*

Pesticides are no longer widely used in Uzbekistan. Although 90 percent of the cotton crop and about 50 percent of the grape crop is sprayed, less than 25 percent of the wheat crop, 10 percent of the melon crop and only five percent of the potato and alfalfa crops are sprayed. Biological controls are promoted but only 10 to 15 percent of farmers employ them. Manual weed control is widespread.

**Seeds.** The national seed monopoly, Uzsabsavotnaveourouglare, is

> *The state enterprise Uzagrochemservice (Uzagrichimterminat) is the only organization permitted to supply mineral fertilizers and pesticides, including potash fertilizer imported from Russia, although it licenses some private entrepreneurs to import as well. Its national branch adds a service fee to the imported price; its oblast branch adds an additional 17 to 21 percent to its purchase price, and the district branch adds its own fee. If the state/collective farm sells the fertilizer to a dehqan or household farmer, it adds its own fee. The total price mark-up of the imported potash fertilizer is between 80 percent and 100 percent. However, imports are increasingly limited due to foreign exchange restrictions and GOU is attempting to establish potash mining in Uzbekistan.*

responsible for producing and distributing seeds under the Ministry of Agriculture. It has been a private institution with no state budget since 1990, but no shares have been issued. It reports to the Ministry of Agriculture and Water Management. Representatives of state agencies and state/collective farms elect its manager.

Uzsabsavotnaveourouglare generally provides a sufficient quantity of seed through domestic production and imports. It contracts seed production to about 20 percent of Uzbekistan's state/collective farms, provides them 25 percent of the expected value of the seed production quota in advance and pays for the remainder upon delivery. Payment is in the form of cash or in-kind (fruits and vegetables). The state/collective farms can sell surplus production to farmers through 127 sales outlets. Uzbekistan now has five cotton corporations and two more will be added. These corporations grow, process (delint, clean and calibrate), package and distribute seeds. The State Breeding Center, Uzdavourouenzoratmarkaz, tests and certifies these seeds.[13]

**Farm Machinery.** Uzbekistan has about half the machinery per hectare of cultivation than found in competitive farm economies such as the United States or Argentina. Its machinery is also generally older, of poor quality and costs more to maintain and operate. The average age of trucks and tractors is 10 years. Shortages of spare parts are endemic. Lack of personal accountability on state/collective farms for maintaining the equipment exacerbates these problems. As a result, equipment use is declining significantly. Machine harvesting of the cotton crop also declined from 40 percent in 1992-93 to four percent in 1997.

The State Committee for Supply and Maintenance, Uzselkozsnobremont, is responsible for supplying and servicing agricultural equipment, as well as designing and manufacturing some equipment. It is a state monopoly with over 60,000 employees and machinery depots in all oblasts and various districts. It has established a joint venture with Case International, Inc. to manufacture, sell, lease and repair farm machinery such as cotton pickers, grain headers, tractors and components. The joint venture is establishing a network of 12 service centers to provide technical support and maintenance of agricultural equipment. It is anticipated that quality, delivery and service will improve over time as this joint venture matures.

**Water Management.** Sustainable agricultural production in Uzbekistan depends on effective water management. The irrigation system currently suffers from critical problems. The primary contributors to water shortages and significant environmental damage are the distribution of water without charge and technical shortcomings in the irrigation system. Economic, political and ethnic disputes have resulted, as well as declining crop yields.

Uzbekistan receives about 110 mm/year of rainfall. Rainfed agriculture is not possible in most of the country; about 98 percent of arable land is irrigated. A major expansion in the early 1970s doubled the area that was irrigated in 1945. Wheat, rice cotton and alfalfa demand large amounts of water (26,400 $m^3$/ha for rice and 11,700 $m^3$/ha for wheat), assuming an efficiency rate of 45 percent. The Syrdarya and the Amudarya rivers supply about 80 percent of Uzbekistan's water needs and 90 percent of needs in the western part of the country. The remaining need is met by tapping underground aquifers.

---

[13]     Uzsavsavotnaveourouglare imports seeds from Russian and the Ukraine. Small private entities have also begun to import seeds, mostly from Holland, and reselling them. The Ministry of Finance and the regional administrations oversee this activity.

Local officials indicate that about half of the districts suffer water shortages, persistent rises of groundwater and excessive salinization of land. This occurs in part because of the enormous waste of water through over-use, high leaching requirements and poor controls. The irrigation system suffers from leaking and seepage due to technical shortcomings, thus transmission losses are also high. About half of the irrigated lands require pumping to irrigate the high lands and to drain water from lowlands. The equipment in the pumping stations needs repair or replacement. Pipes are corroding and canal linings are deteriorating. Furthermore, most secondary distribution systems rely on elevated/open concrete conveyance structures that leak excessively, and piped water is pumped continuously, regardless of the quantity needed by the end-user.

The current structure of water management divides responsibilities and reinforces the free and inefficient use of water. The Ministry of Land Reclamation and Water Resources is responsible for construction and maintenance of irrigation facilities up to the boundaries of the collectives. The district administrations are responsible for the operation of all off-farm irrigation facilities. On-farm irrigation management is supposed to be the responsibility of the farmer. However, the state and collective farms have retained responsibility for distributing water to dehqan and private farms even though the Ministry of Agriculture has been given responsibility for on-farm distribution. The great majority of irrigated lands belong to collective farms, which utilize most of the water. Because of the tradition of free water use and because of the political clout of state/collective farms, water continues to be distributed at no charge, which discourages efficient use.

State and collective farms take advantage of their control over the current shortage of water. They use their distribution power to impose their will on dehqan and household farmers by withholding water to those who do not plant according to the state production quota. They have also used their clout to push most private farmers onto less desirable, unirrigated lands. Thirty-five percent of local officials interviewed stated that water problems are the most urgent ones facing their districts.

The over-use of irrigated water is damaging the environment and reducing crop yields. The situation has been serious enough to curtail water deliveries by 25 percent since 1980 to try to enforce a policy to preserve the Aral Sea, which is fed by the major rivers in Uzbekistan. If the problem of overuse is not addressed, waterlogging and salinity will increase significantly in the next several decades, confining agriculture to salt-tolerant crops and necessitating desalinization of drinking water.

**Processing.** Eighty-four percent of local officials state dissatisfaction with the state of agricultural processing and 73 percent believe that it should be the government's highest investment priority. Although GOU says it has privatized the agro-processing industry, it retains control through monopoly and monopsony in most commodity sectors; state ownership/management control of processors; administrative controls on most aspects of business; and a ban on exports. A small number of genuinely private agro-processors have begun to emerge. Over time they could bring genuine competition throughout the industry.

GOU owns and manages the industry through a 1992 agro-processing law and several subsequent presidential decrees. Most former state processing enterprises have been privatized through the distribution of shares to management, employees and private individuals. However, the State Property Fund retains an absolute majority of shares in all strategic enterprises and between 5 and 30 percent of shares in other processors. Various commodity associations are responsible for administering these shares to all shareholders, and membership in these associations is mandatory. Further, local governments appoint the senior management of these enterprises.

GOU also controls processing through a large number of regulations and practices that determine most important decisions of these enterprises. Such controls include: 1) production quotas; 2) dictated pricing on a cost-plus basis; 3) selection of customers and suppliers; 4) approval of employee salaries; 5) approval of capital investment; and 6) subsidies of inputs, processing and distribution. In many commodity sectors, associations controlled by the Ministry of Agriculture and/or district administrations have been established to monitor compliance, address coordination problems and, in many cases, to conduct the marketing and sales of the products.

The structure of the processing industry ensures that all of Uzbekistan's agriculture remains under government control. Monopoly and monopsony reign. There is usually only one processing facility per district for commodities such as cotton, vegetables, fruits, meats and hides, which covers almost all agricultural production in the country. For example, there are only 179 cotton gins for the country's 163 districts. Most cotton growers must sell to only one processor. The processors exist to impose government control on agriculture. For example, a significant proportion of grapes is grown by the private sector (about 37 percent) with state/collective farms producing the remainder. State/collective farms sell their grapes directly to the market, including the export market. None of their production goes through state associations or trading organizations. However, private grape producers must sell 100 percent of their output through government-controlled associations and trade organizations. Milk is the most liberalized commodity market in Uzbekistan. About 14 percent of private household farm milk production, four percent of total production, is sold directly to bazaars and street markets.

However, government control of processing is weakening. The Ministry of Finance has established the Main Department of Antimonopoly and Price Policy to encourage competitive pricing as well as the Fund for Price Regulation to mitigate the effects of price increases. The have a positive impact on the industry as demonstrated by several new private processing programs (cotton processing and textiles) established with financing by domestic banks. These private enterprises include textile mills, meat processors, bakeries, dairies and fruit processors. Some are extensions of private trading companies while others have been developed by private farms.

The pervasive government control of agricultural processing depresses demand for agricultural products and the prices paid for them. Without a market-oriented processing industry, Uzbekistan's agricultural sector cannot meet its potential and incomes and employment will continue to suffer greatly.

**Credit.** Thirty-one percent of local officials stated that credit is the most urgent problem in rural Uzbekistan. Financing cotton and wheat production accounts for 90 percent of all agricultural lending. A system of "centralized credit" is used in which government-controlled banks lend to input suppliers and trade associations which advance their physical inputs to the producers. Other loans are made with opaque eligibility criteria and a demand that loans are only available when the borrower guarantees a specific quantity of product at a fixed price to a predetermined buyer. Interest rates are normally around 100 percent annually.

Only 15 percent to 20 percent of private farmers have obtained loans and about half of the state/collective farms have difficulty receiving them. Uzbekistan farmers have five major sources of loans—Tadbirkor Bank (35 percent of borrowers), commercial banks (29 percent of borrowers), Pakhtabank (16 percent), Promstrobank (5 percent) and collective banks (8 percent). The Tadbirkor Bank was established specifically to serve private farmers. Therefore, its large market share is a positive development. State/collective farms tend to rely more heavily on Pakhtabank and other commercial banks. The banks make available one, three, five and 10-year loans. The average size of loans is relatively large—as much as 46,000 soums per loan, or about 4,600 soums per irrigated hectare.

Collateral is not well understood or readily available. Farmers are not accustomed to paying interest on loans and bribes are usually required to establish a line of credit. Additionally, with no risk insurance available, farmers are not willing to borrow money they must pay back because of the risk of weather or other catastrophe. Further, many farmers do not have any assets to use as collateral for a loan—land, equipment and even their own production is owned by the government or obligated to it.

GOU is making a significant effort to improve the performance of the financial sector. It seeks to strengthen the legal and regulatory framework for financial intermediation, adopt internationally accepted accounting systems for banks, strengthen bank supervision (with some attention to bank restructuring) and develop capital markets. In April 1996, it adopted the Law on Banks and Banking Activity to begin achieving these goals. As a result of the law, commercial banks are setting aside reserves against non-performing loans and all banks are now required to have annual audits performed by international accounting firms. The payment system has been fully automated, resulting in a major reduction of the time required for settlement. A recent presidential decree called for increased private sector participation and reduced administrative interventions in the banking system.

### Incomplete Implementation of Reforms

GOU has attempted to divest itself from agricultural production; transferred ownership of land to the farmers; deregulated prices and eliminated state orders/procurement on a range of products, with the exception of cotton fiber and cereals; and reduced taxes and forgiven many of the debts of state and collective enterprises to provide more opportunity for these organizations to succeed.

GOU has partially transferred decisions regarding what to produce to state and collective farms and other people who actually manage farm assets (such as family farmers and cattle grazers). It has also approved laws to reform the governance of state and collective farms to make their operations more democratic. State farms first became collective farms, and

| Table 6.9  Private rural land ownership | | | | | |
|---|---|---|---|---|---|
| | 1991 | 1992 | 1993 | 1994 | 1995 |
| Total acreage | 274 | 554 | 571 | 588 | 602 |
| *Annual increase* | - | *102%* | *3%* | *3%* | *2%* |
| Arable land | 226 | 463 | 477 | 489 | 499 |
| *Annual increase* | - | *105%* | *3%* | *3%* | *2%* |
| Source:  Goskomprognozstat, 1997 | | | | | |

now collective farms are being turned into joint-stock farms. By 1997, 2,000 state farms had become collective farming units and "closed-type" joint-stock companies.

The amount of non-arable land a household can hold was increased from 0.1 hectares to 0.5 hectares, and the amount of arable land was increased from 0.1 hectares to 0.25 hectares. Household farms and private farming units (dehqan) numbered 19,828 by February, 1997. Table 6.9 shows one impact of land reform on family farmers. In 1992 the amount of land controlled by private household farmers doubled. The increase resulted from the transfer of some lands from the state and collective enterprises to individual household farmers. By the end of 1996, about 25 percent of agricultural land had been transferred to family-based peasant farms or as household plots under various leasehold or use rights arrangements.

Although the reforms were adopted, they have not been fully implemented. They have significant shortcomings and are not comprehensive enough to be effective. The SA indicates that rural citizens do not believe the reforms are significant, and do not like what they see as the future impact of the implemented reforms on their own well-being. Table 6.10 presents the opinions of rural citizens regarding reforms between 1992 and 1996.

| Table 6.10  Significance of agricultural reforms 1992 to 1996 (percent of household respondents) | | | | |
|---|---|---|---|---|
| Province | Significant | Insignificant | No reform | Difficult to answer |
| Kharazm | 43 | 39 | 10 | 8 |
| Farghana | 35 | 25 | 11 | 29 |
| Samarqand | 29 | 35 | 11 | 25 |
| Qashqa-Darya | 22 | 33 | 35 | 10 |
| Tashkent | 20 | 32 | 17 | 31 |
| Namangan | 19 | 42 | 11 | 28 |
| Total | 28 | 33 | 18 | 21 |

Source:  Social Assessment, Household Survey

Only 28 percent believe that reforms have been significant, while 51 percent believe reforms have been either insignificant or non-existent. Twenty-one percent could not answer the question. Of these, many are probably wary of saying anything critical of the government. Therefore, a significant proportion of this group may also believe reforms have been either insignificant or non-existent. Most citizens also believe the impact of the reforms will be negative (Table 6.11).

Table 6.11  Impact of privatization and farm restructuring on personal well-being  in next five years (percent of household respondents)

| Province | Worsening | Improving | No change | Difficult to answer |
|----------|-----------|-----------|-----------|---------------------|
| Samarqand | 66 | 10 | 5 | 19 |
| Farghana | 60 | 14 | 11 | 15 |
| Tashkent | 54 | 18 | 21 | 7 |
| Kharazm | 49 | 22 | 23 | 6 |
| Namangan | 49 | 9 | 10 | 32 |
| Qashqa-Darya | 48 | 22 | 11 | 19 |
| Total | 55 | 16 | 12 | 17 |

Source:  Social Assessment, Household Survey

Only 16 percent of respondents feel the reforms will be positive.  The remainder feels the reforms will either be negative, of no consequence, or too difficult to understand to assess their impact.  Do citizens dislike the reforms because they are "too much, too fast" or because they are "too little, too slow?"

Those who feel their situation has improved due to receiving land for their own use are more likely to believe land reform is significant.  Those who feel that allocations of land impacted them negatively are more likely to believe the reforms are insignificant.  Where reform is genuinely occurring, opinions are more positive.  Where reform has not occurred, opinions are more negative. The survey indicates that rural citizens believe the reforms have not gone far enough.  Table 6.12 illustrates this belief in the case of land reform.

Table 6.12  Correlation between attitudes toward land reform and impact of land allotment for own use (percent of household respondents)

| Impact of land allotment for own use | View of land reform | | |
|--------------------------------------|-------------|---------------|-----------|
| | Significant | Insignificant | No reform |
| Generally improved | 71 | 58 | 36 |
| Considerably | 16 | 14 | 11 |
| Slightly | 55 | 44 | 25 |
| Remained same | 8 | 10 | 11 |
| Generally negative | 21 | 32 | 53 |
| Slightly | 10 | 10 | 12 |
| Considerably | 3 | 11 | 30 |
| Difficulty replying | 8 | 11 | 11 |
| Total | 100 | 100 | 100 |

Source:  Social Assessment, Household Survey

Rural citizens who oppose the reforms often do so because they believe reform only benefits the privileged.  Thirty-three percent believe the reforms lead to an unfair distribution of privileges and resources to the elite class (the rich, well-connected, managers, and local administrators).  Only three percent believe that ordinary households (cattle-herders, farmers, irrigators, etc.) will benefit and only five percent believe that hard work and initiative will enable individuals to establish successful businesses.

Rural citizens believe that land is too difficult to obtain because its distribution is controlled by the state/collective farms and local administrations.  Further, state/collective farms also control agricultural production, water resource management, farm machinery and social security benefits. Rural citizens believe they use this control to strengthen themselves. Priority in distributing these resources goes to those who fulfill the state orders dictated by the central government. They believe that state farm managers are attempting to use their control of these resources to set up agricultural businesses for themselves and their relatives. The goal of local administrations is to strengthen the production and financial performance of the state/collective enterprises and those dehqans that fulfill state ordered production.

These views are supported by a number of factors:

- Prices and quotas have been deregulated, but not for the two crops that have traditionally been most important to the rural economy: cotton fiber and cereals. Cotton and grain utilize 73 percent of Uzbekistan's sown area, leaving only 27 percent for other crops. GOU still sets procurement quotas and prices on these crops, ensuring that most state and collective farms still grow them at set quantities and prices. Further, each province only has one cotton gin, so cotton producers face a monopsony. Producers of fruits, vegetables and livestock-based products also face monopsonies, reducing incentives in these subsectors as well.

- State farms and collectives now vote to select managers, but GOU controls their management. GOU still approves candidates running for farm chairman. General assemblies of farms can dismiss a farm manager but must base this action "on the opinion of the district and province hakims, as well as the higher order agricultural institutions."

- Most state and collective farms still have not been transformed into joint stock enterprises. Shares in most of these farms have not been distributed to farmers and, as a result, farm managers still answer to district governments and the GOU rather than to shareholders. State and collective farms still remain under the control of GOU in most cases.

- Households are still limited in the amount of land they can manage privately to 0.25 hectares of arable land and 0.5 hectares of non-arable lands. As indicated earlier, household farms accounted for only nine percent of Uzbekistan's arable land in 1995 while private farming units accounted for only four percent. The remainder was controlled by state/collective farms and joint stock companies that operate like state/collective farms.

The state/collective farms control more than just those persons who work on them. Private farmers are also beholden to them because state/collective farms control access to critical resources such as land, irrigation, farm machinery, pensions and medical benefits. Many private farmers also work on state/collective farms to access these resources. Depending on the province, the proportion of household farmers that also cultivates land on state/collective farms ranges from 46 to 72 percent. The income derived from farming state/collective farm land is significant—between 40 percent and 55 percent of income, as depicted in Table 6.13.

| Table 6.13  Annual household income from private plots and collective farms (soums per household) | | |
|---|---|---|
| Province | Private plots/own business | Collective farm land |
| Namangan | 6,435 | 5,305 |
| Farghana | 9,564 | 5,701 |
| Samarqand | 6,509 | 6,163 |
| Tashkent | 15,094 | 10,556 |
| Kharazm | 11,045 | 13,443 |
| Qashqa-Darya | 4,727 | 5,592 |
| Total | 8,150 | 7,349 |

Source:  Social Assessment, Household Survey

Despite the fact that household farmers participate in collective farms to access resources and benefits and earn additional income, household farms are much more productive.  In terms of revenue in soums of income per hectare, soums of income per day and in terms of crop production per hectare, private household farms are much more efficient. Household farms occupy less than 13 percent of Uzbekistan's arable land, yet they account for 45 percent to 60 percent of household income.  In terms of soums per day worked, Table 6.14 details the productivity advantages of private farming and other activities over the state/collective farms.

| Table 6.14  Average income by type of activity (US$ per 8-hour work day) | | | | |
|---|---|---|---|---|
| Province | Collective Farms | Private farming and agribusiness | Non-agricultural Work | Other |
| Tashkent | 0.77 | 3.95 | 1.50 | 1.90 |
| Kharazm | 0.82 | 1.90 | 1.21 | 1.40 |
| Farghana | 0.65 | 1.70 | 1.25 | 1.50 |
| Namangan | 0.70 | 1.50 | 0.70 | 1.01 |
| Samarqand | 0.45 | 1.15 | 1.25 | 1.05 |
| Qasha-Darya | 0.62 | 4.00 | 1.26 | 1.30 |

Source:  Social Assessment, Household Survey

In every case, the collective farm is the least productive organization while private farming and agribusiness is in almost every case the most productive endeavor.  The productivity advantages of private farms and agribusinesses over collective farms are enormous—by factors of two to nearly six—even in Tashkent and Kharazm, where economies of scale should help collective farms.

Private farms are also more productive when measured in crop production per hectare. Land acreage on dehqan farms in 1995 increased 20 percent while grain production increased 320 percent, vegetable production increased 180 percent, fruit production increased 320 percent, grape production increased 450 percent, meat production increased 31 percent, milk production increased 40 percent and egg production increased 49 percent.  These efficiency increases occurred despite the fact that lands dedicated to dehqan farms are usually inferior to those retained by the state/collective farms.  There is clear indication that owning private plots is more desirable than working on state/collective farm lands.

In most provinces it has been difficult to obtain private land or to set up a private farming unit (dehqan). More farmers than not feel it is impractical to establish a private farming unit and succeed in Farghana (59 percent to 39 percent), Samarqand (48 percent to 37 percent), Tashkent (58 percent to 34 percent), Qashqa-Darya (66 percent to 30 percent) and Kharazm (52 percent to 45 percent). More people in Namangan (77 percent to 21 percent) feel it is practical. If reforms were more comprehensive and implemented more fully, household farms would benefit and more strongly support the reform program.

### Participation and Communication in the Reform Process

The rural population in Uzbekistan clearly does not like the current reform program because it does not go far enough; they are caught between the far too prevalent remains of the old planned economic system and a new market system that is incomplete and non-functional in a practical sense. Their negative view is compounded by their perception that they also do not understand the reform program and its various aspects. Lack of participation in developing and implementing the reforms reinforces this perception. Eighteen percent of respondents believe no reforms are taking place, 46 percent say they don't understand the reforms clearly, seven percent say they understand nothing about the reforms, and 11 percent say the question is too difficult to answer. Collectively, 82 percent of respondents do not understand the reforms in some significant way, and only 18 percent understand them clearly. This pattern is consistent across all provinces surveyed. Even among those with full or partial higher education, 63 percent believe they don't understand the reforms, that the question is difficult to answer or that no reforms are taking place.

Understanding is poor because communication of the reforms has been poor (Table 6.15). Television and radio are the most used media of 73 percent of respondents.

| Table 6.15  Sources of information used (percent of household respondents) | | | | | |
|---|---|---|---|---|---|
| Sources | Very often | Fairly often | Infrequently | Never | Difficult to answer |
| Friends/relatives | 11 | 33 | 40 | 12 | 3 |
| Place of Work | 9 | 25 | 23 | 36 | 7 |
| Mahalla committee | 4 | 22 | 36 | 32 | 7 |
| District/provincial authorities | 1 | 9 | 18 | 59 | 13 |
| National government authorities | 2 | 6 | 14 | 63 | 15 |
| State newspapers | 5 | 12 | 23 | 58 | 2 |
| Local newspapers | 6 | 14 | 24 | 55 | 2 |
| Radio programs | 29 | 23 | 18 | 28 | 2 |
| Television programs | 44 | 28 | 14 | 12 | 1 |
| Specialized literature | 2 | 4 | 10 | 80 | 4 |

Source:  Social Assessment, Household Survey

Friends and relatives are used most by 11 percent and the workplace is the source of information about the reforms most used by nine percent of respondents. Government officials and the written word (newspapers and specialized literature) are used least by respondents. This pattern of communication means that the reform process is vulnerable to misinformation—the most reliable and detailed sources of information are used the least, and the least reliable and detailed sources of information are used the most.

Improving the use of media will benefit the reform process, particularly if it is designed as a two-way process. People who understand the reforms are more likely to be optimistic about their impact (Table 6.16) and better able to articulate their needs and interests in the reform process.

Table 6.16 Correlation between understanding reforms and positive assessment of reforms (percent of household respondents)

| View of reforms | No reform implemented | Understand nothing | Don't understand clearly | Understand | Difficult to answer |
|---|---|---|---|---|---|
| Strong pessimists | 20 | 32 | 30 | 12 | 6 |
| Moderate pessimists | 9 | 15 | 32 | 16 | 4 |
| Moderate optimists | 4 | 15 | 51 | 19 | 11 |
| Strong optimists | 7 | 9 | 42 | 40 | 2 |

Source: Social Assessment, Household Survey

While communication from the government to the population needs to improve, so too does communication from the population to the government. There is a disconnect between government officials and the population which is undoubtedly based on experience, as well as information and knowledge. For example, government officials are more optimistic than ordinary citizens about improvements in living standards over the past 10 years. Forty-seven percent of officials surveyed believe that living conditions have improved (only 12 percent of ordinary population thinks so) and 41 percent of surveyed officials feel living standards have declined (72 percent of the general population thinks so). If the responses of officials are understood as a reflection of their objective conditions, they are reasonable; if their responses are understood as a general statement, however, they reflect a significant distortion of reality. In addition, 51 percent of surveyed officials favor the collective farm as the best form of organization for agricultural production, but most ordinary citizens see them as the problem. Communication needs to improve in both directions, particularly to make officials understand the needs and perceptions of the rest of the population.

An extension service can potentially contribute greatly to the communication of both technical and policy information, as well as increase the capacity of rural citizens to take advantage of opportunities that are available. Uzbekistan currently lacks a viable extension service and other means to provide technical assistance. A number of groups could benefit greatly from establishment of an effective extension program. Collective farmers who have not yet become private farmers but want to do so lack direct experience in independent economic activity and knowledge of managing a farm. They would benefit from information and training on organizing a farm enterprise under the new laws, managing a farm, and on agronomy, processing, finance, accounting and marketing. Dehqans, whose main income comes from working private plots but who attempt to develop more substantial farm activities, already have more direct experience in organizing small-scale production and marketing. They require support in more advanced aspects of farm management, agronomy, veterinary medicine, credit and banking, as well as the use of inputs and marketing.

Officially-registered dehqans have experienced the full range of current difficulties. They need the most advanced courses.

Collective farmers who prefer to remain on collectives or other major agricultural enterprises can also benefit from demonstrations of the advantages of private farming as well as technical support in farm management, agronomy and marketing, etc. State/collective farm managers need training that covers the process of transforming collective farms into joint stock companies, managing large-scale agricultural production in a market economy and making their operations transparent and accountable to ultimate owners.

Members of mahalla/qishlaq committees can use support in community organization methods, management and local resource mobilization. Likewise, district administrators assistance in managing municipal/district economies in a market system and in redefining their roles to support the development private sector production and marketing, rather than hinder rural development.

### *Monitoring and Evaluation*

The proposed program is expected to provide long-term, continuous and adaptable support to agricultural development in Uzbekistan. Therefore, the results of initial phases of the program must be recorded, assessed, and used to develop additional solutions to problems. Specific output and impact indicators of the program's progress will be monitored and assessed. Bank staff, GOU officials and local populations should work together to assess and develop these solutions regularly. The monitoring and evaluation program should address the following:

- Changes in policy formulation capacity, including establishment of a permanent institutional arrangement for continuous policy dialogue; such institutions should ensure participation by all stakeholders, including representatives of dehqan farms, household farms, private livestock farms and private agribusinesses;

- The impact of the enactment and implementation of a new land law consistent with concepts of individual accountability and private enterprise;

- The process and outcome of the restructuring of state/collective farms, including the development of business plans, the establishment of commercial bank accounts and implementation of plans;

- Patterns of the development of private agribusiness enterprises, including processors, with business plans and commercial bank accounts, and the constraints they face;

- The process and impact of establishing water user associations on the brigade and farm level, as well as the district level which comprise farmers as well as representatives of agribusiness, the district administration and the Ministry of Agriculture;

- The status of various controls at the tuman level, including settlement accounting, state orders for farm production and price controls on farm produce;

- Changes in farmgate prices for major commodities, including cotton and wheat;

- Changes in the number of commercial loans provided to private farms and private agribusiness; loan size; interest rates, maturity dates; collateral requirements; recovery rates for the loans; and the income levels of borrowers;

- Changes in the availability and quality of production inputs;

- The functioning of the business advisory service; measures may include level of demand, types of services provided, including training and the value added of such services;

- The delivery and impact of technical assistance/training programs by discipline (technical, marketing, finance, and others);

- Changes in the number and use of market facilities that are built or expanded;

- Changes in the number of members in active trade associations and other agricultural and rural-based NGOs; membership rates; and activities undertaken on behalf of the rural population; and

- Changes in yields and overall production of crops, particularly cotton and wheat and high value crops.

## Project Implications

The recommendations presented below directly address decreasing incomes and employment in rural Uzbekistan and the factors that cause them. These recommendations center around three fundamental tasks:

- Privatize government functions,
- Build government and private institutions to address current shortcomings, and
- Involve citizens directly in defining economic reforms and rural development activities.

They address the major weaknesses identified in the SA regarding national and local governments, agricultural input suppliers and producers, processors and other agribusinesses, irrigation and gas providers, and financial institutions. Resolving these problems will significantly improve incomes and employment in Uzbekistan, as well as help the social conditions of women and reduce ethnic tensions.

## *Privatization*

Many government functions should be privatized and institutions must be created in both the private and public sectors in order to address declining incomes and employment in rural Uzbekistan.   The objective of privatization is to enable farmers to make land and equipment more productive; strengthen the delivery of rural services, including social welfare, education and technical assistance/extension services; expand processing capacity for a number of commodities; and increase opportunities and capabilities in marketing and selling agricultural products.   Privatizing state/ collective farms, strengthening the local mahalla/qishlaq committees and privatizing marketing and sales of agricultural products are key elements in increasing incomes and employment.

### Genuinely Privatize Collectives

The transformation of collectives into genuine open joint-stock agricultural enterprises is one of the most important prerequisites for the success of the reforms and rejuvenating income growth in Uzbekistan.  Shares in these joint-stock enterprises must be distributed in a fair, transparent manner to all entitled farmers, dehqans and collectives.  Laws governing this process must be passed by GOU, which should address the distribution of land and agricultural equipment as well as the basic procedures for governing a genuinely democratic joint-stock enterprise.  Specifically, GOU needs to revise the Model Charter of Collective Farms Law to prevent local and district governments from appointing and firing collective farm managers and otherwise interfering in the operations of agricultural enterprises.

An equitable portion of shares should be sold to collective farm members with the remainder sold to household farmers at auctions.  Household and dehqan farmers should have equal opportunities to purchase these shares.  Restrictions on the acreage held by household farmers should be eliminated and district authorities should be prevented from determining the size of dehqan farms.  The amalgamation of land via the free market should be encouraged.

### Liberalize Irrigation and Gas Supplies

Water charges should be introduced and technical shortcomings in the irrigation system which cause water shortages, lead to environmental damage, economic distortions and political and ethnic conflicts should be addressed.  Technical problems of the existing irrigation system must be fixed.  The Pumping Stations Rehabilitation Program now being considered by the Bank will address one major problem.

In addition, irrigation management must be changed to provide economic incentives to use water efficiently and improve agricultural production.  The SA recommends that the Agricultural Enterprise Restructuring Program promote the creation of water user associations on the basis of hydrological units.   On the district level, a new organization should be developed that includes farmers and other major stakeholders—representatives from farming, agribusiness, district administration and the Ministry of Agriculture.  Farm representatives should be drawn from leaders of dehqan, private household farms and other water users. The associations should ensure that all have equal rights to use water on a cost recovery basis.

These organizations should work with the Ministry of Agriculture to provide incentives to use water efficiently by installing individual measuring and control devices, which make volumetric pricing possible. They should also plan for investments in modern irrigation technology and the re-use of drainage water. In doing so, they should establish guidelines for establishing private-sector management of irrigation and how to implement a full-cost recovery system for water-use. Full-cost recovery should be implemented gradually over a period of about a decade. Improvements in the technical performance of the irrigation system should be included in the calculation of full cost so that use of the system can be sustainable economically and environmentally.

Local officials and the rural population believe that gas shortages and high prices need to be addressed urgently, but GOU policy gives low priority to the development of the gas industry. Since 1990 gas prices have increased dramatically and are now at international levels. Now GOU is in a good position to liberalize distribution by facilitating the entry of private enterprise into the fuel supply and distribution market to compete with the current national monopoly, Uzselkhozsnobremonst (the State Committee for Supply and Maintenance).

## Liberalize Input Markets

The supply of inputs such as fertilizers, pesticides, seeds and machinery is often insufficient, over-priced and of low quality. Subsidies to input suppliers mask production inefficiencies. These conditions greatly constrain yields. They are the result of heavy government intervention in all aspects of the sector. GOU must remove established input monopolies and cost-plus pricing (rather than market pricing) and remove currency exchange restrictions to increase the availability of agricultural inputs.

The SA recommends that GOU promote the establishment of private input suppliers in all input industries. Decontrolling prices and phasing out subsidies to existing input suppliers should accompany entry into the market of private companies. Accordingly, inputs should be sold outside the account settlement system. To help private input suppliers establish themselves, GOU should provide limited incentives to encourage the banking sector to provide loans at competitive interest rates to input suppliers that meet established eligibility criteria. It should also facilitate such loans on a seasonal basis to input purchasers.

## Assist Farms in Marketing

GOU should reduce the abuse of power by local officials and state monopolies by eliminating state prices and quotas on cotton fiber and cereals and the purchase and/or distribution of cotton fiber, cereals, water and land by district administrations. In lieu of its current role, GOU should help to establish private service associations/cooperatives to facilitate marketing, procurement, logistics, agronomic services, raw processing, credit contracting and legal aid. This should include construction of physical facilities to market agricultural products, particularly promising products such as meat and potatoes. Where Uzbekistan has a comparative advantage, GOU should encourage the establishment of commodity-focused export marketing associations to increase exports. It should no longer retain the hard currency earnings on any exports, permitting economic entities to export directly and profit from those exports.

### Privatize and Liberalize Processing Industry

The processing industry currently suffers from monopoly, monopsony, government management, administrative controls and a ban on exports. These conditions depress prices and demand for agricultural produce, thereby reducing income and employment opportunities. The SA recommends that state orders and the settlement accounts system be abolished, permitting prices and quantities produced to be determined by the market and permitting farmers to be paid in cash. To infuse competition, GOU should encourage additional private sector participation in processing, particularly those members of state/collective farms now responsible for processing, and provide qualified processors with loans at competitive prices. Processors of cotton, fruits, vegetables and meats should receive special encouragement. Processors should receive technical assistance and other business advisory services. Loans at competitive rates should be made available to improve the overall competitiveness of private processors, including storage, grading and packing facilities.

### *Institution-Building*

### Build Credit Institutions and Markets

The current financial market comprises centralized credit focused on cotton and grain. Use of collateral and sound banking practices is very limited. Interest rates are high and important financial instruments such as risk insurance do not exist. Many transactions are governed by the settlement account system. GOU recently improved the financial system by initial implementation of international accounting standards and practices, closer supervision of bank practices, increasing reserve requirements of banks and permitting more private sector participation. These positive trends should be reinforced and expanded.

The SA recommends that the World Bank finance a line of credit to commercial banks to extend loans to restructured farms and agribusinesses. This program should be structured to prevent control and rent seeking by the local administrative elite. The settlement account system should be abolished. Banks should require applicants to have a suitable business plan that reveals sound understanding of supply and marketing needs. Loans to these private interests should support items such as inputs, machinery and parts, building repair, irrigation, transport needs, etc. Lenders should finance activities that guarantee an early increase in productivity. New financial instruments such as mortgage financing, leasing, use of warehouse receipts and crop insurance should be offered. Financing should be extended at interest rates which permit banks to earn a reasonable profit. Training to facilitate the loan application process should be provided to both lenders and applicants. To help force responsible borrowing, borrowers should be required to risk some of their own capital as part of the loan agreement.

### Strengthen Mahalla/Qishlaq Committees

These local committees are one of the few institutions that has retained, and even increased its status among ordinary citizens. GOU should make the local village committee responsible for implementing the share distribution process. It should also give incentives to these local committees to initiate and complete share distribution–making their budgets depend on the successful sale of collective farm lands, equipment and other assets as well as on taxes on farm production. To ensure adequate local resources, GOU should consider matching with central government funds whatever revenue the local committee can raise with share distribution and taxes on agricultural production.

With access to revenue from agricultural production taxes, enterprise shares distribution and perhaps the national government, the local mahalla/qishlaq committees should also be given more authority and responsibility for developing and implementing social welfare programs. They should nurture community-based institutions to develop these programs, which should include unemployment and healthcare insurance. Additionally, they should fund and oversee social infrastructure such as kindergartens, schools, healthcare clinics, water and gas supplies.

These programs should accommodate young, single women and those working on state/collective farms. The community-based support organizations should be insulated institutionally from local committee administrators to minimize corruption. But the national government should remain responsible for setting aside tax revenue to pay those pensioners that currently are not being paid.

## Public Participation

Mobilizing the public and providing it technical assistance and extension services will lubricate the gears of economic reform, rural finance and agricultural production, processing and marketing—all key ingredients to raising incomes and employment.

### Promote Popular Participation

Lack of popular participation in developing and implementing reforms has slowed progress. The GOU, regional governments and local committees need to establish public forums for: 1) citizens to express their ideas, opinions, criticisms and praise regarding the reforms and other relevant public policy issues; and 2) provide government officials the opportunity to assess reform ideas and provide information to the public on the objectives, philosophy and considerations of their policies and decisions. Such forums include workshops, seminars and surveys. Minorities need to be recruited heavily for these events. As part of this effort, GOU needs to improve its gathering, analyzing and dissemination of objective information GOU and local community organizations may want to consider celebrating cultural days of importance to minorities.

## Provide Technical Assistance

GOU needs to facilitate provision of technical assistance to household farmers, dehqan farmers, collective farm managers and workers, rural businessmen, and national, provincial and local government administrators. Among these groups, women and minorities require special attention. Such assistance should include courses, seminars and workshops covering municipal management, farm management, farm registration, creation of joint-stock companies, privatizing livestock farms, agronomy, veterinary medicine, processing methods, packaging, finance and accounting, credit, contracts, logistics, purchasing inputs, distribution, business plans and marketing. Practical handbooks covering these subjects should also be developed and widely distributed. GOU should also facilitate establishment of consulting centers in rural areas, particularly in provincial and district centers, to provide tailored assistance.

## Chapter Seven

# Sanliurfa-Harran Plains On-Farm and Village Development Project

## Ayse Kudat and Mumtaz Bayram[1]

## Background

### *The Southeast Anatolia Project (GAP)*

Southeast Anatolia, once one of the most underdeveloped regions of Turkey, has been substantially changed through a massive wave of investment during the last three decades. The region has been transformed in its polity, economy, culture, and people by these investments and the wave of rural-urban and inter-regional migration induced by economic and political change, the threat of terrorism, and armed conflict. The Southeast Anatolia Project (GAP) is a multi-faceted and integrated regional development effort aimed at improving agricultural development through irrigated farming, appropriate utilization and management of land and water resources, introduction of more advanced farming practices and cropping patterns, and development of appropriate agricultural mechanization.

GAP was conceived more than two decades ago. At the time it was advertised as a regional development project similar the Tennessee Valley Authority in the United States or the Mezzogiorno development in Italy. In all such projects, the ultimate goal is the integration of the people and the economy of a poor and economically disadvantaged region into the national entity. In the case of Southeast Anatolia the relative poverty of the region was obvious, but integration was targeted for two additional and interrelated reasons. First, the region is ethnically mixed and the majority of the population still speaks various dialects of Arabic or Kurdish in addition to Turkish. Second, the social structure and land tenure system are different from the rest of the country. Its traditional leadership structures, consisting of patrilineages connected over national boundaries, semi-nomadic heritages, and centuries of interactions with the Ottoman and Turkish administration molded the region in ways different than other regions. Thus, its integration into the social and economic entity of Turkey was seen as requiring special effort.

The Southeast appears to have a complex social fabric, but no in-depth studies are available.[2] Within the GAP region itself, there are a large number of sub-regions that have

---

[1]     Ayse Kudat initiated and managed the social analyses for the social assessment process. Mumtaz Bayram, Nilay Cabuk, Feryal Turan, Esma Durugul, Ismet Yalcin, Turan Hazar, and Aylin Baran were among the original authors of the draft SA report based on quantitative studies and stakeholder consultations; they were supported by James Osborn, Bahattin Aksit and Caglar Keyder. This chapter is based on the draft report of October 1998. It contains additional data and interpretation by the authors based on their previous research in the region and consultations subsequent to October 1998. The SA process will continue as Project preparation advances.

[2]     The absence of strong anthropological research tradition, on one hand, and the political instability that characterized much of Eastern and Southeastern Turkey, on the other hand, seem to have discouraged researchers from

different mixtures of socially distinct groups. The ruling tribal families that were nomadic some two centuries ago have become landlords. The tribal leaders used their social position to obtain exclusive rights to land, and the ordinary members of the community became landless tenants or sharecroppers. The low level of mechanization kept the landless dependent on tribal leaders and landowners for many decades. Despite permanent outmigration to other parts of the country, a large portion of the population continues to be landless;[3] this is rare in the rest of the country. The inequity worsened in those sub-regions of Eastern and Southeastern Turkey where mechanization of agriculture started in the 1960s. The landlords no longer required the labor of the landless families on a continuous basis, and the economic situation of the landless became even more precarious throughout 1970s and 1980s. As a result of landlessness and the unequal distribution of income, the poor had no option but to migrate seasonally or permanently in order to find employment and income.

It was against this background that GAP was conceived. It was supposed to provide an economic fix to the social problem of inequality and disenfranchisement; at the same time it was expected to integrate into the national economy a region which was far less developed, economically and socially. The risk with a project which emphasized gains in production was that it could potentially exacerbate the existing inequalities if no measures were taken to counter the unequal distribution of land. Although it was stated at various junctures that land reform was an indispensable component of the project, the political authority never seemed to have the will to enact one. The ceiling imposed on land holdings was easily evaded by landowners through allocating the title to different male members of the family. The failure to confront land tenure inequities through land reform was arguably one of the principal causes of the ethnic unrest which has plagued the region since the 1980s.

Compared to the other regions of Turkey, the GAP region is still disadvantaged in terms of socioeconomic development indicators. Administratively, it covers nine provinces: Adiyaman, Batman, Diyarbakir, Gaziantep, Kilis, Mardin, Siirt, Sanliurfa, and Sirnak. The region has a population of 5,275,013 according to the 1990 census, corresponding to 9.2 percent of the total population of the country. With 69.8 people per square kilometer, the population density in the region is close to the national average of 74 per square kilometer. However, the rate of population growth is 3.6 percent per annum, much higher than the national average of 2.4 percent. The share of urban population is rising; while half the total population lived in urban settlements in 1985, it now exceeds 60 percent. Fertility is higher in rural areas but urban settlements have grown faster because of rural-urban migration. The region extends over a land area of 7.4 million hectares, of which 42 percent are cultivated.

---

gaining an adequate understanding of the social fabric. Given budget and time constraints, the SA was unable to shed much light on this key issue either.

[3]        The outmigration to Europe from this region has been relatively low as compared to other regions (Ayse Kudat. Regional Patterns of Outmigration to Europe. Science Center Berlin. 1976.).

## The Sanliurfa-Harran Plains On-Farm and Village Development Project[4]

The province of Sanliurfa has the largest share of cultivated land in the GAP region (36 percent); the most fertile agricultural land. The Sanliurfa-Harran Plains constitutes one of the most important components of GAP with its potential irrigated area of 152,353 hectares. The area falls within the scope of the Lower Euphrates Project and consists of two irrigation networks. Irrigation water comes from the Atatürk Dam Reservoir and is delivered through the Sanliurfa tunnels. Presently, one of these tunnels is in operation. The main canal network has been completed and phased in for service by the State Hydraulic Works (DSI) and currently includes about 82,000 irrigated hectares.

Extending over a territory of 18,584 square kilometers with Syria to the south and the provinces of Mardin and Diyarbakir to the east, Gaziantep to the west, and Adiyaman to the north, Sanliurfa is the largest province of Southeastern Anatolia. The plains of Sanliurfa-Harran extend over an area of about 1,500 square kilometers. By 1998, about 82,000 hectares of land out of a total of 150,000 hectares had been brought under irrigation. Land consolidation work conducted by the General Directorate of Agricultural Reform in the project area covered 72 percent of the total land of the area. The General Directorate of Rural Services has conducted land leveling work on an area of 32,000 hectares (21 percent) as part of the on-farm development activities. Land consolidation had been completed in more than half of the communities for the proposed World Bank Project and land leveling had been completed in about a third of the villages. Other on-farm development work was in progress, including some of the village roads.

Table 7.1  Population of Sanliurfa by census year

| Census year | Population ** |
|---|---|
| 1927 | 207 487** |
| 1940 | 245 398** |
| 1950 | 298 394** |
| 1970 | 538 131** |
| 1980 | 602 736** |
| 1985 | 795 034** |
| 1990 | 1 001 455** |
| 1995 | 1 243 600* |
| 1996 | 1 302 300* |
| 1997 | 1 363 900* |
| 1998 | 1 428 300* |

Source: Provincial and Regional Statistics, DIE (1994)
* DIE  (State Institute of Statistics) estimates
** Population based on actual censuses

The present population of the province is about 1.5 million (Table 7.1). Cultivated land constitutes 51 percent of the total territory of the province. The average for the GAP region as a whole is 42 percent.  The population of the province was 795,034 in 1985 and exceeded one

---

[4]     Nedret Durutan manages the design of the overall Project. The SA was initiated to assist her in this task. The SA was managed by Ayse Kudat and was carried out by local experts.

million by 1990, with an average annual growth rate of 4.6 percent. In this period, the rate of population growth was above both the national and regional averages (Table 7.2). The population increase was particularly marked between 1995 and 1998 with an average annual growth rate of over 4 percent. The growth rate of the urban population in the same period was even higher.

Table 7.2  Population growth rate (percent increase)

|                    | Province | Region | Country |
|--------------------|----------|--------|---------|
| 1985-1990  Total   | 4.6      | 3.6    | 2.1     |
| 1995-1998  Urban   | 6.3      | 5.7    | 4.3     |
| 1995-1998  Rural   | 2.6      | 1.1    | -0.5    |

Source: DIE 1994 and estimates

In census year 1990 there were 148,521 households in the province; the average household size was 6.74 persons (Table 7.3). The province of Sanliurfa has 11 district centers, 772 villages and 1,646 sub-village settlements. While 551,124 people live in the district centers, 450,331 people live in sub-districts, villages and sub-villages.

Table 7.3  Population, number of households, and average household size

| Census year | Total population | Number of Households | Average household size |
|-------------|------------------|----------------------|------------------------|
| 1960        | 401 919          | 66 759               | 6.02                   |
| 1970        | 538 131          | 84 217               | 6.39                   |
| 1980        | 602 736          | 97 376               | 6.19                   |
| 1990        | 1 001 455        | 148 521              | 6.74                   |

Source: TOBB (Union of Chambers and Stock Exchange), 1997

According to data for the year 1990, 71 percent of the economically active population in the region was employed in agriculture, followed by 20 percent in services, 5 percent in construction works, and 4 percent in manufacturing. While half the population consisted of domestic family labor, 26 percent worked on their own land and 25 percent were wage laborers. In the proposed Project area, there are 178 villages and 12,871 households. According to the 1990 data, the project area has a very young demographic outlook; the age group of 0-14 years constituted 48 percent of the total population. Half of the population was in the age group 15-64 years, and only 2 percent were elderly.

Within the context of GAP, World Bank assistance was sought to help complete critical investments in the sub-region of Sanliurfa where land consolidation is underway. The GAP administration and the Bank agreed to carry out a social assessment (SA) to supplement the economic, financial, and technical assessments required for the design and implementation of the proposed Project.

### Review of Existing Studies

In 1995, the GAP Social Action Plan was prepared covering all nine provinces of the region. The social development objectives for GAP included equity and social cohesion. A 1997 book officially published noted that "population groups of different cultural origin live in the

Region in an interwoven way" and that "if right policies are pursued, new synthesis can be achieved which might contribute to the enrichment of an integrated national culture". It was also noted that "if communal balances are disturbed" conflict might emerge.[5]The government social policy also noted the dominant role of patriarchal tribal social system and the absence of cooperatives and similar democratically organized civil society organizations. Equally important, the government noted that "the uneven distribution of land continues to be an important problem". "Most of the arable land belong to a few big landlords" and most of the people have far too little land to sustain themselves.

Several social survey projects were conducted prior and subsequent to the GAP social action plan. They included studies on trends of social change; population dynamics; status of women and their integration in the process of development. Based on these GAP social policy objectives included the participation of people in design and implementation of development initiatives, enhancing "the partnership of public institutions, local and volunteer organizations", increasing access to social services, and to give priority to women and "youngsters".

A survey carried out in 1992 aimed at depicting the socioeconomic profiles of rural and urban communities in the GAP region and covered 10 urban settlement units and 47 villages in the provinces of Adiyaman, Diyarbakir, Gaziantep, Mardin, and Sanliurfa. Tribal relations were the traditional form of community organization and commitment, and the settlement pattern consisted of many dispersed units and sub-villages (mezras), possibly composed of close kin groups. Noting the difficulties of social infrastructure provision to dispersed settlements, earlier researchers proposed to reduce their numbers through physical and social integration with larger settlements.[6]

In 1992, agricultural production was limited to a large extent to cereals (wheat and barley) and pulses (lentil and chickpea) under dry farming conditions. On irrigated land, which was then only five percent of the total, cotton was the main crop. The share of industrial crops was small, the yield and crop intensity were low in cereals and pulses mainly because of the inadequacy of agricultural inputs. Very few organizations such as cooperatives existed in the region. Livestock production, which was one of the traditional means of subsistence in the region, had already started to decline in importance.

The 1992 research also explored the characteristics of migration to assess the possible impact of GAP.[7] It investigated the skill levels and employment-investment capacities of the settled population should the tendency to migrate be reversed. The research results indicated that 43 percent of the rural communities actually received migrants from other communities. Seventy four percent of the household heads described their work as farming and 40 percent were landless. The highly unequal distribution of assets was reflected in patterns of land tenure. While a third of the households had little land, a small minority held very large holdings. Those with less than 50 donums of land constituted 28 percent of the population and only 3 percent had over 500 donums.

---

[5]     GAP: Social Policy Objectives. Republic of Turkey Prime Ministry. Ankara. 1997.

[6]     M. Sencer. 1993. GAP Bolgesinde Toplumsal Degisme Egilimleri Arastirmasi. Mimeo.

[7]     Bahattin Aksit. GAP Bolgesi Nufus Hareketleri Arastirmasi. Ankara, 1993.

Twenty seven percent of all household heads worked as sharecroppers and 63 percent of them had no land of their own. In 1998, there was little change in the inequitable distribution of land.

The close relationship between economic power and the ability to make a living within the region was evident in the strong tendency among the landless to search for a living outside the area; as land based investments increased, the need for agricultural workers also increased. In 1992-93, about half of all households planning to migrate were landless. Eighty seven percent of these households indicated lack of a job and land, and economic problems as their motives for migration. The SA showed that by 1998, there was a dramatic reduction of the landless seeking employment in other regions because there were sufficient opportunities within the region.

The interdependencies between the economic and political power structures of the region had in the past expressed themselves in high demand for government investments in water and land resources, energy, and infrastructure. These investments have substantially increased the returns to land; those with the largest holdings have benefited the most in absolute terms and this has ensured their political power as well as their ability to make key investments in manufacturing and trade. Irrigation was the single most important benefit brought to the area and to those with land.[8] Therefore, an early assessment of the management and operation-maintenance (M&O) of the irrigation systems was carried out in villages where irrigation infrastructure was in existence and in others where irrigation had been planned to start in 1994.

Several of the key findings of these earlier assessments are particularly worth noting. Economic rationality in farming increased despite continued reliance on subsidies and government investments, and there was a visible increase in the spread of modern technology along with the introduction of irrigation. The most radical change in cropping pattern was the switch to cotton farming. Further, the most common technique of irrigation prevalent in the survey area was labor intensive, with minimal requirements for capital investments.

Most landowners expected the government to carry out investments in irrigation. They saw the State as a benevolent father and felt that farmers' resources were not adequate for such investments and they lacked relevant skills and adequate organization. However, as a result of the findings of this survey, it was suggested that the management and operation-maintenance of the GAP irrigation systems should be transferred to the farmers' organizations with a bottom up approach.

Earlier studies focusing on the status of women in the GAP Region confirmed low appreciation of women despite their active participation in household economy. Three-quarters of all households used only family labor for agricultural activities; women took part in harvesting, seed sifting, hoeing, transporting, and storing. Three-quarters of the rural women were illiterate. Women's access to property was limited by the traditional laws and women themselves believed that this was basically right; asked if they should have a right to inherit from their fathers, 47

---

[8]         Bahattin Aksit. GAP Sulama sistemlerinin Isletme, Bakim ve Yonetimi, 1994, Ankara;  E. Doker and C. Erdogan. Katilimci Sulama Tonetimi ve Su Kullanici Orgutleri.  II GAP Sempozyumu, 1998. TOBB.  Sanliurfa Ekonomik Yapisi ve Uygun Yatirim Alanlari, 1997.

percent of the women strongly disagreed. Their political choices were also dependent upon the opinion of the household heads.

While 60 percent of the households were of the nuclear type, a quarter of all families were extended, a much higher share than observed in other regions. Whether extended families share a household or live in proximity, their existence constituted a strong social control over women's participation in public life. The practice of bride price (baslik), money paid by the bridegroom's family to the family of the bride, was widespread, indicative both of strong traditionalism and the low status of women.

Improvement in the status of women in the GAP region and their integration into the process of development were complex. The research stressed that an integrated set of interventions in different areas and coordination among relevant organizations were essential for success. While the high levels of investments and the accompanying economic changes have brought about important social changes in the region, including a visible improvement in living standards of the landless, significant changes in women's status appear more difficult to achieve.

## The Social Assessment

The first phase of the SA process focused primarily on social analysis and stakeholder consultations in order to design the social goals and provisions, formulate the constructive social mechanisms for development, and identify potential adverse social consequences of economic growth. The process was launched through a rapid assessment in May 1997 and continued with additional rapid appraisals in the summer of 1998 followed by a systematic survey. This chapter reports the first phase of the SA, covering social studies, surveys, individual stakeholder consultations, and a broad stakeholder seminar. Continued dialogue with the direct borrower and other stakeholders is needed before Project components are well defined. Since the Project is currently under preparation, it is too early to discuss SA impacts on its design. However, the most important goals of the Project are social equity and poverty reduction.

Massive technological change from dry farming to irrigated agriculture is taking place in the plains of Sanliurfa-Harran where irrigation started in 1995. The SA raises questions concerning the nature of the social transformation we might expect as a result. Since there has been no significant change in the patterns of land ownership, improvements in income distribution come about indirectly through changing global and national conditions, emergence of new work and employment opportunities, and increasing integration of the region into the world economy. The rapidity of the changes brought about in land productivity through irrigation requires a focus on social or distributional impact of economic and technical change. Other elements of social transformation of the GAP region include globalization and massive changes in the communications structure, intensified formal and informal trade, greater energy capacity and incentives provided to private sector initiatives in the region, adoption of labor-intensive crops made possible by irrigation, greater integration of the region in national politics, social instability and conflict, and immigration to the region from neighboring regions. The outlook of the people of the region has thus changed significantly with all these factors molding their worldview and expectations. This is especially so for the males members of the society who were able to interact with a large number of new actors over the decades whereas the women were confined to their homes.

---

*A User's Perspective*

*The room in Harran was almost empty, there were only a large kilim on the floor and several pillows on which you can lean back, rest your arm or sleep. We were chatting with a young man with shiny black eyes about his family and living conditions in the village. A baby was sleeping in the cradle and another one was playing with a plastic car with a missing wheel. The mother, who had a nice face with smiling eyes, was in long, bright green dress. She moved briskly in and out but did not join the conversation. After awhile, I realized that she did not know Turkish and her husband translated into Arabic for her. Later, she brought a tray with a tea pot, a couple of traditional tea glasses, and some sugar in a plastic container shaped like a sitting chicken. Everything was amazingly clean and shiny and I wonder how she managed this in spite of the lack of running water in the village.*

*The young man was telling us that a small piece of land, handkerchief-sized, was not enough for them. In order to make some money, he traveled to the western part of the country and stayed for about six months. He came back to his family with whatever money he earned. He said that he sold sun glasses to tourists in Kusadasi, a tourist town south of Izmir. He bought the glasses in Sanliurfa, where the prices were relatively cheaper due to border trade, and sold them with a marginal profit.. The room was hot and the baby started to become querulous. The father's voice was full of bitterness. My eyes fixed on the bright red crest of the chicken on the sugar container. I visualized the slim man with shiny black eyes in Kusadasi showing cheap sun glasses to the sun tanned tourists in bikinis, keeping his babies and wife in the back of his mind. I imagined him sleeping in cheap motels, eating as little as possible and earning as much money as possible so that he can come back to the village with enough money to last for the rest of the year. He also brought back small souvenirs for his family--a plastic car for his boy, a sugar container for his wife and a lot of stories reflecting his culture shock. Of course, he did not mention the miserable life he had over there.*

*I thought that the Project can and must change the lives of these people, so that young men with shiny black eyes do not have to live a miserable life away from their familiar environment. They should be home to watch their babies grow. Their wives should not be left to struggle all alone with the challenges of daily life. Their babies should have better futures. The social assessment opened the crust and reflected the frustrations, expectations, dreams and, of course, opportunities of the villagers.*                Nedret Durutan, Task Team Leader

---

### Social Assessment Objectives

The aim of the SA was to explore relevant social development, participation, and institutional issues, identify key stakeholders, define a participation framework, and help design a monitoring and evaluation framework. The broad objective of the survey was to establish a socioeconomic profile of villagers in the irrigated areas, identify the prospective changes and dynamics of Project implementation, and explore the needs, problems, and expectations of the local people. A specific effort was made to determine the quality, speed and direction of the changes introduced by irrigation, land consolidation, and agricultural modernization. The SA explored these impacts by systematic comparison of sub-regions that have received different combinations of development investments.

The SA sought to determine the impact of socioeconomic changes and transformations on different sections of the population. It also investigated emerging problems and possible solutions. Consequently, the SA survey was concerned with forms of land ownership and use, income and living standards, availability and status of technical services related to irrigated agriculture, and the social dimension of related problems. New agricultural technologies (irrigation being foremost), agricultural practices, and crop designs introduced within the framework of GAP affected the relationship between the people and the land, and thus influenced social relations by generating new forms of social organization.

Information from the SA will be used to develop programs that promote equitable sharing of project benefits. Since 60 percent of the people in the survey area are landless, practices and measures that raise the living standards of this population segment are of special importance. One of the primary objectives of the Bank Project is the improvement of the socioeconomic status of the poor and the landless throughout the project area.

### Social Assessment Methodology

The basic hypothesis was that change is inevitable; feudal relationships would gradually dissolve and more inclusionary social relationships would emerge along with a more equitable distribution of social, economic, and political power. To test the hypothesis, the SA investigated relationships between different types of infrastructure investments and emerging social dynamics by reviewing available research, inventories, reports, statistics, and information from public or private sources, conducting in-depth interviews and observations.

The SA survey instrument was designed and pre-tested once this preliminary work was completed. The main survey area consisted of 178 village and sub-village settlements located in the plains to the south of Sanliurfa, an area which was being brought under irrigation. Villages and households were sampled using proportionally-stratified random sampling techniques (Annex 1). Villages in the survey area were ranked according to the presence of irrigation, land consolidation, land leveling, drainage, and on-farm roads and each stratum was randomly sampled in proportion to its specific weight in the area. Households in each sample village were grouped using such criteria as landowners who did not farm their land directly, family enterprises farming their own land, tenants, sharecroppers, and landless peasants. Then the households were sampled randomly with respect to their proportional weight. The survey sampled villages and households to assess the present status and patterns of change with respect to the indicators listed in Annex 1. It gathered and analyzed qualitative and quantitative data on landowners, sharecroppers, tenants, women, and a wide range of institutional stakeholders listed in Annex 2.

The SA used village information questionnaire forms to obtain information on 35 sample communities. The village and household interview forms were tested in the villages of Kisas, Apali, Sevimli and Saglik on 23 July 1998. After necessary corrections were made to the forms, field implementation took place from 29 July to 8 August 1998. Trained interviewers gathered data from the village mukhtar (headman) or members of the village elders council. Also, 450 landed and landless households living in each village were selected by stratified random sampling along with information supplied by the village mukhtar. Standardized semi-structured interview guides were used for in-depth interviews with different groups and categories of people. The findings were re-evaluated by qualitative information coming from observations, focus groups, in-depth interviews, and two symposia.[9]

The analysis of the empirical research findings of the SA process showed that some of the trends observed in the earlier studies had become stronger over time. For instance, the impacts of irrigation had become more marked. Continued difficulties in the development of civil society organizations, weak coordination among governmental institutions, and lack of security and incentives for private sector development continued to slow down the dissolution of feudal relations. However, new trends have also emerged which promise the better integration of the low-income and vulnerable groups into the economy and the society.

## Four Pillars of Social Assessment

### *Key Social Development Issues*

The key social development issue identified by the SA is the challenge of transition from a hierarchical social system, where economic and political power is concentrated in the hands of a few and social inequity is high, to a modern system, where all segments of society benefit from decades of State investment in the region.

> *"The aga lived in a luxuriously built house with a large garden and deep water well, and all the remaining villagers lived close together in a compound settlement. A constant stream of village women and children kept coming to the well of the aga to fetch water. I asked the aga whether there was need for safe water in the community. He responded with a question, "Why would the esir (slaves) need water when they have been getting it free for all these years?." I had no way of telling whether he really meant to refer to the villagers as slaves or as sharecroppers. At home he spoke Arabic and I was unsure whether "esir" had the same degrading connotation for him as it did for me." A local social scientist*

Traditionally, widespread inequities rooted in the tribal system resulted in differential access to land and other productive assets; differences in income and living standard between the landless and landowners were large and those few with very large holdings are influential within the traditional and modern institutions. Kudat has described elsewhere how the tribal social order prevailed throughout the region and how in the transition to the multi-party system in Turkey, social relationships based on lineage and fictive kinship (kirvelik) were used to reserve key roles for influential members of these tribes.[10] The near serfdom that once prevailed in the region can still be seen in the structuring of human relationships; high respect is associated with the term aga (landowner), and esir (slave) is used for the agricultural workers or sharecroppers.

---

[9]  The thirty-five villages which were sampled for the survey represent different combinations of on-farm development services such as land consolidation, irrigation, and leveling. They also represent the survey area with a sampling rate of 20 percent. The total population of all the 178 villages of the project area is 91,115. The average village population is 512 with an average of 72 households per village. The highest and lowest village populations are 2,519 and 29 respectively, and there are 290 households in the largest and only 8 households in the smallest. The 35 sample villages have a total population of 14,042. The SA sampled 2,001 of the 12,871 households in the region. The smallest village in the sample had 15 households and a population of 112; the largest had 110 households and a population of 863. The average number of households per village was 57.

[10]  Ayse Kudat. Kirvelik. Ayyildiz Matbaasi, Ankara. 1974. These relationships are also brought to theoretical articulation by Dogan Avci in his books and essays on "Asian Model of Production" (Asya Tipi Uretim) published in the 1970s.

Gender and intergenerational inequity, rooted in the tribal traditions, were expressed in terms of limited economic opportunities and in the restricted ability of women and younger people to participate in the local and provincial decision making processes. Economic inequity and social inclusion were closely interrelated. The large-scale economic investments made in the area have provided enormous benefits to the economically powerful; the shift to cotton production facilitated by irrigation and producing cotton through labor intensive methods allowed these benefits to trickle down but only slowly to the small holders and the landless. However, in the near future as mechanization increases and as the consolidation of land improves the efficiency of mechanized agriculture, demand for labor may fall. This, in turn, would shift the balance of benefits to the exclusive advantage of the economically powerful groups.

**Social Diversity**

Qualitative research and consultations with village administrators showed that about a fifth of the communities had received migrants from outside within the past five years and 83 percent of the sample villages stated that their villages received seasonal workers from other settlements. Village mukhtars stated that over a third of the migrants came from the city of Sanliurfa and 24 percent came from nearby villages. Seventy-two percent came during cotton harvest. This situation contrasted sharply with the earlier studies from 1992. Clearly, expansion of labor intensive cotton production was not only a major contributor to the trickle down of benefits to the local poor but also generated some work opportunities for the poor from other regions.

For many thousands of years, Anatolia, known as the "cradle of civilization," has been host to a large number of civilizations and ethnic groups. The cosmopolitan nature of the Ottoman Empire, its policies with respect to granting autonomy to all peoples within its territory, and continuous migratory waves of groups have all created a highly diverse society. The social fabric was defined and redefined in a continuous manner. While the major urban centers now accommodate innumerable ethnic groups including peoples from the Former Soviet Union and Eastern Europe who continue to immigrate to Turkey, the situation is also dynamic in traditional regions of emigration. As a result, only a few communities are socially homogenous.

The GAP region has also hosted many different ethnic, language, and religious groups, and its social fabric continues to change. Some of this change is a result of immigration to the region that was once a major provider of migrant workers to other areas. Within the larger GAP region, the proposed Project has a limited geographical scope. Within the narrower Sanliurfa/Harran area, households were asked to name the language other than Turkish that is spoken among family members, as a proxy to ethnic background. The survey of the rural areas in Central, Akcakale and Harran counties revealed that 90 percent of the households were Arabic speaking, about 8 percent spoke Kurdish and 2 percent spoke Turkish exclusively. While Arabic speaking households were equally distributed among all counties, the Kurdish families were concentrated in the Central county and had larger households. All three groups practiced polygamy to the same extent, but 85 percent of the Arabic speaking families and only 69 percent of the Kurdish speaking people paid bride price. More than half of the families in these two groups had members who used to migrate to other regions as agricultural laborers prior to the irrigation of the plains; few do so now (Table 7.4).

Table 7.4  Out-migration prior to irrigation (percent of households)

|  | Language spoken at home | | |
| --- | --- | --- | --- |
|  | Turkish | Kurdish | Arabic |
| No member of the family migrated for work | 71 | 41 | 42 |
| One member migrated | - | 57 | 40 |
| Two or more members migrated | 28 | 3 | 18 |

Source:  Household Survey, 1998

A larger percentage of families who spoke Arabic had low-incomes.  Because the total number of Turkish and Kurdish speaking families was limited, statistical results were not strong; nevertheless, the discriminant analyses revealed a similarity between the Arabic and Kurdish speaking families with regard to land tenure, agricultural income, overall income, and asset ownership.  Adult women from these families do not always speak Turkish; indeed in many of the households that the author visited even younger women tended to speak Arabic only.  This is partly a result of the fact that there is marriage with tribal kin across the borders and the patrilocal norms require brides to reside with their husbands where they speak Arabic with the elder women and men, and continue to have some exposure to the Arabic speaking media as well.  The school age children of both sexes, however, are largely bilingual as long as their economic situation allows them to attend school.  Given the active role of women in crop and particularly in livestock production, the linguistic factor is particularly important.

**Access to Social Services and Infrastructure**

Human development conditions left much room for progress.  Sixty-eight percent of the communities had problems of education infrastructure including the absence or inadequacy of school buildings. While an average village had 98 children of primary school age, 55 of them were not enrolled in school. Some of them are children of seasonal agricultural migrants and thus cannot attend school regularly, but many others whose parents are permanent residents do not attend because their labor is needed or no school is available.

Almost every village headman expressed some concern about health services; 88 percent named the absence of any health center and personnel as their priority problem, but only a few complained about the distance to the health center. Diarrhea and typhoid fever were the diseases most frequently observed. They are of particular danger for children.

Transportation, drinking water, electricity, and communication services are all inadequate in the region. For example, 83 percent of the sample villages have problems with their roads, particularly the lack of tarmac roads and the bad condition of existing roads. In some villages, there were difficulties

> *"You see there is no water. If you have no water in this hot weather, anything may happen: illness, filth....Of course I want to keep my house clean and give a bath to my children...but we can only use discharge water for cleaning purposes. From time to time, our children even drink this water.... First of all, we want drinking water...." A village woman*

associated with excessive mud. The situation with the roads is problematic for children who have to go to other villages for schooling. Forty-six percent of the villages have problems in communication; 40 percent of the sample villages complained about defects and disconnections in telephone service. For 27 percent of the villages their telephone exchange was inadequate and 13 percent stated that they could not pay their bills easily in winter because of difficulty getting to district or city centers.

Drinking water was a priority problem for all village headman and households; 85 percent of the villages have serious problems with drinking water supply and a third have no drinking water. Water is brought in either by tankers or fetched from other villages by children and women. Women often have to fetch the water several times a day, sometimes from a distance of several kilometers. Some also use the water of the canals and discharges, having no other source.

## Social Change

A comparison of SA data with data collected in the early 1990s shows that there have been visible changes in social conditions over the past decade.[11] But the positive impact has been restricted largely to large landlords; landless farm workers who previously sought work in other regions now find seasonal employment in Sanliurfa but have not been able to acquire sufficient land of their own, in general, to become landed farmers. The two symposia held to bring together key stakeholders indicated that overall living conditions have improved in Sanliurfa, where a large proportion of the active population is employed in the agricultural sector. Nevertheless, living standards are still low, and the generally backward conditions of East and Southeast Anatolia characterize the project area. The traditional outlook prevails, gender and age-specific roles are strictly defined within the clan/lineage structure, population growth is high, average household size is 6.6, and the local political structure is closely connected to the economic vested interests.

The farmers themselves are less enthusiastic about changes in living conditions. When asked how GAP had changed their lives, 53 percent of the sharecroppers and 43 percent of landowners indicated no change. Only 20 percent of the landless stated that they could more easily find jobs, and 26 percent of the landowners said they had higher incomes. Even in villages where irrigation, land consolidation, and leveling have been completed, only 29 percent of landowners indicated higher incomes. As income level rises, the percentage of people who indicated they have experienced important changes also rises. Well over two-thirds of high-income groups recognize these changes.

Land tenure continues to be inequitable. While land holdings became smaller in each generation because of land fragmentation caused by inheritance, a small percentage hold large amounts of land and the majority have barely enough to meet subsistence requirements. High fertility rates multiply the numbers of the landless but the potential for intensification of irrigated agriculture provides a clear advantage for capital accumulation for those with land. Not all of the landless are from landless families. Families have a large number of children who marry at an early age, and the males stay within the community to continue farming on the family land owned by their father. While they are landless in a legal sense, they do have expectations to inherit land in the

---

[11]      State Statistical Office. Province and Regional Statistics, 1990; State Statistical Office. Province and Regional Statistics, 1994-96; Union of Chambers and Stock Exchange. Economic Structure of Sanliurfa and Field Investments. 1997; Mumtaz Bayram and H. Erkan. Land Consolidation in the GAP region: Problems and Suggestions. 2nd GAP Symposium. 1998.

future. Often, the smaller holders suffer because their children have far too little to divide among each other. The wealthier families often provide good education for their sons; some of them become professionals and others are provided with capital to establish business. Thus, while the land inherited by the sons of wealthier families might be registered to different names, in essence one of the brothers often assumes its management, sharing an appropriate portion of the revenue with his brothers.

## Social Mobility and Migration

Lack of land and economic problems continue to be the most important causes of migration. Eighty percent of the household heads are native inhabitants, living in the villages where they were born. Only 21 percent have their birth places in other villages or cities, including women who were married into local families but were born elsewhere. Subsistence or economic problems are the leading cause of this intra-rural migration (39 percent). Other reasons include lack of land (25 percent) and changes in employment (15 percent). While these were once the primary reasons to push the poor people out of the region, they now constitute reasons for the poor of other regions to move into the project area. Thus, this type of migration can be seen as the result of positive changes introduced by GAP.

Prior to irrigation, a large part of the landless and some half of those who owned land were agricultural workers in neighboring provinces, even if for short intervals. Review of the previous patterns showed that 79 percent of those who were farming less than 100 donums of their own or someone else's land at the time of the SA survey had been forced to migrate to other regions for work in previous years. A fifth of those who were farming over 100 donums in 1998 previously went to other places for work. More importantly, many families kept moving from one farm to another as a family, using child labor in order to maximize their earnings and many employers preferred women and children to harvest the cotton; one third of the households interviewed stated that they migrated with their families while in half of households only the household head migrated for work.

In the communities that have already benefited from government investments in irrigation, land consolidation and land leveling, 61 percent of the people were previously seasonal migrants. A third of the household heads used to make a living from seasonal migratory work; only 13 percent do so now. Prior to irrigation, Adana and Hatay were the leading provinces where these people went for work. Now migrants from these provinces come to Sanliurfa.

The introduction of irrigation has changed the direction of seasonal migration and the patterns of cropping in the region. As a result, the expertise of the proletarian farmers of Sanliurfa who had been working in cotton production in other regions has become a valuable human resource in Sanliurfa. While 70 percent of the people

> *"We used to go to Adana and Hatay. We lived there in tents in that hot weather. It was like hell in daytime. There was no water and we couldn't take baths. We lived in filth. We also used to live very close to other families and workers whom we didn't know. We always worried about our relatives back in the village. Now we work here, in our own village and live in our own house." The wife of a landless peasant*

interviewed previously went to other provinces for the cotton or citrus fruits harvest, now only 12 percent do so. The great majority of households living in communities that have benefited from state investments in irrigation and land consolidation are no longer forced to make a living as migrants; their skills are highly needed in the region. For instance, 70 percent of heads who were not engaged in farming in Sanliurfa prior to irrigation were seasonal migrants; now only 9 percent are. These persons stated that with the introduction of irrigated agriculture, they have better opportunities for finding jobs such in irrigation, hoeing, cotton harvesting, and thus do not need to migrate.

Some 88 percent of all household heads who either farmed their own or someone else's land responded negatively to the question, "Is there any household member going out to some other place for work?" Even among the landless there are opportunities to work out of a home base; there is no longer a need to live for months in tents in dirt. Sixty-seven percent of the 252 household heads who are landless previously migrated for work; now only 14 percent do so. Instead, they farm as sharecroppers or work on plots owned by their parents or close relatives.

For families who still migrate for seasonal work, only a few are forced to travel long distances and many stay within the region. Thirty-two percent of the seasonal workers in Sanliurfa are from the city of Urfa and 20 percent are from nearby villages and towns where irrigation has not yet started. Most of the seasonal workers from Urfa come from squatter settlements or from settlements that have been submerged by the dam reservoir. People who come from Suruç and Maras are landless peasants who rent land in the area or work on land as sharecroppers.

Some families migrate for other types of work. Some used to purchase goods from the informal "free zone" of Syria and take them to other regions; others work as street vendors or in similar capacities in metropolitan cities such as Izmir. While working in different parts of the country did cause hardship to many families, it was also an important source of social change and one that appears to be particularly visible among the younger men with little or no land.

Clearly, there has been a substantial reduction in out-migration from Sanliurfa as a result of enormous investments. GAP has created employment opportunities for the landless peasants and for people living in areas where irrigation services are not yet available. Families from other regions come during cotton sowing time and stay for eight to nine months, living in small cottage type dwellings allocated by landowners; many also live in tents and move from one farm to another together with other household members using family labor intensively. They provide for their basic necessities during this period by borrowing from the landowners and pay back their debts from their share of the harvest. The poorest of the poor throughout Turkey are these seasonal migrants and their children. Labor intensive cotton production provides continuous incentives for this type of migration and the availability of cheap seasonal migrant workers is an important, if not the most important, element in Turkey's ability to maintain a competitive advantage in textiles.

Labor intensive cotton production has enabled many of the Sanliurfa families to work within their home communities. Against these advantages, the disadvantageous living conditions of families who rely on this type of work are undeniable. It is difficult to target development assistance to this vulnerable group other than through humanitarian assistance. It is also difficult to respond to their needs through the proposed Bank-financed project. While a specific mobile water supply assistance was considered for this group in the context of the social assessment, it was unclear that such an intervention would either be appropriate or feasible. It is expected, however, that some of the proposed improvements in the water supply and sanitation infrastructure will trickle down to these families.

## Social Capital

One of the most pronounced characteristics of the social structure prevailing in the region is the pervasive nature of tribal relations. Membership in tribal lineages is almost universal in the project area. About 61 tribal groups were identified during the SA and these play important roles in shaping the behavior of community members, their affiliations, their membership in civil society organizations including the water user associations, and their participation in local and national level politics. Tribal ties are a protective umbrella and a mechanism for identity building that are highly influential in all spheres of life from political behavior to marriage and family structure.

During elections, the political party of choice is largely determined by tribal decisions; half of the respondents admitted an explicit tribal influence on their voting behavior. One of the expressions frequently repeated during interviews is "here, a person who goes to the ballot surely knows for whom to vote…." At present, tribal chiefs appear as deputies to the parliament, mayors, mukhtars, local leaders of political parties, and heads of water user associations. Half of the landholding families and half of the sharecroppers define tribal loyalty as voting for a specified political party.

The nuclear family is the dominant household form in the area, but people distinguish between social and residential patterns. They act with reference to tribal ties and traditional values and norms. In-depth interviews point out that tribal structure prevents nuclear families from exhibiting individual initiative and autonomous participation. Tribal ties constitute a striking instrument for solidarity and material support; 39 percent explicitly state that this is the major source of support and 43 percent consider it important to have their children marry within the tribe. When it comes to the marriage of girls, the preference for intra-tribal marriage is virtually universal.

Tribal elders and the village mukhtar act as mediators to restore peace when there is a dispute within the village or with other villages. There is a clear overlap between tribal and lineage leadership. Large landowning families often come from prominent tribes and hold key positions within and outside the village communities. Both village elders and the headman have equally important roles in dealing with disputes, with elders playing clearly greater role in family and inter-family issues and the headman focusing more on administrative and inter-community disputes. In disputes with other villages, 78 percent of respondents indicated that the village mukhtar has the authority. Age, gender, political power, administrative position and economic status all define authority within these largely traditional communities. Women and the youth occupy visibly low ranks in the social and political hierarchy and are systematically excluded from access to assets.

An empirical study of tribal influence is needed, but it is beyond the scope of the SA. It would have been difficult to finance under the proposed project given budget constraints. However, without an adequate understanding of this dynamic interrelationship between land tenure and tribal status and status within local and national political and civil society structures, some of the key social engineering required for the democratization of the water user associations and the creation of other civil society organizations will be difficult.

## Social Inequity

**Inter-generational Inequity.** Social norms personalize the nature of relationships in both formal and informal organizations. A person is accorded higher rank or status on account of his/her gender, wealth, lineage, education, employment and age. High status persons provide patronage to those of lower rank, and they expect to extract services in return. For example, the older, wealthy males of lineages who are the village headmen may provide god parenthood and even life long protection (Kudat 1974). In return, they are obeyed and served by the younger villagers. Normatively, patronage—the status of "aga" (feudal lord)—obligates and entitles the patron with a set of responsibilities, including informal assurance of sharecropping arrangements. These cultural patterns originate within the traditional tribal systems and extend to workplace: the farm, water user association or an administration office. Officials who fail to extend traditional patronage to their clients lose face and status. Younger members are undermined in the work place and business rules often make room for social norms. Tribal tradition dominates the public and the private sector, and wealth that can be traced through lineages is given much greater importance than newly acquired wealth.

These deep-rooted traditions govern gender and other social relations; change will be a gradual and indigenous process. Migration is the most important of the many factors that dilute these norms and create countervailing forces since the immigrants know little of the local traditions and tribal relations.

**Gender Inequities.** Women's status is articulated in many different forms. Of these, perhaps the three most important and socially accepted indicators consist of women's participation in the formal labor force outside the agricultural sector, their access to and ownership of property, and their participation in education. In the project area, there is a visible lack of female participation in the formal economy; women are exclusively family laborers. As to property ownership, when questioned about the prospects of his sister inheriting land, a brother says "over my dead body." When questioned "what if her husband insists that she should be given what is legally hers" he answers "we clean these questions with blood" meaning that such a claim would meet resistance. Despite the existence of a civil code which allows women an equal right of inheritance, a mother says: "It is very shameful for girls to ask for land. If they need it, we give them their own share. My husband has land. My father's land is tilled by my brothers."

As a result of the prevalence of patriarchal traditions, few women own property. This is socially acceptable as reflected in the fact that 56 percent of respondents to the SA survey think that it is right to have only male children inherit their father's land. For other types of assets and property there was more support for women's access; half of those interviewed thought that property other than land should be shared equally by each child. But the other half believed that male children should have the right to all property and women should have no claims whatsoever.

With the current practice of land registration and the introduction of land consolidation, women get equal shares according to the provisions of the Civil Code and Legislation on Title

Deed Registry. However, as a result of the process of socialization in which traditional patterns influence thinking, women also have ideas and attitudes which favor their brothers in issues related to succession or inheritance. Land is the main instrument in building wealth in the area, and it is also obvious that partition and control practices favor males. Consanguineous and inter-tribal marriages further consolidate this control. However, there seems to be some flexibility when it comes to property other than land.

Polygamy is still practiced and bride price is widespread. Polygamy was observed throughout the region and traditional norms fully support its practice despite the existence of the Civil Code which strictly prohibits it. Interestingly, 17 percent of all household heads interviewed had polygamous marriages. While it is said that "many wives are for the rich" even those less wealthy practice polygamy; 20 percent of 185 households who owned land and 14 percent of 264 households that made a living as sharecroppers were practicing polygamy.

> *"I don't want my husband to marry another woman. Then there will not be any peace at home. My father had three wives and no peace at home. All of my uncles had two or three wives. I don't approve of this. But in any case, what can I do if he gets married again?" The wife of a landowner*

The examination of the motives for polygamy point to the social acceptability of the custom and the disapproval of divorce. A third of those who had polygamous marriages mentioned the infertility of the first wife as a reason. Another 28 percent stated they regarded polygamy as a tradition to be maintained; loyalty to tradition was somewhat stronger among landowners and higher income groups. In other words, the wealthier could afford to have more than one wife without having to have the excuse of the infertility of the first wife. The share of those who have made a second marriage to have children was 38 percent among sharecroppers and 35 percent among landowners. Another reason for having a second wife was specifically the lack of male children from the first marriage. This reason was cited by 15 percent of the sharecroppers and 12 percent of the landowners.

Polygamy has practically disappeared in the rest of the country; its persistence in Sanliurfa is a remarkable indication of the strength of traditions in the region and the difficulties one should anticipate in attempts to change its social structure, especially the status of women, through the Project. Arranged marriages are getting more rare as the country modernizes, but they are nearly universal in the region; 83 percent of household heads interviewed arranged their marriages by making bride price payments, baslik. Among the remaining 17 percent, a special type of marriage (berdel) was practiced to avoid bride price (baslik). Berdel is the cross-marriage of male and female children of two families, and there are bonding rules for both marriages which continue as long as the marriages last.

Adult women have low educational status but achievements their daughters are major. The relative status of women with respect to education was most clearly observed in comparing the household heads: men were mostly primary school graduates (61 percent). Although the rate of illiteracy of male heads was 29 percent among household heads, 82 percent of their wives were illiterate. Only 16 percent of adult

> *"They do not send girls to school here. I never went to school. I wanted to go, but my mother was ill and they didn't let me. I am 13 years old and I got married three months ago. I didn't want to get married, but my parents said his family was good and wealthy, so it was a good chance for us." The wife of a landowner*

women are primary school graduates.

In sample households, about 40 percent of school age children were not enrolled in any school and 69 percent of households that they came from were poor; only 28 percent came from middle and high-income groups. When asked the reasons for not sending girls to school, some continued to refer to traditional values. However, the younger generations are closing the gender gap. There was no important difference between the literacy rates for boys and girls; 46 percent of the girls and 41 percent of the boys were literate.

Local people's perception of education is not limited to traditional values. Subsistence issues were cited as a major reason for illiteracy by 36 percent of the respondents. Forty-four percent of households in low and 20 percent of households in the middle income groups stated they could not send their daughters to school because of the cost of education. Boys were also kept from school for economic reasons; 55 percent of those who did not send their sons to school stated economic difficulties were the reason. Sixty percent of low-income households did not send their children to school because of economic difficulties. In contrast this reason was given by only 45 percent of middle income families. About 13 percent of the households stated that their children were not going to school because they worked; 71 percent of them farm their own land and draw over half of the needed labor force from their household. Another reason for not sending children to school was the absence of any primary school in the village.

> *"I am working as a sharecropper getting 30 percent of the harvest. I have 70 donums of land and nine children. Since I can not hire labor, I let my children work when they become eight or ten years old. I have difficulty in making a living, but I have my children as labor." A poor farmer*

Female labor is not regarded as productive in spite of its substantial contribution to agricultural practices. Eighty percent of the people view women as wives rather than as farmers; nearly all the male heads of household interviewed described their spouses as housewives. Local women are active in all chores, including tending animals. However, more than half of those who referred to their spouses as housewives called their daughters agricultural workers. During in-depth interviews, women stated that they work in family plots but receive nothing in return. Women are not entitled to keep any of the household revenues from farming.

Health problems are aggravated by poor infrastructure and inadequate services. Poor quality drinking water exacerbates health problems and thyroid fever is common in 29 percent of sample villages. Water is available, but the water quality is quite low because of the rising water table, which originates from excessive irrigation according to the farmers; 64 percent of the households have water outside their houses and 19 percent inside. Sanitation conditions and hygiene are also risk factors; 58 percent of the houses have outside toilets, and in 14 percent of the houses have no toilet at all. Even where there is a toilet, there is no running water in 81 percent of the cases. In addition, 41 percent of the houses have no bathing facilities. Women bear the heavier cost of poor water and sanitation services.

**Economic Inequities**

A majority of the rural households in the area covered by the SA are landless, but public investments have so far been land based, consisting of irrigation, land consolidation, leveling and other on-farm improvements. Therefore, the absolute benefits have accrued directly to the landowners and land values have substantially increased. The landless are unable to receive these benefits directly. Their access to other assets and services is also limited because they lack access to credit.

**Economic Well-being.** Using Basic Components Analysis, the relevant data were combined to create an index of economic development or well-being (EDI). The index was composed of a number of variables: the quantity of cotton grown, size of landholding, size of land farmed as share-cropper, tractor ownership, number of small and large animals, family size, number of school age children not enrolled in school, house ownership, on-farm infrastructures existing in the village, and seasonal migration for work. These variables were grouped under four main factors which explained 57 percent of the original variation. The factors were: basic production factors, labor force, on-farm infrastructure facilities, and livestock ownership. The factors which make up income are: size of land as the basic production factor, household labor force which is an important input for cotton production, accessibility of on-farm infrastructure facilities as indicators of the transition from dry to irrigated farming, and ownership of small animals. The relevant information on groupings was subjected to discriminant analysis and the communities were also ranked in terms of their relative well-being so that the policy makers could give priority to the needs of poorer villages.

Households ranked according to their well-being using EDI values were divided into three major groups with respect to the inflection points of the index. Those that remain in the interval where the EDI has the lowest value form the "low-income" group. Out of 450 households interviewed, 69 percent fall into the low, 25 percent into the "middle" and 6 percent into the "high-income" group (Table 7.5). Most of the high-income group were landowners; most of the sharecroppers were poor. The analysis indicates that the ratio of correct placement into groups is 92 percent. As is shown in Figure 7.1, the number of livestock, size of the household, amount of land under cultivation, and income from crop production are the four key factors in accounting for the economic well being of households. The well-being of the high-income families, as measured by the area under the outer diamond in this figure, far exceeds the well being of low income groups.

Table 7.5  and tenure and income (percent of households)

|  | Low | Middle | High | Total |
|---|---|---|---|---|
| Sharecroppers | 76 | 18 | 8 | 58 |
| Landowners | 24 | 82 | 92 | 42 |
| Total number of households | 311 | 25 | 6 | 448 |

Source: Household Survey, 1998
All cross tabulations presented are satistically significant at 0.0 level

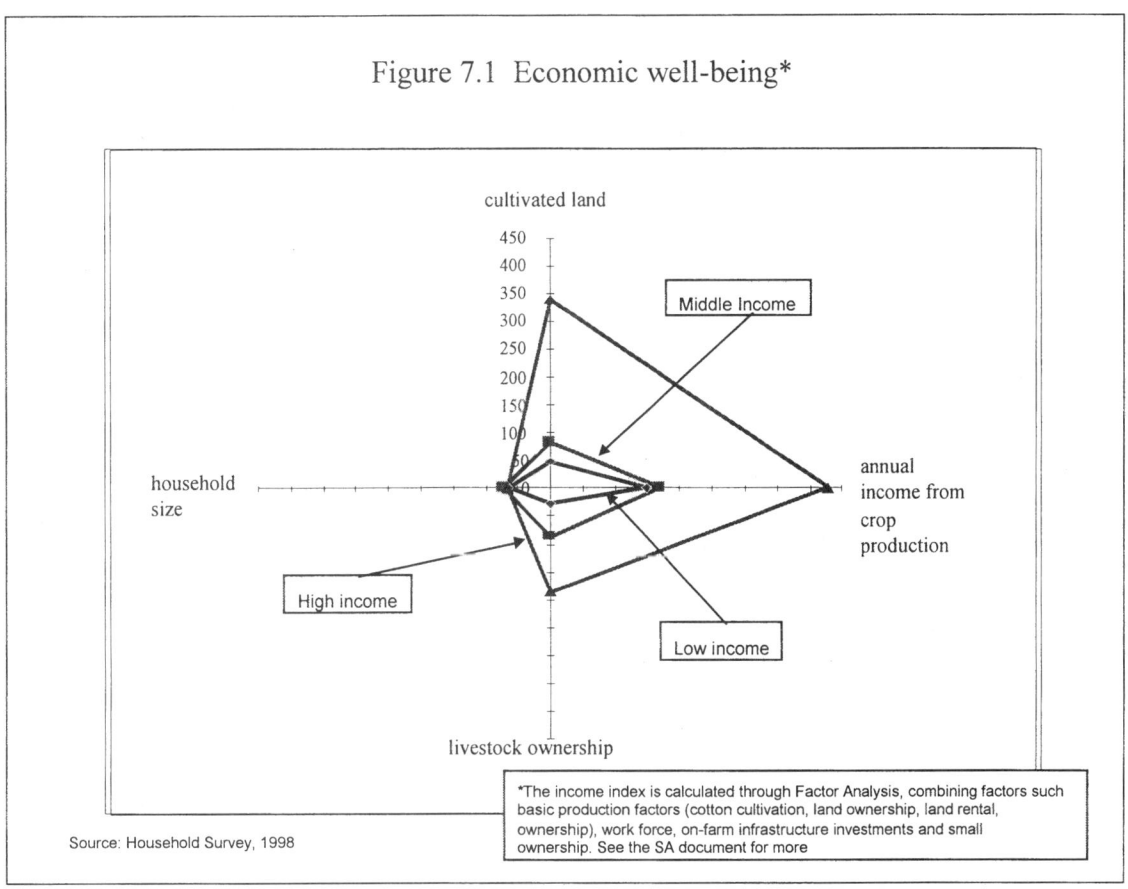

Figure 7.1 Economic well-being*

*The income index is calculated through Factor Analysis, combining factors such basic production factors (cotton cultivation, land ownership, land rental, ownership), work force, on-farm infrastructure investments and small ownership. See the SA document for more

Source: Household Survey, 1998

**Land Tenure.** Property ownership displays a complex picture. Land distribution is skewed. In many cases, official ownership rests with the family elders and local traditions favor the oldest son to assume the management of the inherited land. Fifty-six percent of the households have no land and 45 percent of those who own land have less than 75 donums (Table 7.6).

Table 7.6 Size of landholding and income (percent of households)

| Size of landholding | Sharecroppers | Landowners | Percent of Total |
|---|---|---|---|
| No land | 97 | 0 | 56 |
| 1-30 donums | 3 | 21 | 10 |
| 31-75 donums | 0 | 24 | 10 |
| 76-150 donums | 0 | 28 | 12 |
| 151-300 donums | 0 | 16 | 7 |
| 300 + donums | 0 | 11 | 5 |
| Percent of total | 58 | 42 | 100 |

Source: Household Survey, 1998
n= 450

According to the responses of household heads, 67 percent of all landholders have an official title deed. The remaining 33 percent have land but no title; however their rights would be legally recognized in the process of land consolidation should they be socially recognized. While not all landowners hold a legal title, many of those who are landless do not consider themselves landless because they expect to inherit. The landless sharecroppers report that 65 percent of the land they cultivate is registered to their fathers, 25 percent to their aga (feudal landlord), and 10 percent to some close relative.

Many households farm other people's land, and they typically share 30 percent of the harvest. Some 28 percent of the households are not even sharecroppers but work instead for daily or seasonal wages. Of the remaining 72 percent who farm, the land ranges in size from 1 to 200 donums. Although 47 percent of the landowners have farms smaller than 75 donums, the average landholding is 153 donums, and 80 percent of the landowners farm their own land. The 20 percent of landowners who do not farm let their land be farmed by sharecroppers. On the average, sharecroppers farm 53 donums (Table 7.7).. Small landholders may also rent land from large landowners.

Table 7.7  Land tenure and income groups (donums)

| Land tenure | Sharecroppers | Landowners | Low Income | Middle income | High income | Average landholding |
|---|---|---|---|---|---|---|
| Owner | 0 | 153 | 15 | 113 | 484 | 76 |
| Leaser | 53 | 30 | 33 | 30 | 172 | 42 |
| Total | 53 | 183 | 48 | 143 | 656 | 118 |

Source: Household Survey, 1998

Sharecroppers farm an average of 53 donums while landed households farm 183 donums of land including an average of 30 donums of leased land Looking at income groups, the lower income group farms an average of 48 donums, consisting of 15 donums of owned land and 33 donums of leased land. Since more than 70 percent of this group is composed of sharecroppers, there are strong similarities in the size of land farmed. Also, there are substantial differences in the middle and high-income groups in terms of the size of landholding owned and farmed. Thus, the inequity in land tenure is reflected in large inequities in income and well-being.

**Livestock Ownership.** Livestock management has lost its importance with the shift to irrigated agriculture. Traditionally small animal husbandry (sheep and/or goat) was practiced in the area However, all the available land has been put under cultivation with the introduction of irrigation and little grazing land has been left for the animals. For this reason, very few households currently have livestock and the landless have no opportunity to maintain livestock except in the yard of their homes. As a result, 82 percent of the households have no cattle, and 60 percent have no sheep or goats (Table 7.8). There are important differences between those who own land and those who do not in terms of the number of livestock owned. Prior to irrigation, sheep and goat herding was dominant in the area while cattle breeding was limited. There has been a decrease in livestock after irrigation.

Table 7.8  Average number of livestock owned by households of different income

| | Average | Income level | | | Land tenure | |
|---|---|---|---|---|---|---|
| | | Low | Middle | High | Sharecropper | Landowner |
| Cattle | 0.5 | 0.1 | 0.7 | 3.9 | 0.2 | 0.9 |
| Sheep | 5.0 | 2.7 | 8.6 | 18.7 | 3.6 | 7.2 |
| Goat | 2.0 | 1.0 | 2.2 | 11.2 | 1.2 | 3.0 |

Source: Household Survey, 1998
n= 450

**Ownership and Use of Farm Machinery**. The use and ownership of farm machinery has been gaining momentum since the start of irrigation. Currently, 31 percent of the households have tractors and there is a clear tendency for concentration of these productive assets in the hands of landowners. While 55 percent of households owning land have tractors, only 8 percent of those farming others' land do. Also, the larger the landholding the greater the likelihood of tractor ownership (Table 7.9). More than half of all tractors have been purchased within the last three years. This has been especially attractive as a result of land consolidation and leveling. When a household's land is fragmented with many pieces far apart, it is difficult to use farm machinery. Many landowners now plan to expand their ownership of farm equipment and further increase the return to their land. That this is highly advisable is supported by the data: ownership of a tractor is positively correlated to the level of income.[12] Most sharecroppers use tractors provided by the landowners; as a result they receive only 30 percent of the harvest while the landowners receive 70 percent. The sharecroppers are referred to as "thirty percenters" in the area; elsewhere in Turkey, they usually receive half the harvest. There are no recent anthropological studies to show the changing patterns and dynamics of sharecropping.

Table 7.9  Tractor, plow, and car ownership by household

| Numbers owned | Average | Income level | | | Land tenure | |
|---|---|---|---|---|---|---|
| | | Low | Middle | High | Sharecropper | Landowner |
| Tractor | 31 | 2 | 84 | 96 | 8 | 55 |
| Plow | 15 | 9 | 24 | 35 | 25 | 20 |
| Car | 20 | 9 | 36 | 73 | 9 | 34 |
| Total number of households | 448 | 311 | 111 | 26 | 260 | 190 |

Source: Household Survey, 1998

**Access to Credit**. Access to credit is a particularly serious problem for the landless. Only four percent of the households use credit from credit institutions such as banks and cooperatives, although 14 percent of households borrow money from private lenders and a great deal of informal lending occurs especially prior to harvesting time. While the average amount of credit received from institutions is 1.4 billion TL ($6,800), the average amount borrowed from private lenders is 1.1 billion TL ($5,635). The remaining households stated that they want to use credit but have no access to it (Table 7.10). The Bank of Agriculture requires immovable property as collateral and this requirement forces landless households to seek other sources of finance. Borrowing from within the family and lineage and among close friends also occurs with different interest rates applied among different social categories. Because most people receive over 100 percent annual interest for their deposits even for overnight accounts, the practice of lending with interest has become an integral part of the Turkish culture. Lending in gold, or even in dollars, is also known. Difficulties in getting access to formal credit and the high interest rates discourage most farmers and make it nearly impossible for the landless and small holders to improve their productive activities.

---

[12]     Twenty percent of households have cars. Naturally, the high income group has more cars.

Table 7.10  Use of formal credit by income and land tenure groups

| | Income level | | | Total | Land tenure | |
|---|---|---|---|---|---|---|
| | Low | Middle | High | | Sharecropper | Landowner |
| Credit users  (% of HH) | 0 | 10 | 23 | 4 | 0.1 | 8 |
| Amount  (million TL) | 0 | 1 524 | 2 033 | 1 528 | 4 000 | 1 373 |
| Total number of households | 311 | 111 | 26 | 448 | 260 | 190 |

Source: Household Survey, 1998

**Crop Diversification and Potential Impacts on the Poor.**  Mono-culture is widespread and cotton has become the predominant crop. The SA survey established that 75 percent of the households were engaged in crop farming during the previous year. An overwhelming 93 percent of this farming is cotton; it has a ready market and larger profit margin as compared to cereals and other crops. While 16 percent of the families also grow wheat in addition to cotton, 50 percent grow cotton exclusively and 8 percent grow only wheat. However, during the planning phase of GAP, a different cropping  pattern was envisaged to ensure diversification, encourage horticulture, maintain a certain level of livestock management, and stimulate the transition to high yield production. For instance, cotton was to be grown on 35 percent of the land.. However, the implementation fell far short of expectations.[13]

Diversification continues to be an important challenge along with the need to maintain labor intensity in agriculture. For example, vegetables can be grown in green houses so that the landless and small holders can remain active within the agricultural sector throughout the year. This is a particularly critical social concern in view of the newly accelerated shift to mechanized agriculture; mechanization may result in restructuring of the farms and cause severe reductions in labor hiring and sharecropping arrangements.

Diversification may also enhance returns to the labor of small holders, for instance through vegetable production in greenhouses. Currently, the average cotton yield is 250 kg/donum and the average size of  plot used for cotton production is 83 donums. While ways and means should be found to increase this yield for all income groups, it is evident that high-income groups and owners of larger plots  receive a disproportionately high share from cotton production; they put a great deal more land into cotton production and receive larger returns (Table 7.11).

Table 7.11  Cotton output by level of income and land tenure

| | Income level | | | Land tenure | |
|---|---|---|---|---|---|
| | Low | Middle | High | Sharecropper | Landowner |
| Area (donum) | 46 | 83 | 340 | 55 | 114 |
| Quantity (kg) | 10 989 | 20 332 | 119 907 | 13 067 | 34 821 |
| Total number of households | 183 | 85 | 26 | 158 | 138 |

Source: Household Survey, 1998

---

[13]        In earlier irrigation projects elsewhere, excessive cotton production followed the introduction of irrigation. In the Tarsus and Seyhan Plain Irrigation Projects, cotton culture expanded over larger areas than originally envisaged. However, after a few years, production switched to other crops because of crop diseases, pests, and salination of the soil and cotton production declined.

**Poverty implications of incomplete irrigation infrastructure**. Incomplete investments cause environmental damage as perceived by farmers and confirmed by other stakeholders. The resulting loss of land productivity threatens the landless agricultural workers and the sharecroppers although there is little they can do to rectify the damage. As of October 1998, there was salination on 3,000 hectares and a high water table on 15,000 hectares of land. To remedy this situation requires the urgent introduction of drainage systems. Among landowners, 78 percent state that consolidation of their land is completed and two-thirds have irrigation canals in place. However, two-thirds also point out that their main drainage canals are incomplete and 88 percent indicate that there is no on-farm drainage facilities. Even among the households that have access to irrigation by canals, only 7 percent have a drainage system.

Incomplete land leveling also has negative environmental impacts, but the farmers are reluctant to have leveling done after the irrigation work is completed. Leveling operations would require them to leave the land idle for a period of time, and farmers are unwilling to forego near-term crop revenues.

This situation poses a threat for land and water resources. Much of the soil salinity, contamination of ground water, and erosion can be explained by the failure to complete the investments required by irrigated farming. The EDI value is higher in villages where on-farm infrastructure has been completed both among the land holding and the landless families. This fact clearly demonstrates that infrastructure investments influence household economic well-being positively. Needless to say, for landowners it also increases wealth through appreciation of land values and acquisition of other assets. The EDI value is also higher in localities where irrigation canals and land consolidation has been completed. In addition to irrigation and land consolidation, land leveling also influences well-being. The lowest yields are observed in villages where there is pump irrigation without any on-farm infrastructure. While cotton yield is 286 kg/donum in places where the irrigation infrastructure is complete, it is only 218 kg/donum where pumped irrigation is used without additional improvements. This situation shows the effects of excessive irrigation and lack of drainage.

The importance of drainage for soil conservation and yield increase is recognized by most of the farmers; 62 percent of household heads state that drainage is necessary because it prevents salination and enhances the discharge of excess water. Only 8 percent of the households state that drainage is important for increasing the yields. Since problems caused by the absence or inadequacy of drainage, such as rising water table and pooling, are presently acute, the people overlook its importance in increasing yields. While 69 percent of households indicate they will adopt drainage practices if the relevant works are done by the State, only 27 percent are willing to contribute to the costs of these works. Further, 29 percent of landowners are willing to contribute as much as 10 percent of the costs and another 28 percent are willing to contribute as much as 20 percent of the total cost of drainage services. Given levels of income and the potential returns to investments, a higher cost recovery is possible.

**Agricultural Income**. Agricultural income is derived largely from crops. The share of those who derive income from animal husbandry is less than 5 percent. In terms of EDI, 73 percent of those in the low-income group, 92 percent of those in the middle, and 96 percent of those in the high-income group are engaged in farming. To calculate income in kind, it was necessary to consider the yearly domestic consumption of crops and animals raised by the household and non-agricultural income which has less than a 5 percent share (Table 7.12). As depicted in the table, farming generates income which is on average very close to the gross domestic product (GDP) of Sanliurfa Province. However, sharecroppers make only one third of this average income. The greatest majority of the low-income earners are the landless sharecroppers, followed by small holders. According to crop production information obtained in interviews, 92 percent of the households who farmed others' land a year prior to the SA survey and 59 percent of the landed households earned less than $5,000 last year.

Table 7.12  Annual household income from crops and income in kind (U.S. dollars)

| Household type | Income from crops | Income in kind | Household Size | 1997 income per capita | 1991 income per capita | Rate of increase (%) |
|---|---|---|---|---|---|---|
| Sharecroppers | 2 469 | 2 837 | 6.2 | 457 | 165 | 277 |
| Landowners | 8 769 | 9 663 | 7.3 | 1 323 | 543 | 243 |
| Low-income | 2 733 | 3 002 | 6.4 | 469 | 142 | 330 |
| Middle income | 8 277 | 9 175 | 7.3 | 1 257 | 452 | 278 |
| High-income | 20 815 | 23 715 | 6.8 | 3 487 | 2 047 | 170 |
| Average | 5 570 | 6 155 | 6.6 | 932 | 343 | 271 |

Source:  Household Survey, 1998

The annual income of the low-income group barely reaches a third of the average GDP of the province. While the annual income of the middle group is close to the average GDP, the income of the high-income group is three times the GDP of the province. The comparison of the income levels in 1991, prior to irrigation, with the income levels in 1997 shows that the largest increase in incomes has taken place for the lower income group. As expected, analyses show strong positive correlation between agricultural incomes, irrigation, land consolidation and land leveling. In localities with canal irrigation, average household income from crops is $6,888; it is only $3,340 elsewhere. In places where land consolidation has been completed the figures are $5,956 and $4,299, respectively. As to land leveling, households who have had leveling make $6,943 and those without average $5,014. Low-income areas are those where irrigation is done by pumping. These findings which are in far greater detail documented in the earlier SA report written by Bayram, et al. (1998)[14] clearly show that intended project components consisting of land consolidation, land leveling, and other on-farm services have large pay-offs, even to the lower income groups.

As shown earlier, farming is largely dependent upon the availability of domestic family labor. The relatively low level of agricultural mechanization and the high ratio of landless sharecroppers and agricultural wage workers imply high dependence on domestic labor. The demand for wage labor increases during cotton harvest. Women and children are involved intensively in almost all stages of production and migrate with their families to other regions and communities if the household engages in sharecropping or wage labor. Household size is positively correlated with agricultural income but for different reasons among the different income groups. Large landowning households are larger often because they tend to live in a large compound. High

---

[14]      Mumtaz Bayram, Nilay Cabuk, Aylin Gorgun, Ismet Yalcin and Turan Hazar.  Social Assessment Report: Sanliurfa-Harran Plains on Farm and Village Development Project. Prepared by Oklahoma State University, Su Yapi and the Turkish Rural and Urban Foundation, October 1998.  Including 142 pages of text, graphics, maps and tables.

income and assets encourage larger family size, sometimes also through polygamy. Within the lower and middle income groups, crowded households have more labor to farm and thus earn more.

Women and children are also involved in animal husbandry. Surprisingly, interviews with 450 households indicate that there is no difference between small and large landowners in terms of employing child labor. While the share of school age children not enrolled in school is 35 percent for sharecropper households, the share is over fifty percent for small holders. Economic difficulties is the main reason for not sending children to school and this is clearly reflected in the fact that only 14 percent of large landowners kept their children out of school. Thus, while owners of large land plots can afford to send their children to school, including boarding schools elsewhere in the country, the small holders have a greater tendency to not send their children to school.

**Expenditures.** Sharecroppers spend almost all their income for basic necessities. Even among landowning households 87 percent of the income is spent for basic necessities. Only the high-income households are able to save and make investments. According to statements made by households, low-income households spend an average of $2,263 per year for basic necessities. Relevant figures for middle and high-income households are $4,429 and $11,000 respectively. Per capita spending for low, middle and high-income groups is $353, $607 and $1,617 respectively.

The farmers were asked what they would do if they had an extra $3,500 or $35,000. At least a third of the landless people stated they would spend the smaller amount for basic household necessities. Others would get a house or a second wife. Many joke that with the first extra money one would get an extra wife and if they have ten times as much they would make other investments. Forty-six percent of the landless prefer to buy land and the remainder would start a business or buy a house. Those who own land appear to give priority to starting a business or, to a lesser extent, buying additional land.

Interestingly, in areas where land consolidation, irrigation, and leveling have all taken place, 40 percent of the households stress important changes in their lives. However, in areas where only land consolidation is completed, the share of those indicating improvements drops to 21 percent. Among people who have land, a half indicate substantial improvements in their lives in areas where all services have been completed, but only 36 percent note improvement in areas where services are incomplete.

### Perceived Problems and Priorities for Public Sector Support

Priority problems in farming differ between landed villagers and those who work as sharecroppers although both groups claim that they face many problems. Dispossession or small size of holdings constitute the main problem for one third of rural households, shortages in irrigation water are the main problem for 23 percent, and lack of land leveling is the main problem for 13 percent. There are noticeable differences between problem prioritization of those who own land and others who work as sharecroppers. For those with land, irrigation is the main issue while the landless point to land. In areas that have irrigation but lack land leveling and consolidation, 33 percent of the people indicate land leveling was the main problem in farming; an equal percentage state that limited farm land is the major problem. In settlements where irrigation and leveling services are "officially " completed, people point to inadequacy of the work done. Either the relevant investments are perceived as insufficient or flaws in the organization and management of these services are listed.

The strongest support to land consolidation comes from those who either farm other people's land or those who have medium size holdings. Interviews with the mukhtars of 24 villages where land consolidation has been completed indicate that 59 percent are fully content with this work. In household interviews 71 percent also expressed satisfaction with the relevant investments. Even sharecroppers support land consolidation since dispersed plots make access difficult and cause a loss of time. Consolidation makes it easier for the farmers to plow the land and discharge excess water. The landless people also support consolidation because they hope that private or government land, which is larger than a specific size, may eventually be distributed to landless peasants and to those farmers with very small holdings.

Individuals who have problems with land consolidation state the following reasons for their discontent: negation of their old land, existence of dispersed plots even after consolidation, division of farm land by canals, and receiving less land than expected. Existence of dispersed plots is one of the main reasons for dissatisfaction by the large landowners, but 84 percent of middle sized landowners (76-150 donums) are content with consolidation.

Observations and interviews revealed that the dispersed and divided nature of farm land was a real problem for both sharecroppers and landowners prior to land consolidation. The most obvious problem was the difficulty of using agricultural machinery and tools efficiently. Border disputes were also common prior to consolidation, since the number of neighbors was high. This problem was solved by land consolidation. Finally, consolidation has led to a rise in the value of farm land.

Leveling is seen as an important and useful service which facilitates irrigation and efficiency, but it is lagging behind other on-farm services. The majority (68 percent) express discontent with the present state of progress. Both landowners and those who farm on other people's land assign more or less equal importance to this service. More specifically, while sharecroppers state that the removal of slopes (62 percent) and easier irrigation (48 percent) are the main benefits of leveling, 62 percent of landowners name an increase in yield as the most important factor.

Expansion of irrigation and the improvement of management of the system, availability of credit, improved farm roads, better marketing, improved water supply, better health and education services are also among the priorities. As these are generally discussed above, the following section will focus on the participatory dimension of the services demanded.

## *Institutions*

**Traditional Associations at Network**. Traditional social organization and cultural norms played important roles in structuring the new local institutions and determining participation in political parties. These are, however, largely undocumented. Perhaps partially because of the ethnic composition of the GAP region and the continued dominance of tribal relations, civil society organizations have not yet developed in Harran in the strength that they are observed in other parts of the country. This is especially true of agricultural organizations such as cooperatives. There is a need for better understanding of institutional issues including the composition and management of water user associations, for example, in order to define appropriate institutional arrangements for the delivery of development initiatives. The current level of participation of the landless sharecroppers and small holders is low and alternative institutions that would counteract the influence of the kinship and land-based traditions are needed. The establishment of rural cooperatives, chambers of agriculture, and other civil society organizations could enhance the participation of the landless and the younger generations both among the beneficiaries and decision makers of the change process. To ensure that the Bank-financed Project indeed benefits the poor, well designed social impact monitoring will also be needed.

Two types of issues are important in determining institutional adequacy in implementing the Project. First, for public institutions, do the key actors have the necessary capacity, commitment, and incentives to implement and sustain operations, and will the Project have positive impacts on these institutions? It would be safe to argue that there is sufficient institutional capacity. Since the Project basically aims to complete the on-farm services that the client has already started in other parts of the GAP region, there is a clear capacity to sustain operations with respect to land consolidation and leveling, and little difficulty in completing the farm roads and water supply systems. Although coordination among public institutions is weak, the GAP administration was created to address this concern. As local demand for greater efficiency of public institutions gets stronger, coordination issues will diminish.

The second issue concerns questions as to whether the policymaking bodies and service delivery arrangements will target the poor, and whether "accountability" institutions, including information communication and law enforcement, will ensure that "public" functions are performed. This will depend on several other institutional improvements: the development of civil society institutions, including cooperatives and associations of the landless farmers and the small holders, and the commitment of the political leadership for more active beneficiary participation in regional development. The aim is to enhance the participation of the landless and low-income segments of the population. Given our current knowledge, it is difficult to make a judgement concerning this second institutional issue. Also, it may be difficult for the client to agree that these broader institutional concerns require further attention since implementation of the type of activities proposed for Bank financing has already been completed by the Government under existing institutional arrangements. This will be a challenge for the Project during the next phases of its development.

People depend on social networks, associations, and other forms of social organization to cope with hardship and to access services and resources, including work. Building networks of civic associations, or social capital, extends the range of institutional choices in accessing public services and enables the poor to influence policy formulation. Therefore, enhancing the participation of the landless and the low-income farmers in associations, cooperatives, and other networks is an integral part of a strategy for poverty reduction.

### On-Farm Services

Land consolidation is an important component of on-farm services and has been carried out in 69 percent of the villages. Property disputes among landowners hindered the completion of the work in a number of communities. Some mukhtars stated that they were waiting for land consolidation to begin, and they had even made applications to have the process started urgently. Equally important was the problem of rising water tables because of the inadequacy of drainage facilities and discharge canals. Another important component of on-farm services is land leveling which enhances efficiency of irrigation. In a third of the villages land leveling work had either been completed or was presently in progress; this was felt to be of great importance by 87 percent of the communities.

Although more than half the villages already had irrigation, some of these villages had complaints about the inadequacy of irrigation water and defects in the water distribution scheme. The stakeholder discussions, including those with village leaders, revealed shortcomings of the water user associations with respect

> *Previously we had to go too deep for a well. We could hardly reach water at depths of 20-25 meters. Now there is water at 5-6 meters. Recently, even a new grave has been submerged. There is waterlogging in our plots...no discharge is possible...*
> *Member of an irrigation association*

to maintenance and operation (M&O) of the systems and showed them to have unfair water distribution practices; in particular, the villages located at the end of irrigation canals had insufficient water for their irrigation needs.

While irrigation has been introduced to a large part of the area, farmers report many problems. The major one is the inadequacy of irrigation water. The main reasons for water inadequacy are poor irrigation practices and the absence of night irrigation. Furthermore, inadequate land leveling makes it necessary to resort to labor intensive irrigation during the season. The equity implications of this is important to note; as land leveling is completed there will be less need for labor and this too would reduce the "trickle down" impacts of development inputs such as those to be included in the proposed Project.

### *Participation*

The participation of the local people in the planning, implementation, monitoring, and evaluation process in the development of natural and human resources and in the design of institutional arrangements has been adopted as the basic development policy under GAP. The extent to which this has happened has neither been investigated nor documented. The development of water resources, the construction of the Ataturk Dam, and the decisions concerning water allocation for different purposes were not subject to local level participation but massive public support was assumed.

Once the large-scale water related investments were completed, local participation was expected to increase and it did. Despite continued state guided action, neither the land consolidation nor in-farm irrigation activities would have been possible without community consensus. In areas where land consolidation is carried out, 10 percent of the privately owned plots are taken and used for constructing roads and canals. Although this implementation was not specified in the relevant law, all private landowners consented, saving the state large sums that would have been paid in land acquisition. It led to a 25 percent reduction in the initial capital investments and showed that people were ready to contribute when they were convinced that the final outcome would be to their advantage.

An important mechanism for ensuring that state investments in the area do not exclusively benefit the landowners is to ensure that those who directly benefit from the appreciation of land values participate in the cost recovery. Land consolidation, land leveling, irrigation, on-farm road improvements are inputs that directly benefit those who own land and enhance the value of the land. However, cost recovery levels are low and cover the operations and maintenance costs rather than capital investments. A progressive system of cost recovery can be introduced, those with larger amounts of land who have been shown to disproportionately benefit from these investments. This would allow the financing of development initiatives for the landless and the small holders. The relevant issues are briefly discussed below. However, it is important to note that to introduce new cost recovery systems at this stage of the Southeast Anatolia Project is difficult; the relevant inputs have already been provided to farmers in large parts of the GAP. To institute new cost recovery arrangements for the area proposed for Bank financing would meet with the resistance of the local people and may even raise questions of social divisiveness.

**Drainage.** About a third of the farmers are willing to take part in infrastructure investments if the costs are not too high. Those living in areas where the problem of drainage is more acute all state that they can contribute to the solution of the problem. About a third of landowners are willing to contribute up to 10 percent of the costs and 28 percent state that they can contribute up to 20 percent; the rest prefer to make smaller contributions. These percentages may seem low, but if one considers the high cost of drainage investments, these figures can be taken as reflecting the importance assigned to drainage by farmers. Information and extension services to raise the level of awareness on this issue will obviously affect participation positively. However, current institutional arrangements for cost recovery are inadequate and there are no institutional mechanisms to make other than a one-time payment. If, however, the water user associations are strengthened, the responsibility of collecting farmer contributions for drainage would shift to these civil society organizations and may help to ensure that higher levels of costs are recovered over time. As already mentioned, these associations appear to be controlled or managed by the traditional leaders and large landowners who obviously would oppose participation in cost recovery of public investments in their own lands unless gradual measures are introduced to democratize the associations. Needless to say, tribal leaders and landlords would resist these changes unless they are incremental and based on experience elsewhere in Turkey through study tours and invitations of successful associations operating elsewhere.

**Land Leveling.** Despite high perceived need for land leveling, the high costs associated with it lead farmers to expect relevant initiatives from the state. In addition, once the cultivation on the irrigated land starts prior to any leveling work, it becomes difficult to get farmers' cooperation to implement leveling. Indeed, there is open opposition to equipment operators who try to do "their job". In spite of all these problems, one third of the landowners are willing to contribute to the cost of leveling, "depending on their financial situation". In order to increase their participation they may be asked to be personally responsible from leveling. To this end, equipment used in leveling may be hired by the farmers themselves to complete the work on their land. In this way, disputes between farmers and machine operators may be prevented and the costs involved would be more fully borne by the landowners.

**Farm Roads.** A lack of farm roads creates many difficulties for farmers, including sharecroppers. For example, areas designated for on-farm roads are frequently cropped and this blocks access to subsequent plots. Difficulty in access is also observed in winter and in cases of excessive irrigation or rainfall. Farmers are seriously affected by these problems and state their willingness to contribute to works related to both on-farm and village roads. More specifically, the small holders and the landless express readiness to provide their labor. Some who are better off state their willingness to share in the cost of the relevant works. The cost recovery mechanisms however will have to depend upon the relationship of the required road infrastructure to individual holdings so that an adequate assessment of direct beneficiaries can be made. Once a proposed infrastructure plan is made, it will be possible to assess whether the farmers are willing to contribute their land to the road or if there is a need to consider potential implications of involuntary land acquisition and/or resettlement. Therefore, as the engineering designs advance, further consultations with farmers will be necessary.[15]

**Safe Water Supply**. Willingness to participate in potable water investments is also evident both from willingness to pay surveys and from revealed preferences. Since the majority of those in need are the low-income populations and the landless, water supply investments require priority in poverty focused development initiatives. Half the households feel that drinking water is their priority problem, followed by roads and health facilities. Half of the village mukhtars interviewed state that these problems could be solved with contributions from landowners. Fifty-seven percent suggest a cash contribution while 21 percent prefer a combination of cash and labor. Cost recovery can also be achieved through user charges according to the SA.

As global experience shows, the poor usually pay a disproportionately high cost for services and have a higher willingness to pay for them. However, in the specific context of the proposed Project the justification for continuing the targeting of subsidies to those better off should be questioned. In the Project area, those who are in greatest need of safe water are the landless and other low-income groups. The needs of the temporary migrant workers who come to the region from outside or travel from one community to the other from within the region are the greatest. These people live in tents and have no access to water supply and sanitation. Seeking cost recovery for this basic need from the poor while there is little participation of the landowners in on-farm investments would raise justifiable questions of social fairness.

**Water User Associations (WUA).** Participation problems in the management of water user associations constitute an important source of discontent. Over one-third of the farmers are dissatisfied with the management of irrigation associations because of what they consider unfair and discriminatory water distribution practices; some believe that the more powerful individuals get more water. Interviews with village mukhtars show that more than half of the communities have problems with water user associations. The most frequent problem is the inequitable and

---

[15]     This would have been necessary in case of drainage works as well; however, no drainage component is envisioned for the Bank-financed Project.

inadequate allocation of irrigation water followed by inadequacy of services and poor performance of the workers employed by the WUA. The most frequent suggestion made by village mukhtars to deal with these problems is to have the State ensure fair water distribution and rotational allocation of irrigation water.[16]

This issue was discussed during the 1998 GAP Symposium that brought together key stakeholders. An agreement was reached to improve coordination among the relevant organizations dealing with water resources development. Since a major concern was equitable water distribution, the strengthening of the associations from within was the most important issue. There is insufficient empirical knowledge about the dynamics of these associations and how their management is shaped by the social relationships in the region. Some of the shortcomings are understandable given that they are relatively new and that their managers are engaged in these activities on a part time basis. Even in their present state, WUA are indispensable components of the participatory irrigation system and thus need to be better understood and assisted.

**Innovation and Extension Services**. Given the scale of the public investments made to the region, there is ample room for agricultural innovations and these can be encouraged. Although most farmers state that they are open to innovations and willing to adopt the new technologies, they do not currently benefit from extension services. Nor do they appear to be learning about new approaches from other sources. There are numerous training activities for farmers. However, farmers indicate that these services are not reaching the desired target group because of personnel and equipment shortages. When asked where they learned about irrigated farming, 42 percent of the farmers reply that they resort to trial and error methods; 36 percent learn from other farmers who were more experienced; and only a few use the knowledge provided by formal extension institutions. Farmers have weak ties to these institutions and have little trust for them. Only 13 percent mention that they would approach formal institutions; they prefer to consult with a private agricultural engineers or the water user associations.

Farmers believe that formal extension officers do not take their problems seriously. People are also not informed of what these organizations offer in way of expertise. For example, when asked whether GAP operations are useful, 69 percent of the farmers say they have no idea. Just as rural people are not informed of the operations of GAP administration, they have little knowledge of other governmental organizations such as DSI and Rural Services. Those who are wealthy and own large plots of land hire a private agricultural engineer they trust and know. "When I need advice, I pay for an adviser from Izmir and bring him here," says a rich landowner. However, most do not feel an urge to seek advice and a great majority is unlikely to pay for such advice. Thus, privatization of extension services would result in large savings for the public sector and the private sector would develop to the extent that it generates demand and satisfies the needs of farmers.

---

[16] Two-thirds of the farmers state that the charges for irrigation water are normal and affordable, yet there is little willingness to participate in other than the M&O costs of irrigation. Stakeholders point out that environmental problems have increased with irrigation. These problems partly stem from the incomplete state of the necessary infrastructures; 40 percent of farmers stated that flies and pests have multiplied. This problem is followed by the contamination of drinking water according to 29 percent of the farmers.

The support provided to the private sector for the provision of these services may not adversely impact the poor and the landless; the landowners could pay for the relevant advice and share it with those who work for them. Since this advice is likely to support a more intensive use of the land and water resources, the work potential for the landless would not be reduced either. Whether the landowners would pass over the contributions they make for cost recovery of land consolidation, leveling, drainage, farm machinery, advisory services, etc., to the sharecroppers is, of course, a concern. However, the sharecropping arrangements already provide 70 percent of the harvest to the landowner and consultations indicate that a further deterioration of the shares of the cultivators would be unlikely.

There are many externalities that neither the owners nor the sharecroppers are informed off and/or sensitive to. Among these are many of the environmental issues already noted and those that concern monoculture without rotation. However, there is insufficient research to allow a careful design of extension needs with respect to use of land, water, and technology; this is needed before such services can be designed in a coordinated manner with the private sector playing a supportive role. There is also need to consider government support to extension activities of civil society organizations, including water user associations and other specialized cooperatives.

### Monitoring and Evaluation

As mentioned earlier, this chapter reports on the first phase of the SA process for the Harran Project and thus focuses primarily on some of the key social development issues. The Project is yet to be designed and its monitoring and evaluation (M&E) framework will be defined once other agreements on its components are made. Nevertheless it is important to note that in the context of the SA, M&E is more than simply the determination of criteria or benchmarks for measuring whether project benefits are reaching the targeted stakeholders. It should also become a participatory feedback mechanism so that the needs of various social groups are heard and responded to. There should be specific points in the life of the project when the results of monitoring are used to redirect or reorient the project. Within this process, several social impact monitoring exercises should be defined so that equity concerns raised throughout the SA are adequately addressed. For instance, we should be able to know whether investments in land consolidation and leveling merely help the landlords to increase the efficiency of farming and facilitate their complete shift to mechanization, thus displacing the sharecroppers and the landless. Should this happen, modes of intensification of production through, for instance, greater focus on crop diversification could be considered.

A baseline is needed in order to do an adequate social impact monitoring. Since baseline data are not otherwise available, it is advisable to gather these data during the first phase of the SA. This has been achieved to a certain extent, but there is a great deal of more work to be done on local level institutions, especially on the tribal system and its relationship with other local institutions. If this work is completed, it would facilitate the identification of change scenarios to be monitored. Needless to say, changes will take place without the Project and a great deal has already changed because the government has already carried out investments in other sub-regions of the GAP; thus, the impact measured will not merely be before, during and after project changes along identified variables but rather a comparison of anticipated and actual changes attributable to the Project.

# Project Implications

The social assessment supports a number of recommendations which promote greater equity, including continued land consolidation, integrated investments in land and water resources, agricultural diversification, and greater access to credit. These are highlighted below.

**Land Consolidation Benefits**. While the benefits of land consolidation accrue primarily to landowners, continued support of this process will also benefit sharecroppers. It may also gradually enhance gender equity. There are several distinct social development advantages of land consolidation. First, land disputes and consequent blood feuds have long been problems in East and Southeast Turkey. These have also been important elements in sustaining tribal relationships and marking a special place for the tribal elders. The patrilineal tribal relationships have been instrumental in exclusionary practices vis-à-vis women and have undermined the younger generations. On the positive side, they have provided an important social capital on which people continue to structure their interactions and have harmonious relationships within the lineage. The process of land consolidation opens windows of opportunities, at least for the younger and more educated segments of the area population, to recognize the legal rights of women to land and, at the same time, substantially reduced potential conflict between different tribes and lineages. Now, despite registration of land to women's name, their right to its use or its revenues may be denied or they may be forced to re-register the land to their brothers or fathers. Nevertheless, we can expect gradual changes.

Second, the land consolidation process has strengthened the linkages between formal and informal institutions as the relevant State agencies work closely with communities to ensure full agreement on all land demarcations. Perhaps more importantly, small holders and large landowners also work together, and elders dialogue more closely with the younger generation of landowners. In many ways, the consolidation process makes a contribution to the softening of the hierarchical power structure.

There are other important reasons to support land consolidation. The cultivation of dispersed plots is difficult for landowners and sharecroppers alike and thus it is a desirable development initiative for all concerned. If done properly, based on soil quality considerations, it can reduce erosion, substantially increase productivity, and facilitate introduction of mechanization. Although 70 percent of the accompanying yield increases will go to the owners of land, the increase in the cash value of the sharecroppers' 30 percent share may be higher in relative terms; this has been observed in areas where land consolidation was introduced.

Needless to say, there are social risks as well. From a growth perspective, land consolidation will have positive impacts. If some of the new income generated through this growth is kept within the region and/or re-invested in upgrading agricultural technology and know-how, these impacts will be further enhanced. However, the efficiency gains may reduce the need for sharecropping arrangements and thus diminish the potential positive impacts mentioned above. Similarly, there may be substantial reductions in demand for agricultural labor; this likely scenario too would substantially diminish positive impacts on low-income groups and further increase inequities. To ensure that land consolidation and mechanization do not displace sharecroppers and agricultural laborers, labor intensity in agriculture should be maintained through diversification and shift to high value cash crops. Also, the social development impacts of the project should be monitored.

**On-Farm Improvements Integration**. The integration of on-farm improvements is important to avoid environmental damage to the soil and to enhance work opportunities for the poor. The efficient use of land and water resources of the region require integrated investment and the informed support of farmers. As a result of poor distribution of water and excessive irrigation practices, the problems of drainage and salination are causing damage to the soil and many villages and sub-villages are threatened by a rising water table. The Sanliurfa-Harran Plains have been opened to irrigation since 1994. Such factors as the inadequacy of drainage systems, flat and low topographic structure of the plain, and excessive and non-uniform irrigation practices have led to the rise of the water table, which has further aggravated the drainage problems. The problem of rising water table now threatens many villages and sub-villages in the area; observations in the villages of Tahilalan, Gözele, Ambartepe and Nusretiye showed that many houses had already collapsed or were about to collapse because of a rising water table.

By the end of 1998, land leveling was completed on 31,000 hectares in the project area. In places where leveling has not been completed, there is either no irrigation or irrigation is being practiced at below required standards. Once irrigated agriculture starts, farmers either postpone or neglect leveling and the potential for road building declines. Roads along the main canal and discharges have been built by the state, but most of the on-farm roads and roadways along drainage structures have not been completed. The original aim of land consolidation was to ensure access to each plot. The project area required 3,500 km of on-farm roads of which 1,500 km is completed. Unless the rest are completed, landowners and sharecroppers alike will have difficulties in transporting crops to markets. Another problem is the cultivation of land left as roadways near irrigation canals. With the erosion or movement of the soil supporting the canal structures, the canal may begin leaking or even collapse. From a social perspective, the lack of roads forces farmers to trespass on each other's land; this situation causes disputes.

Land leveling in already irrigated land requires the cessation of cultivation and the consent of the owner; these are not easily obtained. It becomes quite difficult to introduce land leveling once cultivation of irrigated crops has begun. Under these circumstances, the best solution, already practiced in some cases, is the hiring of scrapers and their operators by landowners; in this way the farmers themselves conduct the leveling work. This solution is promising in terms of farmer participation in cost recovery, especially for those who own land. Contributions by the stakeholders to the project at all stages makes the solution of problems much easier.

In areas which have not been brought under irrigation, it is much better to have the Office of Village Services conduct road construction and land leveling work simultaneously in order to use leveling soil for road filling and road excavation soil for land leveling when needed. There is also need to provide information to farmers concerning improved irrigation and drainage practices. Careful targeting of the landless sharecroppers would be especially important; they may have lower motivation for environmental sustainability in the medium and longer terms since some of them cultivate land that belongs to others.

**Public-Private Partnership in Extension**. Public, private, and civil society partnership in the design and implementation of an extension strategy will encourage crop diversification while maintaining labor intensity in agriculture. Support to strengthen livestock management would also help the low-income groups. The dependency on a single crop—cotton—has both high economic and environmental risks; farmers are somewhat but not sufficiently informed of this. Even when cotton cultivation continues, rotation principles should be considered. An important social implication of crop diversification and the development of, for instance, green house cultivation and/or drip irrigation would be to increase or maintain opportunities for income for small holders and sharecroppers who rely heavily on their labor for a living. The initial costs involved in such investments are high. However, a farmer who currently obtains $200 from one donum of land with traditional crops can earn $4,500 by growing tomatoes in a greenhouse on the same plot. Once they are made, even on small plots of land, such investments could support a group of households. However, the lack of marketing opportunities for other crops, including lack of storage and refrigeration facilities for high value crops such as fruits and vegetables, discourage crop diversification and increase undesirable levels of dependency on a single crop.

The Government has made enormous investments in the GAP region. It is not easy to create a well-balanced and dynamic agricultural sector and obtain full returns, but this is the challenge before the Bank and the Government. The current choices made by the farmers are rather conservative. Pursuing alternative choices would require a much better understanding of the social basis of the local relationships than is presently available. It would also require relevant feasibility studies, the availability of long-term and affordable credit opportunities, marketing facilities and advisory services, as well as a well designed information/communications (I/C) strategy targeting different social and economic groups.

Current yields are low because of a poor understanding of irrigation and irrigated farming. Farmers generally believe that they can get higher yields if they use more water and agro-chemicals. They are insufficiently aware of the potential public health and environmental implications. Also, there is no documentation of these practices and their actual impact; thus it is difficult to design an extension strategy for women and men of both landowning and sharecropping families. The coordination of the various local and central governmental organizations is weak. But more importantly, the private and civil society institutions within the agricultural sector appear to be in early stages of their development and do not partner systematically with each other or with the public sector.

Potentially, organizations such as cooperatives or farmers' unions could facilitate the coordination of investments and production activities of small landholders and landless people. They could organize inputs needed by small farmers and facilitate marketing. As is the case in other regions of the country, these civil society organizations would also provide advice and support for credit services to the lower income farmer. Unless these organizations are established, strengthened, and democratized, the new investments made in the region and the new opportunities to be created for credit and advisory services would continue to disproportionately benefit the landowners and those who are already powerful.

Whether provided by the public, private, or the civil society sector, availability of qualified expertise is a key concern. In the past, the east and southeast regions have had great difficulty in attracting or being able to keep such expertise in the region. The best professionals chose to leave for the more developed western provinces or for abroad. Neither monetary incentives nor mandatory systems have helped overcome this tendency, and security concerns add to the reluctance of the top professionals to opt for employment in these regions. The situation does not appear to be much different today. In the future, there must be sufficient and trained personnel for the extension services. Related organizations and agencies must support extension activities. Besides the government, farmers' organizations, universities, and private and voluntary organizations should also participate in the provision of extension services. However, as the region develops and the security issues are effectively addressed, it should be possible to overcome these staffing problems.

Animal husbandry would provide good opportunities to the landless. Although animal husbandry has a long tradition in the region, almost all of the available land has been allocated to crop farming in the process of irrigation and land consolidation, leaving little or no grazing land for small animals in particular. In addition, the exclusive shift to cotton has reduced feed availability. Feed grains such as corn, soybeans, and other fodder crops could be grown as secondary crops to meet this shortage.  However, feed availability alone would not help the landless and small holders to raise livestock. The creation of farmers' associations would facilitate the entry of low-income populations to livestock management; once democratized the water user associations could also facilitate this. Since a certain firm size is necessary in order to produce and market efficiently, households would be encouraged to combine their resources.[17] With proper financing and informational support, livestock enterprises could improve the economic conditions for the landless segment of the rural population.

In addition, Treasury lands are available and can be provided on a long-term rental basis to groups of landless peasants for livestock management. Although the Treasury is planning to sell off the large state farms it now owns to the private sector, given the large share of the landless in the region and concerns of social stability, utilization of all or some of the Treasury lands to help the poor on lease terms might be considered.  Since this issue concerns a much larger group of GAP residents than the limited number of communities within the proposed Project area, it may not be appropriate or feasible for the Bank to include this issue in its dialogue with the borrower; nevertheless it would be important for the policy makers to consider this option.

**Credit Facilities**.  Credit facilities and other inputs should be available to strengthen the position of landless sharecroppers and tenants. Credit, whether for more land, for new technology, or to invest in new crops is difficult not only because of the complicated bureaucracy and high terms of interest, but also because the landless sharecroppers and small holders have no collateral. The farmer associations are the first and the most viable means of overcoming this difficulty. These associations must be strengthened and democratized, especially the water user associations. Otherwise, the opportunities created within them will only benefit those who are already powerful.

---

[17]    In animal husbandry projects developed by the Agriculture Bank, minimum investment costs are $65,000 for 10 cows and $218,000 for 50 cows. This is beyond the means of any low-income household, especially since the terms are far too short. Therefore, a cooperative bringing together many households would facilitate the entry of the poor into the livestock management process.

Another way of ensuring that the landless benefit equitably from State investments is to provide them land either with long term credit or on a long term lease basis. As mentioned above, the Treasury has 15,000 donums of agricultural land which should be rented long term or given to the landless, and there is a large State farm that will be privatized. As recent experience shows, unless a different policy decision is taken, these lands will end up in the hands of a few influential locals at unreasonably low prices. Their distribution to the landless in the region, however, might have high economic and social returns provided that the repayment period is sufficiently long.

**Developing Civil Society Organizations**. Support should be provided to develop civil society organizations so that the productive capacity of the landless and the small holders can be strengthened and gender and inter-generational inequities are reduced over time. The identification of mechanisms to do so would require additional analytical work based on detailed case studies and stakeholder consultations during the subsequent stages of the SA. Nevertheless several concrete suggestions can be made with respect to water user associations. First, of great importance, a better understanding of the social dynamics of the existing local institutions including the village administration and the water user associations is needed as these currently play important roles in the lives of communities. Since resources are limited during the project preparation, such work could be built into the monitoring and evaluation component of the project.

GAP is a special entity that does not exist elsewhere in Turkey. It has been supported by a special administration set up to ensure strong coordination among institutions so that the investments made to the region achieve their objectives. Nevertheless, consultations carried out by a large number of stakeholders and the farmers clearly point to weak institutional coordination. An important step in this direction would be the commissioning of a specific study to better understand the reasons for this weakness. Equally important to understand, and more difficult to do so in the context of the M&E activities during project implementation, are the factors associated with the relatively low level of development of such civil society organizations as cooperatives, professional chambers as well as of the private sector. A better understanding of these issues would have high returns in terms of long-term social sustainability of investments in the region but would require commissioning independent researchers for specially designed social impact monitoring activities.

The landless and small farmers cannot obtain adequate access to credit, technology, and information unless these associations are strengthened and geared towards such objectives. Voluntary unions must be encouraged to enhance solidarity and cooperation among farmers especially in the provision of inputs and other farming activities. The strengthening of associations, cooperatives, and other forms of civil society organizations would also ensure that government financed initiatives that primarily benefit those with larger assets are built on cost recovery. For instance, even though there is low willingness to participate in infrastructure investments and widespread expectation that the government should assume the primary responsibility for such investments, a higher level of cost recovery targeting direct beneficiaries is possible and more equitable. There are institutional difficulties in arranging longer term repayment arrangements, but these difficulties could be overcome through strengthening of the water user associations.

At the same time there is need to limit the use of WUA management as a source of power and prestige, and the tribal control mechanisms that appear to play important roles in WUAs should also be loosened. This can be accomplished through legislative actions discussed below and by issuance of written guidelines for the overall principles of WUAs, their management plans, organizational structure, and functions. The guidelines would create an effective and democratic self-control mechanism within each union. Elected union managers must be accountable to the farmers. These procedures must be incorporated into the guidelines by those in charge of developing union regulations. The GAP administration together with the State Hydraulic Organization could support this process.

One of the most important tasks for the public sector is the legislative support to the operation of WUAs as there are major gaps in the current legal framework. For example, there is no clarity concerning real estate acquisition entitlements of WUAs or the role of the regulatory bodies. Despite similarities between WUAs and cooperatives, the existence of a regulatory framework for the latter, and sufficient experimentation with the former, delays in the finalization of the legislative continue.

Another important mechanism for the strengthening of the WUAs and their democratization would be the establishment of WUA federations. This has proven to be the case in other parts of the country. The primary reason for this is the weakening of the domination of local agas on individual WUAs and the shift to a more competitive and technically sound decision making process through linkages between WUAs.

It is possible to increase the level of sharecropper and small holder participation through information building by effective extension systems that are jointly designed and implemented by the local units of the Ministry of Agriculture, faculties of agriculture at the universities, and non-governmental organizations. Such a service is essential to deal with dissatisfaction and conflicts regarding the distribution of irrigation water. The problem of water distribution may be reduced by demonstrating that there will be enough water for every farmer if correct practices are followed. With reader friendly and periodic brochures and guides to be published by governmental and non-governmental organizations, as well as through training activities described above, water user associations may be re-structured in a more democratic way and they can develop more equitable models for those who benefit from irrigation systems.

The exchange of information and experience among associations may be a more effective mechanism to launch the process of democratization of the WUAs. For instance, the farmers of Çukurova, where participation has been successful, may be brought together with the farmers of the region. With the organization of programs such as those mentioned above, there can be cooperation, exchange of information, communication, and technical assistance among WUAs. The Bank could play an important role in facilitating this and in financing study tours. Face to face contacts and exchange of information between the farmers of both regions will raise awareness and the level of participation. Such programs may be developed under the coordination of the GAP administration with the participation of other key stakeholders, including the Provincial Directorate of Agriculture.

The recruitment of qualified agricultural engineers would also help improve current practices. If a single union can not afford to employ such personnel, unions can combine resources and engineers. Irrigation services also may be requested by WUAs from local firms working in the field of agriculture. Agricultural engineers working in the WUAs may find easier access to governmental organizations, thus forming a bridge in conveying information to the farmers. In order to enhance the performance and efficiency of WUAs, their record keeping, accounting, and auditing systems must be made uniform. Governmental organizations must organize training in these fields.

Bank-financed rural development attempts ought to consider supporting crop diversification in directions that would encourage labor intensity in agriculture; supporting water and transport infrastructure through use of local and possible rural labor, including the labor of women; and carrying our further consultations to identify investments in rural industries to help generate work for the landless. The SA process to date has shown that the major factor that facilitated the trickle down of benefits of land-based interventions from landowners to the landless sharecroppers and agricultural laborers has been the labor-intensive nature of the agricultural production, particularly cotton picked by hand. We also note an accelerating tendency for high income landowners to acquire farm machinery, signaling a potential loss of work for the landless. Making sure that large scale construction firms that have their headquarters in major metropolitan areas of the country and that tend to use their own personnel with heavy overheads do not become the exclusive beneficiaries of funds allocations for road and water supply infrastructure or for leveling;

## Conclusions

The Southeast Anatolia Project has brought important economic benefits to what was once a backward region. A concerted government effort, coordinated by a specially established administrative structure, has dramatically changed the face of the region. GAP administration has played a strong role in coordinating the activities of external donors and facilitating the exchange of experiences among key stakeholders through such activities as conferences and symposium. What was once a feudal and highly hierarchical social order has gradually given way to a more participatory society. Some of the land based investments trickled down to sharecroppers and landless farmers, and the development of the water and energy infrastructure has led to the emergence of services, trade, commerce, agribusiness and other non-agricultural sector activities. The highly unequal distribution of assets was somewhat modified in 1960s through a land reform which imposed a ceiling on individual land holdings; however, communities responded by registering the land in the names of the male members of extended families and lineages.

More recently, GAP has redistributed benefits in two ways. First, extended family structures are strong even when the head of the family is the sole owner of land, so the benefits received by the landowning family heads are distributed to the sons whether or not they reside in the same household. This is especially so among the aga families where social and economic relationships are organized around patrilineal extended family principles. The head of family is almost always a male, either the father or the eldest brother with whom the widowed mother would reside. Second, intensification of agriculture, the shift to cotton production, and the high yields/cash obtained from irrigated agriculture created work opportunities for the landless and the poor. The social and economic transformation of the region was accelerated through land consolidation, a pre-condition for the expansion of the irrigation system. Although the impacts on living standards in measurable economic terms are identifiable, those on the social fabric and institutions have received little attention; this hinders the ability of policy makers to "design the content of induced change and chart the collective action path towards accomplishing it" (Cernea 1996: 11).[18]

The proposed Project has several components: land consolidation, land leveling, farm roads, potable water supply, institutional strengthening, and advisory services. The SA supports these interventions and recommends a higher cost recovery for on-farm investments. It also supports institutional and capacity building improvements including support to civil society organizations such as the water user associations to ensure that benefits trickle down, cost recovery is achieved, dependency on the state is reduced, and the participation of the landless and the sharecroppers in regional development is increased. The SA also suggests that maintaining labor intensity in agriculture is important as is supporting labor intensive public works programs and rural industries.

Given insufficient knowledge of the interrelationship of social and formal institutions and of community dynamics, social impact monitoring should be a key component of the Bank-financed project. This SA provides a baseline against which social monitoring efforts can be built. However, qualitative and longer term research is needed to carry out in-depth institutional analyses. Also of key concern would be the monitoring of Project impacts on existing inequities, including gender and inter-generational inequity, as well as on the landless. Many of the social impact assessments could be contracted to independent research institutions or firms to order to ensure objectivity and high scientific quality. In order to ensure adequate response to the M&E activities, the project should be designed in a flexible manner. The result would not only be critical to guide future development initiatives in the region but would also provide valuable information on the transition from one system of production relations to another.

---

[18]     Michael Cernea. 1995. Social Organization and Development Anthropology: The 1995 Malinowski Award Lecture. The World Bank, Washington, D.C.

# Annex 1
## Selection of Sample Units in the Survey

Sampling from the region was made in two stages. In the first stage, 35 villages were selected from a region consisting of 178 villages, a sampling ratio of 20 percent. At the second stage, 450 households were selected from the total of 2001 households, 22 percent sampling ratio. In defining sample units, stratification was applied in line with the objectives of the survey. The determining factors in this definition were land consolidation, irrigation, and leveling. Settlements in the region were divided into seven strata as follows:

Sampling Strata and Number of Villages

| Strata | Land consolidation | Irrigation | Land leveling | Number of villages | Number of villages sampled |
|--------|--------------------|-----------|---------------|--------------------|----------------------------|
| A | completed | completed | completed | 53 | 11 |
| B | completed | completed | no | 27 | 5 |
| C | completed | no | no | 36 | 7 |
| D | completed | no | completed | 4 | 1 |
| E | no | completed | completed | 6 | 1 |
| F | no | completed | no | 17 | 3 |
| G | no | no | no | 35 | 7 |
| | | | TOTAL | N = 178 | N= 35 |

In the first stage 35 villages were selected from the region by "stratified proportional sampling." In the second stage, "stratified random sampling" was applied to obtain the sample of households according to their status (those who farm their own land and others who rent land or work on other people's land as share croppers and get 30 percent of the production). This process gave 35 villages and 450 households as sample units. A skilled team equipped with survey information and questioners who knew the area conducted the interviews using standard questionnaire forms. In order to obtain information relevant to village development, the village information questionnaire was applied to villages, whereas the household interview form was used in interviews with 450 households.

Qualitative information was obtained through in-depth interviews with those who farm their own land, those who rent land or work as share croppers, and women. Observations and informal talks were also recorded and considered in writing this report. Information and data obtained through these channels included the following items: village infrastructure, demographic characteristics, land ownership, social structure, economic activities, irrigation practices, crop farming, animal husbandry activities, credit facilities, agricultural inputs, extension, health, and education. All these served to formulate an understanding of people's living standards, conditions, perception of problems, related behavior, and attitudes. Socioeconomic analyses made with these data will be presented in the later stages of our work.

## Annex 2
## Key Stakeholders and Institutions Included in Consultations

### GAP Regional Development Administration (GAP-BKI)

Established in 1989, GAP-BKI is in charge of regional land use planning, organization of agencies involved in infrastructure works, monitoring and evaluation of projects covered by GAP, and overall coordination of relevant activities in the region. GAP-BKI has its headquarters in Ankara and a Regional Directorate in Sanliurfa. It is also responsible for the management of the Koruklu GAP Agricultural Research Station which provides services for the region.

### General Directorate of Rural Services (KHGM)

KHGM is in charge of conducting activities and extending services in such areas as the development of land and water resources, rural settlement planning, land leveling, on farm services, and construction of village roads. KHGM was organized in 1984 by combining the Land-Water Management and Road-Water-Electricity Management Units. There is a research unit of KHGM in the project area.

### Rural Services Research Institute

With a staff of agricultural engineers, the institute conducts research on the problems confronting farmers. Previously engaged in research on water and soil composition, the institute has added research on crop adaptation, fertilization, and irrigation. A senior level manager from the organization has stated that the land leveling work is not satisfactory, thus causing problems with irrigation. There are also some areas in the project where salination has already started. Therefore, there is an urgent need to introduce drainage systems in these areas and take relevant measures in other areas to prevent the emergence of the same problem.

### State Hydraulic Works (DSI)

Attached to the Ministry of Energy and Natural Resources, DSI is in charge of planning and project development for large-scale water works. Planning, project development, construction, operation and maintenance of dams, and large scale irrigation and drainage networks are all under the responsibility of DSI. DSI is also engaged in relevant consulting and training services to help water user associations overcome their current problems. Since an additional 20,000 hectares of land will be opened for irrigation, DSI has assigned importance to raising awareness in the water user associations to ensure efficient water use.

### General Directorate of Agricultural Reform (TRGM)

TRGM is in charge of land consolidation, land use planning, land distribution, provision of Treasury land to landless peasants, provision of equipment, training for farmers, and formation of farmers' organizations in areas delineated by the Council of Ministers. In the GAP Region, TRGM conducts surveys to identify priority areas for land consolidation and it does the consolidation work. TRGM conducts land consolidation programs on State land, nationalized private lands, and private lands. With land consolidation, agricultural enterprises will become more economical and farming will be more efficient. Training farmers is an important activity of TRGM; a farmer training center was established in 1987 on an area of 79 hectares adjacent to the Sanliurfa-Akçakale highway.

## Provincial Directorate of Agriculture (TIM)

The primary functions of this organization are research and training on issues related to crop farming and animal husbandry. It has units in charge of organization, support, and development of crop farming, plant protection, and pest control. TIM in Sanliurfa has been conducting various training activities for farmers in irrigated areas, but they have not reached a high percentage of the rural people because of shortages in personnel and equipment.

## Water User Associations (WUAs)

There were seven water user associations on 39,000 hectares of land in 1995. With the establishment of four more WUAs in 1997 and 1998, there are now 11 WUAs on an irrigated area of 82,000 hectares. They all use surface water for irrigation. Recently, the 11 water user associations combined resources and started issuing a newsletter to inform farmers. Also, three unions joined together to construct a management building. The WUAs are a non-governmental and grass roots type organization based upon farmers' participation in ensuring sustainable agricultural development in the region. Village mukhtars are natural members of these organizations. Each elected representative stays in office for a period of three to five years. Each union is represented by the mukhtar and two representatives from the villages it covers. The higher council has seven members. These unions are making efforts to obtain financial resources for their personnel, machinery, tools, and administration buildings by applying for credit to various institutions, including the World Bank. The World Bank provided a grant of $42,000, which was used to buy wireless radios and motorcycles. The plans of the unions over the next five years include procurement of materials and equipment worth $6.5 million. The World Bank loans will play an important role in the development of the water user associations. However, the water user associations need support and consolidation to resolve problems such as the persistence of feudal relations, insufficient solidarity, need for democratization in management, and shortage of personnel. The major problems confronted by water user associations include high personnel costs, low level of training with respect to the farmers, excessive and poorly distributed use of water, low level of investment for equipment, and shortage of funds for the maintenance of the irrigation systems. The major problems faced by agricultural engineers and technicians are the difficulties of conducting repair and maintenance operations and controlling water distribution due to the absence of roads near tertiary irrigation canals.

## Organizations Consulted, Sanliurfa

GAP Regional Development Administration
Regional Directorate of Rural Services
Regional Directorate of Agricultural Reform
Regional Directorate of DSI
Rural Services Research Institute
Provincial Directorate of Agriculture
Agriculture Bank
Harran University
Chamber of Agriculture
Agricultural Equipment Office
Directorate of Organized Industrial District
Chamber of Industry and Commerce

Provincial Directorate of Environment
Kisas Municipality
Sanliurfa Municipality
Soil Products Office (TMO)
TEMAV
Central WUA
Kisas WUA
Haktanir WUA
Onortak WUA
Koruklu WUA
Cabiresan WUA
Tek Tek WUA
Suayb WUA
Firat WUA
Provincial Directorate of Health
Toros Fertilizers Co.
Göktepe Co.
DSI Operation and Maintenance Directorate of 15th Regional Directorate
Harrani Textile
Turyap Real Estate

# Annex 3
# Definitions

Area under irrigation:
Rural settlements located in areas which have been brought under irrigation through canals provided by the State Hydraulic Works (DSI). The source of water is the Atatürk Dam.

Land development:
On-farm services carried out in order to ensure the highest yield and reduce costs.

On-farm services:
Land leveling, land consolidation to bring dispersed plots together, in-farm roads or paths, and other services make it possible to increase agricultural output by efficient irrigation.

Land consolidation:
Bringing divided and dispersed agricultural plots together so that economic management, soil conservation and efficient irrigation are possible. Establishing a single contiguous block of land of equal size to the dispersed plots and taking relevant measures for technical, economic, and social improvements to raise the living standards of individual households.

Land leveling:
Smoothing agricultural land to ensure uniform distribution of irrigation water.

Extended family:
Married children and their parents living in the same household.

Household:
Persons living in the same house and subsisting from the same budget.

Landed:
Persons owning agricultural land whether they farm it themselves or not.

Landless:
Share croppers, tenants, and agricultural workers who work on land owned by someone else. Others who are engaged in non-agricultural works.

Share croppers:
Commonly observed in irrigated farming and in cotton farming in particular. The share cropper provides labor such as ploughing, hoeing, watering, tending, and harvesting and gets 30 percent of the yield. The landowner provides the land, seed, machinery, fertilizers, chemicals and pays for the irrigation water and demands 70 percent of the yield.

Daily or seasonal agricultural worker:
Works directly for cash or in kind wages in agricultural enterprises. Not a form of agricultural enterprise, but a labor intensive participation in an enterprise.

Water user association (WUA):
Local organization in charge of operating and maintaining the irrigation schemes. Active since 1995 in areas under irrigation. The village mukhtars and local water users are natural members of these associations which are managed by a board.


Donum:  A local land measurement equal to 1000 square meters.

Aga: landowner, feudal landlord

Baslik: bride price

Berdel: the cross-marriage of male and female children of two families

Esir: sharecropper, serf

Kirvelik: fictive kinship

Mezra: sub-village

Mukhtar: village headman

Salma: flood irrigation

Tava: check flood irrigation

# Distributors of World Bank Group Publications

Prices and credit terms vary from country to country. Consult your local distributor before placing an order.

**ARGENTINA**
World Publications SA
Av. Cordoba 1877
1120 Ciudad de Buenos Aires
Tel: (54 11) 4815-8156
Fax: (54 11) 4815-8156
E-mail: wpbooks@infovia.com.ar

**AUSTRALIA, FIJI, PAPUA NEW GUINEA, SOLOMON ISLANDS, VANUATU, AND SAMOA**
D.A. Information Services
648 Whitehorse Road
Mitcham 3132, Victoria
Tel: (61) 3 9210 7777
Fax: (61) 3 9210 7788
E-mail: service@dadirect.com.au
URL: http://www.dadirect.com.au

**AUSTRIA**
Gerold and Co.
Weihburggasse 26
A-1011 Wien
Tel: (43 1) 512-47-31-0
Fax: (43 1) 512-47-31-29
URL: http://www.gerold.co/at.online

**BANGLADESH**
Micro Industries Development Assistance Society (MIDAS)
House 5, Road 16
Dhanmondi R/Area
Dhaka 1209
Tel: (880 2) 326427
Fax: (880 2) 811188

**BELGIUM**
Jean De Lannoy
Av. du Roi 202
1060 Brussels
Tel: (32 2) 538-5169
Fax: (32 2) 538-0841

**BRAZIL**
Publicações Tecnicas Internacionais Ltda.
Rua Peixoto Gomide, 209
01409 Sao Paulo, SP.
Tel: (55 11) 259-6644
Fax: (55 11) 258-6990
E-mail: postmaster@pti.uol.br
URL: http://www.uol.br

**CANADA**
Renouf Publishing Co. Ltd.
5369 Canotek Road
Ottawa, Ontario K1J 9J3
Tel: (613) 745-2665
Fax: (613) 745-7660
E-mail: order.dept@renoufbooks.com
URL: http:// www.renoufbooks.com

**CHINA**
China Financial & Economic Publishing House
8, Da Fo Si Dong Jie
Beijing
Tel: (86 10) 6401-7365
Fax: (86 10) 6401-7365

China Book Import Centre
P.O. Box 2825
Beijing

Chinese Corporation for Promotion of Humanities
52, You Fang Hu Tong,
Xuan Nei Da Jie
Beijing
Tel: (86 10) 660 72 494
Fax: (86 10) 660 72 494

**COLOMBIA**
Infoenlace Ltda.
Carrera 6 No. 51-21
Apartado Aereo 34270
Santafé de Bogotá, D.C.
Tel: (57 1) 285-2798
Fax: (57 1) 285-2798

**COTE D'IVOIRE**
Center d'Edition et de Diffusion Africaines (CEDA)
04 B.P. 541
Abidjan 04
Tel: (225) 24 6510; 24 6511
Fax: (225) 25 0567

**CYPRUS**
Center for Applied Research
Cyprus College
6, Diogenes Street, Engomi
P.O. Box 2006
Nicosia
Tel: (357 2) 59-0730
Fax: (357 2) 66-2051

**CZECH REPUBLIC**
USIS, NIS Prodejna
Havelkova 22
130 00 Prague 3
Tel: (420 2) 2423 1486
Fax: (420 2) 2423 1114
URL: http://www.nis.cz/

**DENMARK**
SamfundsLitteratur
Rosenoerns Allé 11
DK-1970 Frederiksberg C
Tel: (45 35) 351942
Fax: (45 35) 357822
URL: http://www.sl.cbs.dk

**ECUADOR**
Libri Mundi
Libreria Internacional
P.O. Box 17-01-3029
Juan Leon Mera 851
Quito
Tel: (593 2) 521-606; (593 2) 544-185
Fax: (593 2) 504-209
E-mail: librimu1@librimundi.com.ec
E-mail: librimu2@librimundi.com.ec

**CODEU**
Ruiz de Castilla 763, Edif. Expocolor
Primer piso, Of. #2
Quito
Tel/Fax: (593 2) 507-383; 253-091
E-mail: codeu@impsat.net.ec

**EGYPT, ARAB REPUBLIC OF**
Al Ahram Distribution Agency
Al Galaa Street
Cairo
Tel: (20 2) 578-6083
Fax: (20 2) 578-6833

The Middle East Observer
41, Sherif Street
Cairo
Tel: (20 2) 393-9732
Fax: (20 2) 393-9732

**FINLAND**
Akateeminen Kirjakauppa
P.O. Box 128
FIN-00101 Helsinki
Tel: (358 0) 121 4418
Fax: (358 0) 121-4435
E-mail: akatilaus@stockmann.fi
URL: http://www.akateeminen.com

**FRANCE**
Editions Eska; DBJ
48, rue Gay Lussac
75005 Paris
Tel: (33-1) 55-42-73-08
Fax: (33-1) 43-29-91-67

**GERMANY**
UNO-Verlag
Poppelsdorfer Allee 55
53115 Bonn
Tel: (49 228) 949020
Fax: (49 228) 217492
URL: http://www.uno-verlag.de
E-mail: unoverlag@aol.com

**GHANA**
Epp Books Services
P.O. Box 44
TUC
Accra
Tel: 223 21 778843
Fax: 223 21 779099

**GREECE**
Papasotiriou S.A.
35, Stournara Str.
106 82 Athens
Tel: (30 1) 364-1826
Fax: (30 1) 364-8254

**HAITI**
Culture Diffusion
5, Rue Capois
C.P. 257
Port-au-Prince
Tel: (509) 23 9260
Fax: (509) 23 4858

**HONG KONG, CHINA; MACAO**
Asia 2000 Ltd.
Sales & Circulation Department
302 Seabird House
22-28 Wyndham Street, Central
Hong Kong, China
Tel: (852) 2530-1409
Fax: (852) 2526-1107
E-mail: sales@asia2000.com.hk
URL: http://www.asia2000.com.hk

**HUNGARY**
Euro Info Service
Margitszgeti Europa Haz
H-1138 Budapest
Tel: (36 1) 350 80 24, 350 80 25
Fax: (36 1) 350 90 32
E-mail: euroinfo@mail.matav.hu

**INDIA**
Allied Publishers Ltd.
751 Mount Road
Madras - 600 002
Tel: (91 44) 852-3938
Fax: (91 44) 852-0649

**INDONESIA**
Pt. Indira Limited
Jalan Borobudur 20
P.O. Box 181
Jakarta 10320
Tel: (62 21) 390-4290
Fax: (62 21) 390-4289

**IRAN**
Ketab Sara Co. Publishers
Khaled Eslamboli Ave., 6th Street
Delafrooz Alley No. 8
P.O. Box 15745-733
Tehran 15117
Tel: (98 21) 8717819; 8716104
Fax: (98 21) 8712479
E-mail: ketab-sara@neda.net.ir

Kowkab Publishers
P.O. Box 19575-511
Tehran
Tel: (98 21) 258-3723
Fax: (98 21) 258-3723

**IRELAND**
Government Supplies Agency
Oifig an tSoláthair
4-5 Harcourt Road
Dublin 2
Tel: (353 1) 661-3111
Fax: (353 1) 475-2670

**ISRAEL**
Yozmot Literature Ltd.
P.O. Box 56055
3 Yohanan Hasandlar Street
Tel Aviv 61560
Tel: (972 3) 5285-397
Fax: (972 3) 5285-397

R.O.Y. International
PO Box 13056
Tel Aviv 61130
Tel: (972 3) 649 9469
Fax: (972 3) 648 6039
E-mail: royil@netvision.net.il
URL: http://www.royint.co.il

Palestinian Authority/Middle East
Index Information Services
P.O.B. 19502 Jerusalem
Tel: (972 2) 6271219
Fax: (972 2) 6271634

**ITALY, LIBERIA**
Licosa Commissionaria Sansoni SPA
Via Duca Di Calabria, 1/1
Casella Postale 552
50125 Firenze
Tel: (39 55) 645-415
Fax: (39 55) 641-257
E-mail: licosa@ftbcc.it
URL: http://www.ftbcc.it/licosa

**JAMAICA**
Ian Randle Publishers Ltd.
206 Old Hope Road, Kingston 6
Tel: 876-927-2085
Fax: 876-977-0243
E-mail: irpl@colis.com

**JAPAN**
Eastern Book Service
3-13 Hongo 3-chome, Bunkyo-ku
Tokyo 113
Tel: (81 3) 3818-0861
Fax: (81 3) 3818-0864
E-mail: orders@svt-ebs.co.jp
URL:
http://www.bekkoame.or.jp/~svt-ebs

**KENYA**
Africa Book Service (E.A.) Ltd.
Quaran House, Mfangano Street
P.O. Box 45245
Nairobi
Tel: (254 2) 223 641
Fax: (254 2) 330 272

Legacy Books
Loita House
Mezzanine 1
P.O. Box 68077
Nairobi
Tel: (254) 2-330853, 221426
Fax: (254) 2-330854, 561654
E-mail: Legacy@form-net.com

**KOREA, REPUBLIC OF**
Dayang Books Trading Co.
International Division
783-20, Pangba Bon-Dong,
Socho-ku
Seoul
Tel: (82 2) 536-9555
Fax: (82 2) 536-0025
E-mail: seamap@chollian.net

Eulyoo Publishing Co., Ltd.
46-1, Susong-Dong
Jongro-Gu
Seoul
Tel: (82 2) 734-3515
Fax: (82 2) 732-9154

**LEBANON**
Librairie du Liban
P.O. Box 11-9232
Beirut
Tel: (961 9) 217 944
Fax: (961 9) 217 434
E-mail: hsayegh@librairie-du-liban.com.lb
URL: http://www.librairie-du-liban.com.lb

**MALAYSIA**
University of Malaya Cooperative
Bookshop, Limited
P.O. Box 1127
Jalan Pantai Baru
59700 Kuala Lumpur
Tel: (60 3) 756-5000
Fax: (60 3) 755-4424
E-mail: umkoop@tm.net.my

**MEXICO**
INFOTEC
Av. San Fernando No. 37
Col. Toriello Guerra
14050 Mexico, D.F.
Tel: (52 5) 624-2800
Fax: (52 5) 624-2822
E-mail: infotec@rtn.net.mx
URL: http://rtn.net.mx

Mundi-Prensa Mexico S.A. de C.V.
c/Rio Panuco, 141-Colonia
Cuauhtemoc
06500 Mexico, D.F.
Tel: (52 5) 533-5658
Fax: (52 5) 514-6799

**NEPAL**
Everest Media International Services
(P.) Ltd.
GPO Box 5443
Kathmandu
Tel: (977 1) 416 026
Fax: (977 1) 224 431

**NETHERLANDS**
De Lindeboom/Internationale
Publicaties b.v.--
P.O. Box 202, 7480 AE Haaksbergen
Tel: (31 53) 574-0004
Fax: (31 53) 572-9296
E-mail: lindeboo@worldonline.nl
URL: http://www.worldonline.nl/~lindeboo

**NEW ZEALAND**
EBSCO NZ Ltd.
Private Mail Bag 99914
New Market
Auckland
Tel: (64 9) 524-8119
Fax: (64 9) 524-8067

Oasis Official
P.O. Box 3627
Wellington
Tel: (64 4) 499 1551
Fax: (64 4) 499 1972
E-mail: oasis@actrix.gen.nz
URL: http://www.oasisbooks.co.nz/

**NIGERIA**
University Press Limited
Three Crowns Building Jericho
Private Mail Bag 5095
Ibadan
Tel: (234 22) 41-1356
Fax: (234 22) 41-2056

**PAKISTAN**
Mirza Book Agency
65, Shahrah-e-Quaid-e-Azam
Lahore 54000
Tel: (92 42) 735 3601
Fax: (92 42) 576 3714

Oxford University Press
5 Bangalore Town
Sharae Faisal
PO Box 13033
Karachi-75350
Tel: (92 21) 446307
Fax: (92 21) 4547640
E-mail: ouppak@TheOffice.net

Pak Book Corporation
Aziz Chambers 21, Queen's Road
Lahore
Tel: (92 42) 636 3222; 636 0885
Fax: (92 42) 636 2328
E-mail: pbc@brain.net.pk

**PERU**
Editorial Desarrollo SA
Apartado 3824, Ica 242 OF. 106
Lima 1
Tel: (51 14) 285380
Fax: (51 14) 286628

**PHILIPPINES**
International Booksource Center Inc.
1127-A Antipolo St, Barangay,
Venezuela
Makati City
Tel: (63 2) 896 6501; 6505; 6507
Fax: (63 2) 896 1741

**POLAND**
International Publishing Service
Ul. Piekna 31/37
00-677 Warzawa
Tel: (48 2) 628-6089
Fax: (48 2) 621-7255
E-mail: books%ips@ikp.atm.com.pl
URL: http://www.ipscg.waw.pl/ips/export

**PORTUGAL**
Livraria Portugal
Apartado 2681, Rua Do Carm o 70-74
1200 Lisbon
Tel: (1) 347-4982
Fax: (1) 347-0264

**ROMANIA**
Compani De Librarii Bucuresti S.A.
Str. Lipscani no. 26, sector 3
Bucharest
Tel: (40 1) 313 9645
Fax: (40 1) 312 4000

**RUSSIAN FEDERATION**
Isdatelstvo <Ves Mir>
9a, Kolpachniy Pereulok
Moscow 101831
Tel: (7 095) 917 87 49
Fax: (7 095) 917 92 59
ozimarin@glasnet.ru

**SINGAPORE; TAIWAN, CHINA MYANMAR; BRUNEI**
Hemisphere Publication Services
41 Kallang Pudding Road #04-03
Golden Wheel Building
Singapore 349316
Tel: (65) 741-5166
Fax: (65) 742-9356
E-mail: ashgate@asianconnect.com

**SLOVENIA**
Gospodarski vestnik Publishing
Group
Dunajska cesta 5
1000 Ljubljana
Tel: (386 61) 133 83 47; 132 12 30
Fax: (386 61) 133 80 30
E-mail: repansekj@gvestnik.si

**SOUTH AFRICA, BOTSWANA**
For single titles:
Oxford University Press Southern
Africa
Vasco Boulevard, Goodwood
P.O. Box 12119, N1 City 7463
Cape Town
Tel: (27 21) 595 4400
Fax: (27 21) 595 4430
E-mail: oxford@oup.co.za

For subscription orders:
International Subscription Service
P.O. Box 41095
Craighall
Johannesburg 2024
Tel: (27 11) 880-1448
Fax: (27 11) 880-6248
E-mail: iss@is.co.za

**SPAIN**
Mundi-Prensa Libros, S.A.
Castello 37
28001 Madrid
Tel: (34 91) 4 363700
Fax: (34 91) 5 753998
E-mail: libreria@mundiprensa.es
URL: http://www.mundiprensa.com/

Mundi-Prensa Barcelona
Consell de Cent, 391
08009 Barcelona
Tel: (34 3) 488-3492
Fax: (34 3) 487-7659
E-mail: barcelona@mundiprensa.es

**SRI LANKA, THE MALDIVES**
Lake House Bookshop
100, Sir Chittampalam Gardiner
Mawatha
Colombo 2
Tel: (94 1) 32105
Fax: (94 1) 432104
E-mail: LHL@sri.lanka.net

**SWEDEN**
Wennergren-Williams AB
P. O. Box 1305
S-171 25 Solna
Tel: (46 8) 705-97-50
Fax: (46 8) 27-00-71
E-mail: mail@wwi.se

**SWITZERLAND**
Librairie Payot Service Institutionnel
C(tm)tes-de-Montbenon 30
1002 Lausanne
Tel: (41 21) 341-3229
Fax: (41 21) 341-3235

ADECO Van Diermen
EditionsTechniques
Ch. de Lacuez 41
CH1807 Blonay
Tel: (41 21) 943 2673
Fax: (41 21) 943 3605

**THAILAND**
Central Books Distribution
306 Silom Road
Bangkok 10500
Tel: (66 2) 2336930-9
Fax: (66 2) 237-8321

**TRINIDAD & TOBAGO AND THE CARRIBBEAN**
Systematics Studies Ltd.
St. Augustine Shopping Center
Eastern Main Road, St. Augustine
Trinidad & Tobago, West Indies
Tel: (868) 645-8466
Fax: (868) 645-8467
E-mail: tobe@trinidad.net

**UGANDA**
Gustro Ltd.
PO Box 9997, Madhvani Building
Plot 16/4 Jinja Rd.
Kampala
Tel: (256 41) 251 467
Fax: (256 41) 251 468
E-mail: gus@swiftuganda.com

**UNITED KINGDOM**
Microinfo Ltd.
P.O. Box 3, Omega Park, Alton,
Hampshire GU34 2PG
England
Tel: (44 1420) 86848
Fax: (44 1420) 89889
E-mail: wbank@microinfo.co.uk
URL: http://www.microinfo.co.uk

The Stationery Office
51 Nine Elms Lane
London SW8 5DR
Tel: (44 171) 873-8400
Fax: (44 171) 873-8242
URL: http://www.the-stationery-office.co.uk/

**VENEZUELA**
Tecni-Ciencia Libros, S.A.
Centro Cuidad Comercial Tamanco
Nivel C2, Caracas
Tel: (58 2) 959 5547; 5035; 0016
Fax: (58 2) 959 5636

**ZAMBIA**
University Bookshop, University of
Zambia
Great East Road Campus
P.O. Box 32379
Lusaka
Tel: (260 1) 252 576
Fax: (260 1) 253 952

**ZIMBABWE**
Academic and Baobab Books (Pvt.)
Ltd.
4 Conald Road, Graniteside
P.O. Box 567
Harare
Tel: 263 4 755035
Fax: 263 4 781913